The uncertain quest

Note to the reader from the UNU

The economic and social development of a nation have become directly dependent on the strength of its scientific and technological base. In order to take advantage of the full range of available technologies – from traditional to highly advanced – for the purpose of development, concepts such as the management of technological pluralism and of technology blending have emerged as key components in the design and implementation of scientific and technological policies for development.

The United Nations University has established an extensive network of scholars representing different disciplines, cultural values, and perspectives in approaching the issues of science, technology, and development. Considerable research has been carried out and knowledge accumulated through country studies and regional comparative studies. Building on this experience, the UNU invited a selected group of specialists mainly from its network to develop a sourcebook on science and technology policies to be published in several languages. The book provides a scholarly assessment of the role of science and technology in the development process and a critical analysis of their social, economic, and political dimensions.

The uncertain quest: Science, technology, and development

Edited by Jean-Jacques Salomon,
Francisco R. Sagasti, and Céline
Sachs-Jeantet

 **United Nations
University Press**

TOKYO • NEW YORK • PARIS

The views expressed in this publication are those of the authors and do not necessarily reflect the views of the United Nations University.

United Nations University Press
The United Nations University, 53-70, Jingumae 5-chome, Shibuya-ku, Tokyo 150, Japan
Tel: (03) 3499-2811   Fax: (03) 3499-2828
Telex: J25442   Cable: UNATUNIV TOKYO

Typeset by Asco Trade Typesetting Limited, Hong Kong
Printed by Permanent Typesetting and Printing Co., Ltd., Hong Kong
Cover design by Apex Production, Hong Kong

UNUP-835
ISBN 92-808-0835-4
United Nations Sales No. E.93.III.A.6
04300 P

Contents

v

Contents

Foreword

Heitor Gurgulino de Souza

Rector of the United Nations University

Science and technology have long been recognized as essential driving forces in the development process. Yet the decade of the 1980s brought with it many disappointments on the part of the developing countries in their attempts to actually use the potential of rapid technological change to their advantage. For this reason, it became obvious that in light of global change and the struggle for a new world order in many areas, innovative theories, new paradigms, and creative approaches regarding science and technology policies were long overdue.

The United Nations University (UNU) has established over the years an extensive network of scholars representing different disciplines and cultural values and perspectives in approaching the linkages between science, technology, and society. Thus, considerable research had been carried out in the past through country and comparative studies. Building on this experience the UNU invited a selected group of specialists engaged in these past endeavours to write an authoritative sourcebook on science and technology policies for development to be published in several languages.

The objective of this UNU sourcebook on science, technology, and development is to provide a scholarly assessment of the role of science and technology in the development process and a critical analysis of the social, economic, and political dimensions of science and technology. All contributors have themselves been attached to science and technology policy units, either at a university or research institute or in other high-level administrative positions related to this field throughout their careers.

In order to establish a direct bridge between this scholarly work and the UNU's postgraduate training activities for professionals from developing countries, some of the authors of our sourcebook were also involved as lecturers in related training activities carried out by the UNU in cooperation with Unesco and several non-governmental organizations in Latin America, Asia, and Africa. Choosing priorities in science and technology, identifying strategies appropriate to existing national resources and needs, and establishing adequate institutions and strategic alliances of governmental, academic, and private sectors are essential stepping-stones in the management of science and technology. The UNU's training activities were designed to assist these processes.

Science and technology capacity-building in developing countries continues to be of great concern to the UNU, and new institutional and programmatic arrangements have been made by us over the past years to strengthen the position of science and technology in such fields as natural resources, development economics, environment, energy, and new technologies, particularly biotechnology and microelectronics, to mention just a few. The consolidation of the World Institute for Development Economics Research (UNU/WIDER) and the establishment of the UNU Institute for New Technologies (UNU/INTEC), the UNU International Institute for Software Technology (UNU/IIST), the Programme for Natural Resources in Africa (UNU/INRA), and the Biotechnology Programme for Latin America and the Caribbean (UNU/BIOLAC) represent major progress in this direction. Furthermore, a new series of international seminars on "The Frontiers of Science and Technology" was established by our University Centre in Tokyo, in cooperation with major Japanese universities, to explore some of the most recent trends and their potential implications.

New paradigms are evolving in development theory and practice. The "human development" aspect now emphasized by the United Nations Development Programme (UNDP), the higher importance attributed to the role of the private sector by all major multilateral and bilateral aid organizations, the recognition of ecology as a vital factor in development, and the inevitable long-term trend towards a "one-world technology" – implying that there will not be one technology for the developing and one for the developed world, but global technologies – are all part of this process.

I wish specially to thank the editors, Jean-Jacques Salomon, Francisco R. Sagasti, and Céline Sachs-Jeantet, and the authors of our sourcebook for their efforts, as well as the International Development Research Centre (IDRC) and Unesco for the moral, intellectual, and financial support they have given to this project.

Preface

The idea for this book came about in the course of discussions between the International Council for Science Policy Studies (ICSPS) and the United Nations University. In the industrialized countries, the number of institutions and research projects concerned with the study of the economic, political, and social issues raised by science and technology has been growing steadily over the last 30 years; under the labels "science studies," "science policy studies," "science, technology, and society," or "science of science," these issues have acquired academic recognition and university status, making it possible not only to improve understanding of these topics but also to provide relevant training. By contrast, very few developing countries have organized teaching or research on these subjects, although some have earned an excellent reputation, especially India and countries in Latin America. In recommending that the United Nations University should arrange regional seminars to provide training related to these issues, the ICSPS stressed that certain themes should be given top priority and that it was essential to make a state of the art as regards the problems and the literature in this field.

Jean-Jacques Salomon, then President of the ICSPS, was behind the publication of the book *Science, Technology and Society: A Cross Disciplinary Perspective* (ed. Ina Spiegel-Rösing and Derek de Solla Price, 1977), which was very well received and made a substantial contribution in all the industrialized countries, in both the West and the East, to the rise of disciplines concerned with studying science and technology policies. A similar exercise, this time examining the specific situation of developing countries, was therefore proposed by the ICSPS, taking into account the progress made in the field since the publication of the Spiegel-Rösing and de Solla Price book and also the special problems faced by the developing countries in the last decade. We are particularly grateful to Professor Heitor Gurgulino de Souza, who had only just been appointed Rector of the United Nations University in 1988, for his immediate enthusiastic reaction to the project and for his constant support ever since.

The project was rather more ambitious than the earlier one: for one thing, the well-established tradition of studies in this area had generated a vast corpus of books and academic theses that all the specialists in the industrialized countries have long since read and absorbed. In addition, whatever the differences among the industrialized countries – for instance, among the OECD member states or between countries run on free-market principles and those with managed economies – they had all invested massively in science and technology since the Second World War, they had all encouraged research and development (R&D) in similar areas, and hence they had a pool of shared experiences and debate. Naturally, neither the corpus of literature nor the experience was to be found on the same scale in the developing countries, especially given that their choices of development strategies had been extremely diverse, sometimes conflicting and rarely directed towards major R&D efforts. In many ways, it is still useful to refer to the book edited by Spiegel-Rösing and de Solla Price – which shows how much their pioneer work contributed, intellectually and academically – for a review of the contexts and fundamentals that continue to shape the study of the links between science, technology, and society. In putting together our own volume, we have taken for granted that the earlier survey is still an essential work of reference, and we have not felt it necessary to discuss again some of the topics (especially as regards the growth of institutions, professionalization, and standards) where its coverage is still valid and relevant to the present situation.

We went about the task of producing this volume in the same way as was true for the last. At an initial meeting of all the contributors in Paris in June 1990, we tried to draw up the overall structure of the book in the light of the proposed chapter outlines. The editors and the authors then sent each other many kilos of correspondence, and read several drafts, before a four-day meeting at the Saline Royale d'Arc et Senans in June 1991, where we tried to ensure that the various chapters constituted a coherent framework, which itself had inevitably been modified in the course of time. After another, three-day meeting in Paris, this time just among editors, in January 1992 to review the revised contributions, the editors had to check all the chapters, make sure that they were neither too long nor too short, take some drastic decisions in order to eliminate as much repetition as possible, and prune the bibliographies to manageable proportions.

From the very outset, we were aware of the limits of our project. We knew that we could not deal with all the issues relating to science and technology as applied to development, and we never even thought that we could. For one thing, this field is vast and has no well-defined frontiers, and the problems are constantly changing over time and in response to changes in the economic, political, and social contexts, both nationally and internationally. There is no comprehensive economic and social theory that clearly explains the links between science, technology, and development – hence part of the uncertainty of the quest. The best one can do is to stress the complexity of these links, to summarize existing knowledge, and to highlight some of the partial lessons that result from the many studies of these links. Secondly, the developing countries themselves are quite disparate, with their own characteristics, "styles," and constraints that make it impossible to establish satisfactory typologies. Under the same heading, they are different social organizations: there are differences in degrees of external exposure, in terms of trade, access to external funds, maturity of production, patterns of social conflict, etc. Lastly, and most importantly, authors were asked to examine the issues through small "windows" and from just a few angles, and hence could be accused of bias, since the view from Latin America, for example, could obviously not be the same as from South-East Asia or Africa. If we had tried to cover more topics, and to analyse them more thoroughly, with more case-studies and illustrations, we would have ended up like the builders of the proverbial Tower!

Hence this book – with its limitations and its deficiences that we are the first to acknowledge – seemed to us the best solution to the problem of how to treat the matters that we had set ourselves as the aim of the endeavour both concisely and in a useful way for teaching purposes. We have tried to present a survey of the state of the art; to emphasize some key issues, but not all; to make available studies that, along with the bibliography of relevant publications, will provide a sound basis for teaching and learning and an analytical framework for reflecting upon the role of science and technology in the development process. We expect our audience to be researchers, academics, research administrators, and decision makers concerned with all aspects of devising and implementing policies on science and technology. We should like to stress that if science and technology are essential components of any development strategy, the policies relating to them should be integral to all the other aspects of a thoughtful and consistent development policy, ranging from the economy to education, from agriculture and industry to the environment, from business to health, etc. Lastly, we have tried to highlight certain challenges for the future, and point to areas and directions where research and discussions might be pursued. It is clear that if there is to be a follow-up to this volume – and it is up to others to undertake the task – it should take the form of a series of sector studies, with case-studies on the history and the various disciplines in each region, and even perhaps the different experiences of each country within the region.

It has also to be said that events are moving ever faster, especially as a result of scientific and technological progress, so that we live in a world of constant and extremely rapid change. When we first thought of this book, the communist bloc was still in existence, even if there were cracks in the foundations. As time passed, and the communist economies collapsed and the Soviet Union imploded, we wondered (and others asked us) whether we should not have covered the industrialized countries of the Second World that had suddenly become "new developing countries" by analogy with the "newly industrialized countries." This would have meant making the book even larger; but in any case, there are very good reasons for making a clear distinction between the former communist economies and the developing countries. Besides, there is a risk that the tendency of the West to take a special interest in the problems of the ex-communist countries might mean that it would be even more neglectful of the problems of the "old" third world. To have contributed to this trend would have been contrary to the ideas underlying the conception and the production of this book.

If we had prepared the book 25 years ago, the title as well as the contents would have been more optimistic with regard to the potential positive influence of science and technology on development in both industrialized and third world countries. Now, the damage to the environment from industrial activity and the dangers of nuclear weapons and nuclear energy mean that progress as such can no longer be taken for granted. We must be wary of the sidetracks, the adverse effects, and the costs of change resulting from scientific and technical advances. This quest is all the more uncertain today in relation to development. The title of this volume reflects not our doubts about what can be achieved through science and technology but our conviction that this is less than ever an inevitable process, all of whose promises can be kept.

Acknowledgements

We are delighted to record that this project benefited from the outset from wide-ranging support from the United Nations University, the engine of the project, but also from the International Development Research Centre, the United Nations Educational, Scientific and Cultural Organization, and the International Council for Science Policy Studies. We should like to thank in particular the officials of each of these institutions with whom we dealt, who were unstinting in their faith in us and in their support: the Rector of the UNU, naturally, but also his Senior Adviser, Sogo Okamura, and Dieter Koenig, Scientific Affairs Officer; Marc Chapdelaine, Head of the Science, Technology and Society Unit at Unesco, his successor, Vladislav Kotchetkov, and Kotchetkov's consultant, Folin Osotimehin; Brent Herbert-Copley, Programme Officer at IDRC; Everett Mendelsohn, Professor at Harvard University and President of the ICSPS; and Georges Ferné, Secretary of the ICSPS, without whose help we should never have been able to carry through this project. We should like also to thank those who contributed during the earlier stages of the preparation of the volume: Roy MacLeod, Professor at the Australian National University; Geoffrey Oldham, Director of the Science Policy Research Unit of the University of Sussex; and Henrique Rattner, Professor at the University of São Paulo. Finally, we would like to acknowledge the support provided throughout the project from the Conservatoire National des Arts et Métiers, and the Centre Science, Technologie et Société, in particular from Nadine Glad, and from the World Bank.

We would also like to say how grateful we are to the contributors, not merely for the chapters that they wrote but also for their valuable comments, discussions, and criticisms. If the book gives the impression of being a collective endeavour, conceived and brought to publication in a real spirit of international cooperation, it is thanks as much to them all as to us. Last but not least, we would like to express our gratitude to Ann Johnston, who edited the original English version of the manuscripts, most of them written by authors who are not native English speakers. Moreover, without her tenacity, professionalism, and encouragement, this volume would never have been accomplished.

All parts and all aspects of science belong together. Science cannot develop unless it is pursued for the sake of pure knowledge and insight. It will not survive unless it is used intensely and wisely for the betterment of humanity, and not as an instrument of domination by one group over another. Human existence depends upon compassion and curiosity. Curiosity without compassion is inhuman; compassion without curiosity is ineffectual.

Victor F. Weisskopf

Introduction: From tradition to modernity

Jean-Jacques Salomon, Francisco R. Sagasti, and
Céline Sachs-Jeantet

A "science" of some sort has existed in every society at all periods of human history. There can be no action, whether on natural or social phenomena, without a certain amount of rational empirical knowledge of the physical, living, and social world. Such knowledge has always played an important role in the development of societies, in their material as well as in their institutional and cultural achievements. However, it is in modern industrial societies that science and technology became the critical factor in the process of long-term economic growth and development. Many civilizations and societies have ignored or simply not paid attention to the notion of progress, but nevertheless have witnessed some degree of technical change that occurred over the very long term.

Expectations about prospects for improvement in the standard of living are a rather recent phenomenon, and they rose extremely slowly in the pre-industrial era. The idea of progress emerged in the context of the Judeo-Christian civilization and developed mainly with the Scientific Revolution in the seventeenth century, the Enlightenment in the eighteenth century, and the Industrial Revolution that is still with us. Subsequently, economic growth became – for better or for worse – the basis of every society's hopes for the future, and science and technology became more and more instrumental in the fulfilment of these expectations. It is in this framework that policies for and through research and development (R&D) activities became more and more indispensable to the conception, elaboration, and implementation of broader policy and political objectives. Max Weber considered that the modern state is defined by bureaucracy, so that

1

any current policy-making process can be defined as bureaucracy plus science: most political decisions today have recourse to scientific disciplines as regards methods, proofs, results, and even promises.

The importance of science and technology

Science and technology do indeed matter, and nowadays more and more. This should be self-evident, and yet in many developing countries, there is so little appreciation of this fact, among decision makers as well as the general public, that people either do not know or do not realize the benefits that a consistent and deliberate development strategy can derive from scientific and technical resources. Furthermore, people often overlook the fact that science and technology function successfully only within a larger social/political economic environment that provides an effective combination of non-technical incentives and complementary inputs in the innovation process. Science and technology are not exogenous factors that determine a society's evolution independently from its historical, social, political, cultural, or religious background.

As a recent report of the International Council for Science Policy Studies has emphasized:

technological change and innovation cannot have their socially beneficial effects if the cultural and political contexts are not prepared to absorb and incorporate them, and to achieve the structural transformations which will be required – a process which is much more difficult and complex than a mere transfer of resources (in this case, science and technology rather than capital) from the rich to the poor as a way of correcting imbalances. Science and technology have had an enormous impact on reducing the burden of physical work and improving social welfare. These contributions have only been made possible by the enormous methodological power of scientific reasoning which extends human ability to imagine and to develop alternatives. This being said, however, the development of science and technology is much more than the application of objective logic. It is built on a social consensus about goals and values. Science and technology exists only through human beings in action in certain contexts, and as such cannot be entirely value-free and neutral. [7, pp. 16–17]

Unquestionably, scientific and technological progress has provided many benefits over the long term for the industrialized countries and in more recent times for developing countries. The most striking evidence of this in the industrialized countries is per capita income, which has increased almost tenfold in the space of two centuries.

2

What is more, this purely quantitative indicator gives no idea of the individual and collective benefits that have accompanied this enormous rise in income: longer life, lower infant mortality, eradication of certain diseases, higher level of education, more rapid means of communication, better living and working conditions, greater social protection, more leisure opportunities, etc. Whatever inequalities persist, and however large (and sometimes growing) the pockets of poverty still to be found in the "rich" countries, the general level of material improvement is manifestly positive. This is all the more a reason to try to improve the current situation of most developing countries, whose conditions are such that the benefits of scientific and technological progress do not contribute to their development in the same way, at the same level or speed.

This reading of technical progress – the only one that is objective – is derived from the figures selected by economists for the purpose of calculating rises in gross national product and productivity. They can lead to irrefutable conclusions regarding the quality and standard of living from an economic standpoint, and this is already a decisive achievement. But such an assessment does not go beyond the quantitative facts concerning production, consumption, the working week, health and hygiene, life expectancy. As soon as one takes a broader view, the balance sheet of progress is more ambiguous and becomes a matter of subjective reactions and convictions. Our economic indicators are quite incapable of gauging the social costs and drawbacks (e.g. for the environment) associated with economic growth and technical progress. But they are also incapable of allowing for all the new knowledge and technical know-how – largely the products of progress – that have enabled human beings to extend their knowledge of nature and themselves, to reduce the level of superstition, and to act more rationally to achieve a better life. There are, of course, darker sides in this balance sheet of science and technology, from the arms race and the creation of a nuclear arsenal capable of "overkilling" mankind to the global environmental issues resulting from a process of industrialization that threatens the future of the whole earth. Nobody today can share the positivist optimism of the Enlightenment's concept of progress; the straight road to greater knowledge and material progress does not lead by the same token to the less direct road to "happiness" and "moral progress."

"Whether like the sociologist, Herbert Marcuse, or the novelist Simone de Beauvoir, we see technology primarily as a means of human enslavement and destruction, or whether, like Adam Smith, we see it primarily as a liberating Promethean force, we are all involved in its advance. However much we might wish to, we cannot escape its impact on our daily lives, nor the moral, social and economic dilemmas with which it confronts us. We may curse it or bless it, but we cannot ignore it." This was how Christopher Freeman began his book on *The Economics of Industrial Innovation* [5, p. 15]. Indeed, whether one likes it or not, the final trade-off is between poverty and growth. Where Freeman was concerned only with technology, we are concerned here with both science and technology.

In rejecting modern science and technology, Simone de Beauvoir is consistent in her deliberate preference for poverty. But most economists have tended to accept with Marshall that poverty is one of the principal causes of the degradation of a large part of mankind. Their preoccupation with problems of economic growth arose from the belief that the mass poverty of Asia, Africa and Latin America and the less severe poverty remaining in Europe and North America, was a preventable evil which could and should be diminished, and perhaps eventually eliminated. [5, p. 15]

Freeman continues:

Innovation is of importance for increasing the wealth of nations not only in the narrow sense of increased prosperity, but also in the more fundamental sense of enabling them to do things which have never been done before at all. It is critical not only for those who wish to accelerate or sustain the rate of economic growth in this and other countries, but also for those who are appalled by narrow preoccupation with the quantity of goods and wish to change the direction of economic advance, or concentrate on improving the quality of life. It is critical for the long-term conservation of resources and improvement of the environment. The prevention of most forms of pollution and the economic re-cycling of waste products are alike dependent on scientific and technological advance. [5, p. 16]

We quote at such length from Freeman, who was offered the first chair of science policy in the world and led with great success the Science Policy Research Unit at the University of Sussex, not only to pay tribute to his pioneering work but also because we share his conviction – which is the guiding principle of this volume as a whole – that there is no substitute for rational thought. We can learn to make better use of science and technology, but we cannot escape from them – unless of course we are prepared to give up all attempt to cope with the difficulties, tensions, and challenges of the world in which we have to live. Freeman added:

The famous first chapter of Adam Smith's *Wealth of Nations* plunges immediately into discussion of "improvements in machinery" and the way in which division of labour promotes specialized inventions. Marx's model of the capitalist economy ascribes a central role to technical innovation in capital goods – "the bourgeois cannot exist without constantly revolutionizing the means of production". Marshall had no hesitation in describing "knowledge" as the chief engine of progress in the economy.

From Schumpeter to Samuelson, most economists today come to the same conclusion. The central importance of science and technology for economic progress is equally the main concern of this book.

Science, technology, and society

The social and cultural factors – the attitudes and the beliefs attached to economic, political, and social organization – influence the role that science and technology play in a given society. In their turn, the spread of new knowledge, products, and processes derived from scientific and technological progress transforms social structures, modes of behaviour, and attitudes of mind. The role of technical change in the process of economic growth is recognized by all theories of development. But what precisely is that role? In particular, what part did science and technology play in the economic and social transformations that accompanied the Industrial Revolution from its beginnings? Answers to these questions can be neither easy nor, consequently, swift, requiring as they do a subtle analysis, a long-term historical perspective, and reference to examples drawn from different branches of social science [2, 14].

Today the ways in which technical change transforms attitudes, institutions, and societies cannot be reduced to a simple linear relationship that is automatic, i.e. deterministic. Technology is one social process among others: it is not a question of technical development on the one hand and social development on the other, as if they were two entirely different worlds or processes. Society is shaped by technical change that, in turn, is shaped by society. Conceived by man, technology eludes his control only in so far as he wants it to. In this sense, society is defined no less by those technologies that it is capable of creating than by those it chooses to use and develop in preference to others [15].

Indeed, the present situation is very different from the expansion of mechanization encouraged by the development of machine tools and the steam engine in the nineteenth century. The spread of the "new technologies" (electronics, computers, telecommunications, as well as new synthetic materials and biotechnologies) creates far greater disparities than those that were possible between European countries at the beginning of the Industrial Revolution. Moreover, it involves much greater challenges than those tackled by nineteenth-century European societies (which were pre-industrial rather than purely agricultural), which achieved success thanks to their long preparation in basing their interpretation of natural phenomena and their handling of techniques on, among other things, mathematics, experimentation, measurement, calculation, and proof [16]. On the one hand, in fact, the geopolitical situation in the world today is more complex, with events and actors constantly in motion on a continental scale, further augmented by the explosion in the means of communication themselves. On the other hand, the very tools (both conceptual and practical) that allow us, at least partially, to understand the world in which we live and to manipulate it, have continued – in large measure thanks to the spectacular progress of science and technology – to become ever more "sophisticated" and therefore difficult to master without specialist skills and qualifications.

It is against this background of the increasing complexity of problems as much as of methods that the "shock" of the new technologies has struck both developing and industrialized countries. For the latter – given the economic difficulties of the early 1980s, the very moderate rates of growth and the persistence of high unemployment – the adjustment to the new technical system that is just beginning to spread poses problems that are not very different from those that gave rise to the various stages of mechanization in the course of the nineteenth century. Whatever the social costs in terms of redundancies and job displacement, and however substantial the pockets of poverty that remain (and that sometimes even grow as a result of the crisis and uneven development), we are nevertheless dealing with societies where basic needs are by and large satisfied, and furthermore the resources available to train and retrain the labour force are considerable. It is not for nothing that they have been called "post-industrial" societies, characterized by the dominance of the service sector, the very rapid growth of information-related activities, and the large scale of investment in education and research.

By contrast, for most of the developing countries, the most basic needs for survival – food, health, shelter, and education – are far from having been met, so that the things that are perceived by the rich countries as essential can seem to the poorer countries like a display of luxury or a gimmick of a consumer society. In addition, they face the double pressures of the population problem, which seems unlikely to see major improvement before the end of the century, and the debt problem, which has become so dramatic that some countries can barely cope with payment of the interest charges. Against this background some people question the claim that the new technologies are what many developing countries should seek as a high priority in order to meet their real needs. And yet – given both the growing interdependence of economies and the internationalization of trade on the one hand and the undeniable opportunities to modernize and "catch up" that are offered by the new technologies on the other – it seems inconceivable that any country should choose to deprive itself of the products and the infrastructures that increasingly define the "nervous system" of the contemporary world and determine its functioning [8]. In this connection one cannot underestimate the relevance and the value of "technology blending," i.e. the application of new technologies economically deployed to upgrade, modernize, or develop traditional activities (or to exploit natural resources that would otherwise remain untapped) while causing minimal social and economic disruption.

The rapid spread of a new technology does not of itself imply rapid social change. Other factors are involved, such as economic, social, and educational policies, the negotiations and agreements between interest groups, the well-established customs of daily life and social institutions, the society's values and traditions. Once again it needs to be stressed that science and technology are not independent variables in the process of development: they are part of a human, economic, social, and cultural setting shaped by history. Nothing is more revealing from this standpoint than the case-studies of technology blending, which indeed show precisely that the application of new technologies in traditional sectors is not simply a technological issue but more so an institutional, social, and political one [1]. It is this above all that determines the chances of applying scientific knowledge that meets the real needs of a country. It is not the case that there are two systems – science and technology on one side and society on the other – held together by some magic formula. Rather, science and technology exist in a given society as a system that is more or less capable of osmosis, assimilation, and innovation – or rejection – according to realities that are simultaneously material, historical, cultural, and political.

All in all, there is no inevitability in technical change: neither its pace nor its direction is predetermined (even though one cannot underestimate the strength of certain industrial and national lobbies in imposing their factories or products), and the success of an innovation is never certain. Technology influences economics and history, but it is itself the product and the expression of culture. The same innovations can therefore produce very different results in different settings, or at different periods within the same society. Technical change and technology itself thus make up a social process in which individuals and groups always make the determining choices in the allocation of scarce resources, an allocation that inevitably reflects the prevailing value system [14]. At the same time, science and technology are not "black boxes" with principles and effects that leave unchanged the social structures of the societies that adopt them. They cannot be shipped like commodities: the process is never neutral, straightforward, or permanent; it demands levels of skill and often also perseverance, without which it constitutes a tool without a handle or a box of tricks without a key.

It is from this angle that the links between science, technology, and society in developing countries should be addressed. Beyond a certain threshold of resources, capital accumulation is never by itself a guarantee of growth. On the contrary, it is first and foremost the organization of society – which in turn determines the organization of production – that allows a country to create and exploit its scientific and technical resources. These factors define the extent to which science and technology can operate to initiate and stimulate the process of development, and not vice versa. If science and technology are not external to this process, it is because they cannot themselves be either developed or used other than in a given economic and social framework. Extreme underdevelopment is in this sense the stage of development that puts no pressure on the social structure to become involved in scientific and technical research. And, lacking a favourable economic and social structure, even countries above this level may find themselves unable to take advantage of science and technology. If there is a lesson to be kept from history, and especially from the history of science, it is that the routes and institutions by which knowledge develops and is transmitted across a society, as much as across cultural frontiers, are never linear nor mechanistic.

The institutional and policy requirements

In what follows, we intend to highlight, first, the crucial importance of scientific and technical resources for social and economic development; second, the variety of situations facing the developing countries, especially as regards their endowment in terms of science and technology, and thus the fact that there is no single model for defining and implementing strategies; third, the contradictory, if not disappointing, results achieved by development economics, and the indispensable effort that needs to be made in order to integrate science and technology policy in an overall policy for economic and social development. Any attempt to make general statements on the subject runs the risk of failing to capture what is actually happening, for two reasons: national circumstances are too diverse for a single model to fit them all, and science and technology today are too complex to be dealt with in general terms. These words of warning apply equally to general discussion of the so-called "third world." And particularly with regard to technical innovation: "It is not possible to come to grips with the complexities of technology, its interrelations with other components of the social system, and its social and economic consequences, without a willingness to move from highly aggregated to highly disaggregated modes of thinking" [12]. In other words, any analysis of the interaction between social organization and technical change must always be refined to take account of each country's characteristics, especially its relative level of scientific and technical assets, the nature and quality of those assets (higher education and training institutions, laboratories, etc.), and the use made of them in the framework of its specific economic, political, and social conditions.

Whatever its pace and level, development is a journey between tradition and modernity. In this dynamic process, quantitative indicators are always relative: development is never finished and certainly is never achieved once for all, nor is the process measurable only quantitatively. Neither "take-off" nor increasing industrialization can ever be a reliable guarantee against slipping back, as the example of eastern Europe shows. In addition, although the available data provide points of comparison, we are not dealing with a scale of values derived from a single, comprehensive and unassailable theoretical model. The journey takes time, incurs costs, requires the making of choices, and so demands a resolute collective determination not simply to cope with the risks arising from change but to try, from a long-term perspective, to guide change in a particular direction.

As Gunnar Myrdal [10] has emphasized, the terminology used by the social sciences is not neutral. We now talk about "developing countries" rather than "underdeveloped countries" because we want to play down the realities of structural imbalance and stress instead the chances of catching up. The courteous language of diplomacy suggests that there is merely a short time-lag separating the industrialized countries from those that are not there yet: all that is needed in order to bridge the gap is to adopt the "right" economic policy. The term "developing country" is illogical, according to Myrdal, because it conveys the idea that there are countries that are not developing. Besides, it gives no indication of whether a country wants to develop or is taking practical steps to foster its development. In this sense, the first requirement – not just in terms of chronology but above all of principles – can be summed up in the determination to try to develop, not so much with a view to breaking with the past (or at least not with all earlier traditions) as to acquiring the means to modernize. These means are partly but not entirely economic; institutional, social, political, and cultural factors also count. The development process is a package in which success depends on many different elements in combinations that can never be determined by economic indicators alone.

The most general lesson is that technical change does not transform societies independently of other factors that are not related to technology as such. The Industrial Revolution witnessed the start of a new type of growth, which was connected with a succession of technical innovations that speeded up the pace of change, although their origins and development depended on a wide range of non-technical factors. In Europe, capitalistic competition encouraged technical developments geared to increasing labour productivity. These developments happened and were able to spread only because the economic, institutional, and social circumstances were favourable. In their turn, these circumstances were altered by the progress of science and technology and then influenced the rate and direction of technical innovation. The process was extremely complex, as Landes [9] stresses in the conclusion to his history of the Industrial Revolution: "There is a wide range of links, direct and indirect, tight and loose, exclusive and partial, and each industrializing society develops its own combination of elements to fit its traditions, possibilities, and circumstances. The fact that there is this play of structure, however, does not mean that there is no structure."

In this delicate and uncertain "play of structure," which is affected by the historical and cultural background of each country, the institutional and political prerequisites for making good use of the scientific and technical resources available mainly relate to these non-economic factors. The growing interdependence of nations and the emergence of the world economy have not abolished the individuality of cultures and societies. The journey from tradition to modernity raises the same question for all developing countries, but they are *the only ones* in a position to reply, in line with the decisions that they take *themselves* about science and technology as about everything else. This question is double-edged: how to modernize without sacrificing tradition? how to preserve tradition without compromising modernization? More than ever, the hurly-burly of politics that we are witnessing as we approach the end of the twentieth century warns any developing country to be sensitive to the implications of this question.

The new international context

The decade of the 1990s will hold as many surprises and shocks as that of the 1980s. A new and as yet fluid world order is in the making as we approach the transition to the twenty-first century, and in this volatile context extrapolation of past trends into the future is a risky enterprise. Predictions are less effective than attempts at mapping uncertainties and identifying desired outcomes, for the latter are likely to be brought about by the purposeful actions of governments, international institutions, private enterprises, non-governmental organizations, religious groups, research and academic institutions, and mass media, among a growing number of actors in the international and national scenes.

The uncertain world in which we live has many dimensions: a rapidly shifting political setting, changes in the patterns of world economic interdependence, growth and diversification of social demands, emergence of environmental concerns, and major transformations in the cultural landscape.

The political setting
The end of the Cold War has undermined the ideological, military, and political foundations of the international order that prevailed

11

during the last half-century. The world is in transition to a "post-bipolar" political and economic order, whose nature is in the process of being defined but which will require a profound re-examination of the means for providing national, regional, and international security as a precondition for development. Some of the elements of this new order include the virtual elimination of the threat of an all-out nuclear war, an increase in the number and intensity of regional conflicts, the likelihood of a more cooperative approach to conflict resolution among key political and economic players, and a larger role for international institutions in fostering and maintaining international security.

The range of possible outcomes for these various elements of the emerging political order is wide. The demise of East-West rivalry has complex implications for national security in developing countries. Conflict and insurgency based on Cold War ideology, once generously financed by the superpowers, have all but vanished, as has the possibility of playing one camp against the other. But Soviet and American disengagement could encourage other countries to build and exercise military power, with the enthusiastic support of aggressive arms merchants.

Ethnic and religious tensions within countries have contributed to this possibility, since they can attract support from neighbouring states. New regional conflicts over natural resources such as water, oil, or tropical forests, and over environmental spillovers could also encourage military aggressiveness. These tensions and conflicts may be kept in check by concerted actions by the major military powers, by regional and international organizations, or a combination of both. So far, despite diminished global superpower rivalry, there is no evidence of a decline in regional disputes, or in organized violence by ethnic groups, secessionist movements, terrorists, or drug traffickers.

At the same time, states are becoming less important as political units in the sense of being able to control whatever phenomena – economic, social, environmental, or technological – take place in the world at present. This is hard to get used to, for political systems are geared to focus on states as the locus of power and decision-making and as the main unit of political, social, and economic analysis.

The pre-eminence and sovereignty of states is being eroded in many aspects of foreign and economic policy, as is highlighted by the renewed importance of the United Nations in conflict prevention and resolution, by the proliferation of regional trade and economic agreements, by the growing economic power of international corporations, and by the conditions established by international financial institutions for obtaining access to resources under their control. The movement towards supranational action is likely to proceed by fits and starts, with temporary reversals and renewed bouts of nationalism, but will probably gain momentum as the new century approaches.

Political pluralism, popular participation, and democratic movements are becoming a fact of life everywhere: East, West, North, and South. It is now almost unthinkable to accept – at least without outrage, loud protest, and international sanctions – any government's imposition of a repressive regime on its citizens. By the early 1990s eastern European countries had their first open elections in half a century, almost all of the countries of Latin America had democratic regimes, a military coup failed in Russia, the Central Asian states of the former Soviet Union were struggling to become modern nations, White rule was ebbing in South Africa, and there were pressures to abolish one-party rule in many countries of Africa. However, as the civil wars in the former Yugoslavia and in Somalia have shown, advances towards democracy and peaceful coexistence are by no means guaranteed.

As a consequence of these changes, the exercise of power and authority in the management of resources for development – usually referred to as "governance" – has become a legitimate subject of concern, particularly by international organizations and development cooperation agencies. In addition, non-governmental organizations of all types (trade unions, professional associations, environmental and human rights advocacy groups, grass-roots movements, church organizations) have also become extremely active and show that civil society is finding multiple ways of expressing itself at the local, national, regional, and international levels.

The international economy

The major transformations taking place in the patterns of world economic interdependence include the rapid growth and globalization of financial markets, changes in trade patterns, and new situations in key countries that affect the world economy. International financial markets now comprise a tight web of transactions involving global securities trading, arbitrage in multiple markets and currencies, portfolio investments through a bewildering array of international funds, and massive transborder capital movements. Financial transactions have acquired a life of their own and are becoming uncoupled from the production and distributions of goods and services: in 1989 the combined daily average of trade in the foreign exchange markets of Japan, the United States, and the United Kingdom reached US$430 billion, six times the 1979 level and 50 times the average daily volume of international trade in goods and services [4].

After a decade of rapid and substantial increases in commercial bank lending to developing regions during the 1970s, the debt crisis that started in the early 1980s reduced private bank flows to zero by the end of that decade: of the approximately US$60 billion of net debt-related flows to developing countries in 1980, US$32 billion came from commercial banks. In contrast, by 1990 total net flows fell to about US$20 billion, and the amount obtained from commercial banks fell to near zero. As a consequence, direct foreign investment acquired much greater importance as a channel for resource transfers to developing countries [22].

However, not all developing countries have been able to benefit from the rapidly expanding flows of foreign direct investment, and towards the end of the 1980s, only five developing economies accounted for about 80 per cent of foreign direct investment flows: China (24 per cent), Brazil (18 per cent), Mexico (17 per cent), and Egypt and Malaysia (10 per cent each). The reasons why most of the poor countries in Africa, Asia, Latin America, and the Middle East have not been able to attract direct foreign investment are many and varied and include their remote geographic location in relation to the main export markets, the relatively small size of their domestic markets, deficiencies in physical and institutional infrastructure, lack of a skilled workforce, and inadequate investment incentive regimes. An indirect result has been the inability to benefit from the transfer of technology, marketing, and managerial capabilities that are associated with direct foreign investment.

There have also been changes in the direction and content of international trade, such as the emergence of the North Pacific as the world's largest trading area (with the North Atlantic taking second place), the halting movement towards worldwide trade liberalization (best exemplified by the on-again off-again GATT negotiations), the rise of regional trading blocs (Europe after 1992 and the North American Free Trade Agreement), and the shift in the content of international trade against primary commodities (exported primarily by developing countries) and in favour of high technology services and manufactured products (typically industrialized country exports). A new web of commercial linkages between transnational corporations

– covering manufacturing, finance, trade, and services – has now emerged, of which strategic alliances in pre-competitive research and development are a prime example.

In addition, we have seen completely new situations in several key countries and regions that affect the world economy significantly. During the 1980s, for the first time in recent history, the United States became a net debtor; Japan has become a dominant economic and financial actor on the international scene; Europe is rapidly moving towards economic, and maybe some form of political, unity; the USSR has dissolved and its republics are undergoing a painful transition towards market economies, a path followed earlier by central and eastern European countries; Latin America has weathered the debt crisis of the 1980s, initiated policy reforms, and appears poised for renewed economic growth after a decade of stagnation; the worsening situation in Africa has reversed the precarious gains of the preceding three decades; continuous instability and strife plague countries in the Middle East; and in Asia a few newly industrialized economies continue to grow rapidly, India and China are experimenting with economic policy reform and liberalization, while other countries in the region begin a difficult process of reconstruction after decades of war.

The range and diversity of possible outcomes in practically all aspects of the international economy appear much larger during the 1990s than at any time during the last three decades. Growing interdependence has created an international economic environment that transmits disturbances and magnifies disruptions. Technological advances in telecommunications and information sciences have contributed to this (witness the impact of computer trading in stock markets), while the absence of effective international rules and institutions to regulate financial and trade flows – and the limitations of economic policy coordination among the world's leading economies – have helped to increase uncertainty.

Social demands
The explosive growth in social demands in the developing regions has been largely triggered by population increases during the last 30 years. Coupled with a significant slow-down in population growth in the industrialized nations, this has led to a highly skewed worldwide distribution of social needs and of the capabilities to satisfy them.

The dynamics of population growth strongly condition the demand for food, education, employment, housing, and other social goods

[6]. Food and nutrition demands have multiplied many times over, particularly in the poorest countries, and although world aggregate food production is sufficient to provide each and every human being with adequate nourishment, existing political, social, and institutional arrangements – at both the national and international levels – have proven incapable of doing so. Armed conflicts, droughts, and natural disasters have conspired to make it even more difficult to ensure access to food in many developing countries.

Demand for basic health care and elementary education expanded at a rapid pace during the last three decades, as developing countries made efforts to improve the provision of these services to growing populations. Migration and accelerated urbanization created huge demands for housing, sanitation, transportation, and energy supply – a situation that adds unmet urban needs and widespread urban poverty to the deprivation that characterizes rural populations throughout the developing world.

Unemployment has emerged as perhaps the most troublesome and persistent problem in developing countries. This is also a growing issue in industrialized countries, where technological change seems now to depend so heavily on capital that unemployment appears to have become one of the new structural characteristics of economic growth for the foreseeable future, if not forever. If there are reasons to be anxious about the outlook for employment in the industrialized countries, there are few grounds for optimism about most developing countries. Here the jobs created by the new technical system are generated against a background of non-employment, and the promises of a production system that will be more and more based upon robots and fully automated factories may increasingly conflict, in view of demographic trends, with the job expectations in developing countries. The spread of the new technologies is already transforming the very nature of work and leisure, creating jobs that are less and less like traditional tasks, although it is precisely these traditional tasks that still offer the highest number of employment opportunities in developing countries. The inability of the modern sectors of their economies to absorb new entrants into the labour force has led to a variety of "informal" arrangements for workers to earn their means of subsistence. Developing countries face the difficult challenge of raising labour productivity while at the same time absorbing the growing number of entrants into the labour force.

A significant drop in the population growth rate of industrialized

16

countries is to be expected during the 1990s, from an average 0.5 per cent per year in the 1980s to only 0.3 per cent in the 1990s. This implies a rapid rise in the number of elderly people (particularly in Japan and Germany), a significant increase in the ratio of dependents (children and old people) to workers, and a further shift in the balance of world population. Ageing in industrialized nations will have a major impact on the demand for social services, as well as important consequences for the patterns of consumption, employment, and savings and for the direction of technical progress.

In developing countries the rapid pace of population growth is expected to continue through the 1990s, although at a moderately slower pace than in the 1980s – from the present rate (i.e. 1992) of 2.0 per cent per year to 1.8 per cent per year during the next decade. As a consequence, youth will remain by far the largest segment of the population in most of these countries, whose economies must expand at rates significantly above those of population in order to satisfy the growing demand for work.

Population imbalances could pose the problem of uncontrolled mass migration from developing to industrialized countries, threatening social cohesion and international solidarity. In some western European countries there is already a backlash against "foreigners," although the fear of massive inflows of workers from the east has failed – as yet – to materialize. In Asia, migration pressures are likely to build up as a result of the growing demographic imbalance between Japan and the poorer, overpopulated countries of the region. Despite the increased participation of women in the labour market, Japan will experience a decline in the labour force after 2000, and labour shortages will be compounded by moves to reduce the number of working hours [11].

The role of human capital and technological capabilities will become even more important as a major determinant of long-term growth in the developing countries in the 1990s. The level and quality of investments in human resources will have to rise significantly during this period in order to deal with the rapid rise in the number of young people, and also to enable the labour force of developing countries to utilize new technologies that increase productivity.

Environmental concerns
During the 1970s and 1980s environmental concerns have risen to the top of the international public policy agenda. There is now greater

17

awareness of the limits that the regenerative capacity of natural ecosystems imposes on human activities, as well as of the dangers of the uncontrolled exploitation of natural resources (the sea, forests, land, rivers) and from overloading the capacity of the earth to absorb waste (air and water pollution, acid rain, toxic and nuclear wastes). The 1980s witnessed the emergence of truly global environmental problems, such as depletion of the ozone layer and global warming, that underscored the possibility that unforeseen ecological instabilities could cause irreversible environmental damage.

The problems of environmental sustainability and resource use are closely related to population growth and poverty in the developing countries, and to the often wasteful consumption habits of rich nations. Major changes in lifestyles will be essential in both groups of countries to address successfully the problem of environmental sustainability in the transition to the twenty-first century. According to the World Bank,

the most immediate environmental problems facing developing countries – unsafe water, inadequate sanitation, soil depletion, indoor smoke from cooking fires, and outdoor smoke from coal burning – are different from and more immediately life-threatening than those associated with the affluence of rich countries, such as carbon dioxide emissions, depletion of stratospheric ozone, photochemical smog, acid rain and hazardous wastes. [22, pp. 2–3]

The Earth Summit in Rio de Janeiro endorsed "Agenda 21," a wide-ranging world programme of action to promote sustainable development, but the negotiations exposed the divergence of perspectives between industrialized and developing nations on approaches to sustainable development [21]. Questions of lifestyles, national sovereignty, barriers to trade and financial assistance, in addition to access to less-polluting technologies, are now at the centre of the debate on sustainable development (see the contribution of Ignacy Sachs in this volume).

As a consequence of the greater importance of environmental concerns, access to development assistance during the 1990s will be increasingly linked to the attainment of environmental objectives. Another result is that some industrialized countries – notably Japan and Germany – are positioning themselves to compete in what will be one of the most dynamic markets of the future: that of environmentally sound technologies. Being able to deliver "green" technologies could soon become a source of competitive advantage in the global search for new markets.

Cultural transformations

Three powerful cultural forces are shaping the international scene in the transition to the twenty-first century: the growing importance of religious values and the rise of fundamentalism as a main driving force of economic and political actions in many parts of the world; the tensions between cultural homogenization pressures brought about by the pervasive influence of mass media and the desire to preserve cultural identity; and the emergence of moral and ethical issues at the forefront of choices about inter- and intra-generational equity, particularly in relation to the environment, income distribution, and poverty reduction, and the new biomedical technologies. It is not a coincidence if, in most industrialized and in some developing countries, special commissions have been instituted, often at the level of the legislative branch but also in the framework of advisory bodies independent from the executive, in order to anticipate and assess the impact of technical change and sometimes even that of scientific discoveries: offices of technology assessment, commissions on biomedical ethics, on freedom and information sciences, on the prevention of technological risks, etc. An increasing number of fields call for mechanisms of regulation so as to correct, limit, and if possible avoid the negative or unanticipated effects of scientific and technological activities. These institutional innovations on the political scene reflect a change of values in societal reactions, at the national and international level, to progress.

The revival of religious and spiritual concerns has been a characteristic of the 1980s and 1990s, which have witnessed the renaissance of Islamic values in North Africa, the Middle East, and Central Asia; a revival of the Orthodox Church in eastern Europe and the former Soviet Union; the spread of evangelical churches in Latin America and other developing regions; a surge of popularity of the Pope; the growing influence of Christian fundamentalism in US political life; and the renewed interest in mysticism and Oriental religions, often associated with "New Age" movements that eschew rationality. This revival points to the fact that, because of the overriding concern with improving material well-being and standards of living, the spiritual dimensions of human development have been neglected during the period after the Second World War.

As a consequence of the globalization and pervasive influence of mass media – a direct result of technological advances in communications during the 1970s and 1980s – two contradictory cultural forces can now be seen at play: pressures towards the standardization of aspirations and cultural values throughout the world, and the desire to reassert individuality and preserve cultural identity. These two contradictory forces create cultural tensions and emotional stresses, particularly in developing countries, where the images of affluence brought by television programmes from industrialized nations contrast sharply with the harsh reality of mass poverty – and with the fact that those worlds of plenty are simply unattainable for the vast majority of the population.

Moral and ethical questions, once the province of academics and religious activists, are finding their way into public debates on the rights of future generations in relation to sustainable development and on issues such as racism, abortion, corruption, crime, and drugs. A renewed concern with human rights throughout the world has led to a questioning of the principle of non-intervention in the internal affairs of states where governments do not respect basic human rights. Finally, reversing the trend that prevailed during the 1980s, equity considerations are finding their way onto the political agenda of many industrialized and developing countries, at the same time that the moral and ethical aspects of technological change and economic behaviour have begun to receive greater attention.

Against this background of fundamental changes in the international context, North-South cooperation is likely to remain a peripheral concern of industrialized countries, especially as they focus their attention on their own internal problems, on coordinating economic policies, on improving competitiveness, and on easing the transition of the former Soviet Union and of eastern Europe towards market economies.

As prospects for greater resource flows to developing countries appear doubtful, policy reform, structural adjustment, and the mobilization of science and technology for development objectives will take place in a resource-constrained environment. This will test the political will of governments to embark in the uncertain and long-term enterprise of building science and technology capabilities, particularly when facing a multiplicity of urgent short-term needs.

Modernity and the uncertain quest

In 1963, C.P. Snow [19] wrote an essay on the "two cultures," calling attention to the differences that exist between scientists and literary intellectuals, deploring the lack of communication and understanding between them, and making a strong plea for the emergence of a more integrated culture in which the humanities and the sciences would contribute equally and grow through mutual interaction. However important the differences and lack of communication between Snow's "two cultures" – which may indeed have increased as a function of the growing impact of scientific methods and activities on all societies – they have been overshadowed by the even more profound and disturbing material differences between the rich and the poor nations of the world. Indeed, Snow made reference to these glaring inequalities and attributed their existence in part to the inability of the West, with its divided culture, to grasp their magnitude and to understand the need for urgent and profound structural transformations of a social, economic, political, and cultural character.

It is obvious that the end of the twentieth century and a great part of the new one will be dominated by the growing gulf between the industrialized and the developing countries, in so far as one can speak of "two civilizations" rather than of several worlds. The concept of the third world emerged as both a third element and a buffer more or less manipulated by and manipulative of the two rival blocs, communism and capitalism, that confronted each other after the Second World War. Now that communism has admitted defeat, the notion of the third world is all the more meaningless in that most of the former communist countries have created a new category, that of industrialized countries that have become in their turn newly developing countries. Moreover, developing countries do not constitute a homogeneous category, and the need to distinguish various levels of development and even underdevelopment is more pertinent than ever.

The world is still divided into two civilizations that interact strongly, although the interaction is one-sided: the second civilization is dependent and deeply affected by the first and lacks the capacity of influencing it to the same degree. The first civilization is based on the growth of science as the main knowledge-generating activity, the rapid evolution of science-related technologies, the incorporation of these technologies into productive and social processes, and on the emergence of new forms of working and living deeply influenced by the *Weltanschauung* of modern science and science-related technologies. The second civilization is characterized by the lack of a capacity to generate scientific knowledge on a large scale and by a passive acceptance of scientific results generated in the first; by a technological base that comprises a substantive component of traditional technologies and a veneer of imported ones; by a productive system

21

whose modern segment is dependent on the expansion of production in Western industrialized nations and on the absorption of imported technology and whose traditional segment vegetates and is based on an often stagnant traditional technological infrastructure; and by the coexistence of disjointed and even contradictory cultures.

The first civilization, corresponding to the developed, or highly industrialized, countries, has an endogenous scientific and technological base. This base is still present, in spite of its current difficulties and disruptions, in the former communist countries of eastern Europe, and one of the most important agreements signed in 1992 between the United States, the European Community, and Japan was intended to help these countries to keep this base alive. This comes down to helping somebody not to sink when he or she already knows how to swim. The second civilization is not swimming, but struggling to stay afloat, with the exception of a handful of countries that have recently succeeded in catching up with some of the best swimmers in the first civilization. The great majority of the countries in the second civilization are not only lagging behind but lack, above all, most of the basic ingredients – in terms of resources, institutions, manpower, and cultural background – indispensable if they are to benefit from scientific knowledge and new technological innovations. The historical reasons for this situation deserve to be carefully studied in these countries by local scholars and should be made a part of science studies and research programmes, which would help policy makers and society at large become more aware of the internal and external conditions that have jeopardized – and still jeopardize – the development, if not the emergence, of a scientific and technological capacity. It is hoped this sourcebook will contribute to a better understanding and thus a greater mastery of all these conditions.

Development is an uncertain quest in which the seekers rely heavily on science and technology. The quest is uncertain not only because there is no prior guarantee of success (nor that it will be lasting), but above all because it raises questions about the *price of modernity*: the *benefits* that a country can expect to derive from it, in political, economic, social, and cultural terms, as well as the *sacrifices* that it is prepared to make on its behalf. Development is not a neutral process with no impact on the social structures that are involved; science and technology do not always bring about improvements to those areas that they affect. In short, despite what was promised by the rationalism of the Enlightenment and even more by the positivism of the nineteenth century, scientific and technical progress does not necessarily coincide with social or moral progress.

Since the beginning of the Industrial Revolution, economic progress has meant upheavals. Schumpeter agreed with Marx at least in this regard, and stressed the "revolutionary character" of industrial capitalism, which leads to the obsolescence, destruction, and renewal of economic and social structures. This is what is involved in innovation, and now that innovation is worshipped as the driving force of international competitiveness, it is important to recognize that it always has a price attached: technical change is accompanied by social change. As Schumpeter rightly said, "No matter how many more stagecoaches you have, you will not thereby acquire railways," and he emphasized that economic growth is a process of change that is constantly revolutionizing economic institutions from within, destroying the parts that are out-of-date and creating new ones in their place. What happens is not that more stagecoaches are added to the existing stock, but they are replaced with railways in a process of "creative destruction" [18].

Economic development has growth (i.e. a sustained increase in national income) as a corollary, but growth in quantitative terms does not necessarily mean development. From the start of the Industrial Revolution, and especially since the pace of technical change has quickened thanks to the growing cross-fertilization of science and technology, people in the industrialized countries have been pondering the gap between wisdom and strength. The issue of how to bridge that gap is constantly raised by modernity, and the economic implications naturally have philosophical dimensions.There has to be choice at least regarding the importance attached to tradition, to its structures, hierarchies, codes and rites, as against rationalization, with its constraints, order and disorder, its capacity to transform and destroy. As Alain Touraine [20] has pointed out, scientific and technical thinking threatens to reduce human beings to purely instrumental rationality, while attacks on rationality from the viewpoint of particular faiths, traditions, or communities threaten to retard or even prevent any change by searching for compensations for the present in a mythical past. To bring together the economic vision and the cultural one involves the same difficulties as making a bridge between the particular and the universal, or between facts and values.

The developing world has forced the industrialized countries to recognize not only that their cultures are extremely diverse, but that that diversity is perfectly legitimate. Both sides have learned, too, that development cannot take place without dialogue between cultural heritage and instrumental rationality, even if the two cannot be entirely reconciled. In the upheavals marking the end of the twentieth century, especially after the collapse of totalitarian ideologies and regimes, the whole world is in quest of new paths and alternatives leading to a better social order. And just as the developing countries are having to take on board some of the aspects of modernity that they used to criticize, so the industrialized countries are having to restore some aspects of tradition that they used to challenge. No society, clearly, can ever again impose its own values or development model on any other. Science and technology can contribute a great deal to development, but they cannot do everything, and above all they do not offer a ready-made solution to the problem of values that is raised by the clash between tradition and modernity. Modern societies have realized that they can no longer place their trust in progress as people thought in the Enlightenment. But while nobody can believe any longer that growth necessarily brings with it greater democracy and happiness, everybody knows now that development requires growth and a certain degree of rationality: not any longer confidently relying on technical or administrative efficiency alone, but rather on an awareness and a mastery of the consequences of scientific and technical change.

References

1   Bhalla, A.S, and D. James, eds. *New Technologies and Development: Experiences in Technology Blending*. Boulder, Colo.: Lynne Rienner, 1988.

2   Braudel, Fernand. *Civilisation matérielle, économie et capitalisme, XVè-XVIIIè siècles*, 3 vols. Paris: Colin, 1979.

3   Bugliarello, G., and D.B. Doner, eds. *The History and Philosophy of Technology*. Chicago: University of Illinois Press, 1979.

4   Fardoust, Shahrokh, and Ashok Dareshawar. *Long-Term Outlook for the World Economy: Issues and Projections for the 1990s*. International Economics Department, Working Paper Series, no. 372. Washington, D.C.: The World Bank, 1990.

5   Freeman, Christopher. *The Economics of Industrial Innovation*. Harmondsworth: Penguin Books, 1974.

6   Gowariker, Vasant. *Science, Population and Development: An Exploration of Interconnectivities and Action Possibilities in India*. 7th ed. Pune: Unmesh Communications, 1992.

7   ICSPS (International Council for Science Policy Studies). *Science and Technology in Developing Countries: Strategies for the 90s, A Report to UNESCO*. Paris: Unesco, 1992.

8   Johnston, Ann, and Albert Sasson, eds. *New Technologies and Development*. Paris: Unesco, 1986.

9   Landes, David S. *The Unbound Prometheus: Technological Change and Industrial Development in Western Europe from 1750 to the Present*. Cambridge: Cambridge University Press, 1969.

10   Myrdal, G. *Asian Drama: An Inquiry into the Poverty of Nations*. New York: Pantheon Books, 1969.

11   OECD. *OECD Forum for the Future: Conference on Long-Term Prospects for the World Economy, 19–20 June 1991*. Paris: OECD, 1991.

12   Rosenberg, Nathan. *Perspectives on Technology*. Cambridge: Cambridge University Press, 1976.

13   ———. "Technology, Economy, and Values." In: Bugliarello and Doner, eds. *See* ref. 3.

14   ———. *Inside the Black-Box: Technology and Economics*. Cambridge: Cambridge University Press, 1982.

15   Salomon, J.-J. *Le destin technologique*. Paris: Balland, 1992.

16   Salomon, J.-J., and A. Lebeau. *Mirages of Development*. Boulder, Colo.: Lynne Rienner, 1993. Originally published in French as *L'écrivain public et l'ordinateur*. Paris: Hachette, 1988.

17   Schumpeter, J. "The Communist Manifesto in Sociology and Economics." *Journal of Political Economy* 57 (June 1949): 199 – 212.

18   ———. *Capitalism, Socialism and Democracy*. London: Allen and Unwin, 1950.

19   Snow, C. P. *The Two Cultures: A Second Look*. New York: Mentor Books, 1963.

20   Touraine, Alain. *Critique de la modernité*. Paris: Fayard, 1992.

21   UNCED. *Final Report of the United Nations Conference on Environment and Development: Agenda 21*. Rio de Janeiro, July 1992.

22   World Bank. *Global Economic Prospects and the Developing Countries – 1992*. Washington, D.C.: World Bank, 1992.

25

Part 1: Science, technology, and development

Part 1 sets the scene. Jean-Jacques Salomon first reviews the emergence of modern science: its successive institutionalization, professionalization, and industrialization. The fact that this process tended to happen in a different order in developing countries raises particular problems for them. In recent years, in industrialized countries the expansion of modern science and technology has gone hand in hand with the rise of science policy – that is policy *for* science and policy *through* science – as a result of increasing concern about the impact of advances in science and technology on society. Science is linked to the state, and in the context of the Cold War, there was a full-scale mobilization of scientific research. It is impossible to underestimate the importance of the innumerable innovations generated by economic competition and by defence-related R&D during this period, and especially the role they played in the conception and development of the new technologies that characterize the "new technical system" now flourishing. In an era of increasing international competitiveness, innovation rests on a much wider range of actors, institutions, and issues, raising a lively debate on the role of the state: how far should it intervene, under what circumstances, and on what criteria? The chapter ends with a discussion of the universality of science and the coexistence and complementarity of rationalities, which may challenge Western science as a unique model, but not its operational effectiveness.

What is development? Nasser Pakdaman traces the evolution since the Second World War of the ideas, theories, and practices that have lain behind the efforts of the third world countries to emerge from "underdevelopment." The patchiness of their success – indeed, the frequent failures – has led some commentators to refer to the rise and fall of development economics, as if the subject were bound to disappear so that others could rise, like a phoenix, from its ashes, relying increasingly on an ever wider range of social sciences (sociology, anthropology, history, etc.). There is now a better understanding of the factors leading to economic growth, but there is still no clear definition of what constitutes economic development, beyond the fact that it involves a process of gradual transformation over the long term, and the ingredients are never exclusively economic. The current preference for "sustainable development" arises out of an awareness that the pure "economic paradigm" has its limits, whether inspired by the Left or the Right, and that economic theory and practice must abandon the illusions of rapid "take-off" or "catching up" and instead fit in with the historical realities that shape the specific characteristics – and constraints – of each country.

Though it is difficult to achieve international comparability using existing R&D and innovation indicators, Jan Annerstedt attempts to provide from the existing statistics a comprehensive picture of measurements of science, technology, and innovation, stressing the uneven relationship in R&D spending: in 1988–1989, the third world had a little more than 4.5 per cent of total R&D funds, with considerable differences among developing countries. A proposed worldwide science and technology–related typology identifies countries (a) with no science and technology base, (b) with the fundamental elements of a science and technology base, (c) with a science and technology base well established, and (d) with an economically effective science and technology base, notably in relation to industry. Finally, the author argues that to develop policies that could avoid further marginalization in foreign investment and technology transfer, the developing countries need much more detailed and statistically grounded analyses of the role of science and technology in the globalization process, and he reviews the innovation indicators in the making.

1

Modern science and technology

Jean-Jacques Salomon

The emergence of modern science

It has been said that all the old scientific movements of all the different civilizations were rivers flowing into the ocean of "modern" science [31]. Modern science has its roots in a past that is extremely diverse in both time and space, ranging from the earliest civilizations of Asia, Mesopotamia, Egypt, to the "Greek miracle," through the Judeo-Christian, Arab, and scholastic traditions. However, science as we understand the term is a relatively recent phenomenon. A major advance occurred in the seventeenth century, an advance so different from all previous ones that it can be called an unprecedented "intellectual revolution."

Gaston Bachelard [1] has labelled it an epistemological breakthrough and Thomas Kuhn [19] a paradigm shift. Either way, this turning-point was of even greater historical significance because it began in Europe and developed almost exclusively there for several centuries. The economic and social transformations coming in the wake of the invention of printing and the enormous stimulus to curiosity provided by the "great discoveries" and accompanying this scientific revolution helped to ensure, strengthen, and speed up the expansion of Western civilization relative to all the others. It is not surprising that the history of Western science has often been written as a history of conquest, and oversimplified in such a way that science has featured as an agent of European colonialism or as a residual feature of post-colonial imperialism. Yet history is no less complicated than is the concept of a scientific revolution [7].

Modern science did not happen in a single day – it took time to make an impact on people's thinking and on institutions, with added difficulties because, when experimental science started, most facts were still so uncertain that speculation had a field day. Furthermore, some of the most innovative thinkers (such as Kepler and Newton) in many respects belonged to the old order, half in the modern era through their radical contributions to astronomy, but half in the past because of their links with hermetics, mysticism, or astrology. In a system of thought that had not freed itself from alchemy nor from the bookish tradition handed down from Aristotle, the spread of new ideas was hindered by strong resistance, resulting from a combination of prejudice, dogma, and habits. The scientific revolution of the seventeenth century has generated a huge literature, which is constantly being reinterpreted and reassessed [24].

"Nature is expressed in mathematics": Galileo's famous phrase appeared in his *Saggiatore* in 1623; it marks symbolically the break with the ancient notion of Nature as an ensemble of substances, forms, and qualities and suggests instead a completely different conception in terms of quantitative phenomena that can by definition be measured and therefore potentially controlled. This "intellectual reform" led not only to the transformation of science – which gradually developed into a range of many and varied sciences, each of them in turn splitting up into more and more specialized subdisciplines – but also to one of perceptions, structures, and institutions. The break between arts and crafts and science reflected a break in the social order and hence a class distinction; technology, until then reserved for the "servile class," becomes the indispensable collaborator of speculative science, which had been reserved for the "professional class." This nearing of theory and practice is a revolutionary turn at both the intellectual and the social level. For the old saying, "to know is to contemplate," a new one was substituted: "to know is to act, to manipulate, to transform" – *knowledge is power*, in Bacon's phrase. And by the same token, the technician's know-how is to be closely associated with the scientist's theoretical way of thinking and doing.

The process of the creation, expansion, consolidation, and success of modern science has had three distinct phases: institutionalization, professionalization, and industrialization. In all the industrialized countries these phases occurred in the same historical sequence and took several centuries, whereas in the developing countries – most of which became independent nations only very recently – they have often occurred in a different order, with professionalization starting

before institutionalization, or even industrialization before professionalization. The problems of the scientific and technological systems in many of these countries, like the lack of social recognition of
their scientists and research institutions, can often be largely attributed to this hasty development, which frequently occurs without the
benefit of any previous scientific tradition and within a few decades in
circumstances very different from those of the industrialized countries.

The institutionalization of science

Bacon, in his utopia *New Atlantis* (1627), already envisaged scientific
research as a public service, taking in most of the functions that it in
fact acquired between his day and ours: research would become a
profession, managed by administrators, the subject of political
decision-making, requiring funding and choices to be made; it would
yield usable and useful results; it would be responsible for informing
and educating at all levels, drawing on a wide range of specialists,
from researchers to administrators, even to scientific attachés, whose
brief would be both to make known a country's discoveries abroad
and to monitor – if not spy on – developments elsewhere. The link
that modern science established between theory and practice creates
a power to act inseparable from its power to explain.

Institutionalization began in the scholarly communities of the
Academies, the first ones appearing in Italy: they distanced themselves from both Aristotelian science (grammar, rhetoric, and logic)
and from other institutions (political, religious, philosophical), which
did not share their exclusive concern with "perfecting the knowledge
of natural things and of all useful arts . . . by experiment," to quote
the charter of the Royal Society (1662). Herein lies the origin of both
the secularization of the modern world – the differentiation of the
sphere of scientific proofs and facts from that of faith and conviction
– and the reductionist, positivist, or even "scientistic" leanings of
some scientists. One can also see in the stance of the Academies the
beginnings of the conflicts that science has had ever since Galileo
with authorities who thought they could impose their beliefs, contrary
to scientific theories and scientifically established facts. Indeed, little
has changed since Galileo wrote to Christina of Lorraine that to in-
terfere with the work of researchers "would be to order them to see
what they do not see, not to understand what they understand and
when they seek, to find the opposite of what they find."

Nevertheless, from the outset, the scientific establishment has

31

been linked to those with political power, demanding their protection and support and in return providing useful and usable results. The style of institutionalization naturally varied according to the national context. The Académie Royale des Sciences in France was created by Louis XIV's minister Colbert and kept under tight royal control; its members received salaries and the state treasury allocated 12,000 livres per year for equipment and experiments; and certain foreign scholars (such as Huyghens and the Cassinis) were hired abroad for huge salaries – an early example of an organized "brain drain." By contrast, the Royal Society in London enjoyed purely formal official support and until 1740 had an annual budget of less than £232, mainly contributed by the Fellows, and only two official appointments. Both, however, were eager to gain recognition through services rendered to the state, e.g. by solving the problem of calculating longitude at sea, a major strategic concern for which the maritime nations offered substantial rewards [27].

The process of institutionalization spread throughout the seventeenth and eighteenth centuries. The laboratories attached to the academies provided a new setting, outside the universities, for the activities of researchers and the development of new ideas. But institutionalization did not yet mean professionalization, even though the members of the Paris or Berlin academies received salaries. The membership was still limited to a tiny élite, many of which were active in politics, the army, or the church rather than engaged in scientific research. Institutionalization helped to foster the "role of the scientist as researcher," but this role was just starting to develop and was far from achieving social recognition [2].

The professionalization of science
A profession is a legally recognized occupation, usually offering a lifetime career path as well as a livelihood. Scientific research began to achieve this status in the early nineteenth century, but did not do so fully until the eve of the Second World War. The Ecole Polytechnique in France started the process: it provided for the first time technical training involving both a research laboratory and teaching by specially appointed professors (e.g. Monge). However, Polytechnique soon concentrated on teaching rather than on "science in the making," and its graduates became senior civil servants rather than research scientists. The German chemist Liebig, a graduate of Polytechnique, introduced the model to his university in Giessen, whence it spread throughout the Continent. Research became the

purview of (professional) university teachers rather than of (amateur) academicians. Humboldt's reform of the university was very much along these lines, and made scientific research an integral part of the university's responsibilities. Merely to possess knowledge and transmit it was not sufficient; the university must also create knowledge.

These developments were reflected in the changing membership of the Royal Society: the number of academic scientists more than doubled between 1881 and 1914, when they made up 61 per cent of the total, while other categories such as "distinguished laymen," soldiers, and clergy were drastically reduced. First used by Whewell in 1840, the term "scientist" came to replace "natural philosopher" or "savant," first in the English-speaking countries, a century later elsewhere. Indeed, the language and activities of science had become incomprehensible to anyone who had not had the appropriate training. New specialisms, disciplines, and subdisciplines proliferated and generated their own networks of institutions, journals, and meetings. The number of researchers grew enormously: not just scientists, but engineers and technical experts, increasingly working in teams or groups, often outside the universities in public or industrial laboratories, or for defence establishments. As in any profession, the growth in numbers led to fierce competition for recognition and hence resources and survival. James Watson gives a very personal and vivid account of the discovery of the genetic code in *The Double Helix* [59], describing the ruthless behaviour often required to be recognized as one of the top research teams in the world and to achieve the ultimate accolade, the Nobel Prize. The American catch-phrase, "publish or perish," is another example of the distortion of the scientific ethic brought about by competition within a worldwide scientific community, where the "credit" attached to results produced and published to gain fame also determines the financial "credit" that all research programmes require to survive.

The process of professionalization implies membership in a community, with its own rules and initiation rites and tests for entry and continued acceptance. The scientific community in fact has a double role: communication and regulation. It is responsible for disseminating the results of work in progress, as well as publicizing and promoting science, both within its own ranks and outside, to decision makers and the general public. It also looks after scholarly exchanges, sanctions qualifications and research projects, sees to the promotion of researchers and honours them with prizes and grants. In institutional terms, these functions are carried out by the Academies, learned societies, "peer review committees," boards of examiners, and juries. The basic qualification for the researcher is the doctorate, which originated in Germany in the mid-nineteenth century and is now the standard entry requirement for the profession.

In basic research, unlike technological research, scientists are expected to share their results freely with the rest of the scientific community. Progress occurs through and depends on publishing results and on cooperation that by definition transcends national and ideological boundaries: it is indeed a matter of "public knowledge," where the norms set the conditions for working in the field, just as they do for advancing knowledge and know-how [64]. In return, scientists expect to receive additional resources in order to continue their work, perhaps leading to further and more substantial recognition. There are indeed certain similarities with the process of canonization by the Church, except that the candidates are alive and the cursus of honours (publication in prestigious journals, membership of learned societies, national and international prizes, etc.) helps them to advance in their careers. Kuhn [19] has shown that professionalization in the natural sciences is inseparable from this regulatory role of the scientific community. If science is able to advance, it is precisely because the learning process depends on the publication of current research efforts in a given field. A scientific revolution occurs when a new "paradigm" is adopted, obliging the community to throw away the books and articles produced on the basis of the previous paradigm. There is no equivalent in scientific education of the art museum or the library of classics. Whereas in the arts or social sciences one cannot ignore the work of the great names of the past – the writings of Plato or Weber are still a fundamental element of discussions in philosophy or sociology – a modern student of physics is not required to read Newton, Faraday, or Maxwell.

Finally, the process of professionalization not only leads to recognition of status in the abstract, but also (perhaps above all) involves socially sanctioned rewards in terms of income and resources directly linked to the activity of research. This social legitimation occurred earlier in the United States than in Europe, just after the First World War. As Ben-David [2] has pointed out,

The requirement of a Ph.D. made suitable candidates scarcer, and raised thereby the market value of those who possessed the degree. But its principal effect was to create a professional role that implied a certain ethos on the part of the scientist as well as his employer. The ethos demanded that those who received the Ph.D. must keep abreast of scientific developments, do research, and contribute to the advancement of science. The employer, by employing a person with a Ph.D., accepted an implicit obligation to provide him with the facilities, the time, and the freedom for continuous further study and research which were appropriate to his status.

In Europe in the interwar period, scientists had great difficulties in convincing governments to recognize their role as researchers. In fact, research activities still appeared there to be an end in themselves – a calling rather than a productive function – in the context of a university culture, insulated by its institutions and context from community problems and mundane affairs; they were kept on the fringe of university functions and remained there for such a long time that Jean Perrin, Nobel prizewinner in physics, could say, as late as 1933, that "the use of university grants for scientific research is an irregularity to which the authorities are prepared to turn a blind eye" [52].

It was only after the Second World War that the function of scientists devoting themselves full time to research came to be fully recognized in most of the capitalist industrialized countries, with negotiable salaries. In the United States, this negotiation takes place on the basis of individual contracts, whereas in countries such as France, it is part of the standard negotiations with trade unions and professional organizations relating to conditions in the public service. Whatever the system, however, research has joined the general category of professions that provide their members with their livelihood. This stage would probably not have been reached as fast or on the scale that it has without the stimuli of developments in industry and of deliberate policies for science and technology launched after the Second World War.

The industrialization of science
The industrialization of science should not be confused with industrial research. The latter dates back to the mid-nineteenth century and merely brings together the laboratory and the factory. Industrialization means the development of big equipment and the application of industrial management methods to scientific activities themselves. This stage of "big science" [43] occurred only between the world wars and increased rapidly after 1945. In fact, science and technology had relatively little contact with one another until the middle of the nineteenth century; and technology contributed to science (via scientific instruments) rather than vice versa. As is well known, the Industrial Revolution was not closely linked to science at the outset,

but rather was produced by craftsmen and engineers, often trained on the job. The most famous example is the steam engine, which was invented almost a century before the principles of thermodynamics were understood.

The turning point came again thanks to Liebig, who brought about the creation of "applied science" in Germany with the exploitation of advances in organic chemistry in the dyeing industry between 1858 and 1862. Von Baer's team, working on the synthesis of indigo, was given direct support by the Badische Anilin und Soda Fabrik, which invested almost £1 million in both research and development, i.e. establishing the chemical reactions required on a large scale prior to commercial production. Similarly, Menlo Park, created by Edison in 1876, was the first R&D laboratory in electromechanics and one of the first instances of substantial venture capital being invested by banks hoping to profit from future inventions. Edison did not so much mark the end of the heroic age of great inventors as the beginning of science-based technology. A self-taught experimenter rather than a scholar himself, he brought to Menlo Park scientists and technicians trained in the best European institutions.

Industrial research soon spawned a new type of entrepreneur, entrepreneurs with science degrees from universities and engineering schools, who were employed by industrial firms or who themselves started new industries. It is important to realize that these developments depended on special conditions whose absence in developing countries often explains their difficulties in properly integrating scientists and laboratories into the production process. For industrial research to flourish, there must already be a layer of relatively mature and varied industries, and the industrialists themselves need to have an adequate scientific background that they can bring to bear on both management and production. There must also be a pool of scientists willing to undertake "directed" research on the problems facing firms, with the aim of producing commercially viable results within a reasonably short time [6]. In some specific cases of scientific research (elementary particles, fusion, astronomy, space research, genome) no progress is conceivable without a critical mass of manpower, equipment, and institutions. These prerequisites could not be satisfied in Europe until the beginning or even the middle of the twentieth century. The Industrial Revolution was accompanied by essential transformations of higher education: the combination of research and teaching, the creation of new specialisms, the modification of university structures in line with changes arising from scientific progress, but also the introduction of university-industry contracts and the increasing recruitment by industry of university-trained scientists.

The industrialization of research – and even of science itself – is the most recent development, dating back to the aftermath of the First World War. The system for supplying weapons, transport, food, and health care (the first vaccines) set up in order to wage the war provided a model for the rational management of technology in terms of organization, discipline, standardization, coordination, separation of line and staff, etc. [47, 48]. The First World War did not so much create new weapons as adapt existing civilian technologies for military purposes (automobiles turned into armoured cars, aeroplanes into bombers, etc.). It was the first war where the outcome was determined by success in maintaining a constant supply of *matériel*, of machines as much as munitions, and also the first where military operations started to be mechanized and submitted to scientific management. The basic principles underlying the American and European industrial systems with regard to machine tools, spare parts, standardization, and mass production were then extended from the military to the civilian economy via the armies' suppliers: Taylorism and Fordism thus had their first applications [25]. The changes begun in the interwar period, most vividly illustrated by the creation of enormous industrial laboratories such as Bell Laboratories or Du Pont de Nemours in the United States, were considerably strengthened during and just after the Second World War, which was the immediate stimulus for new weapons systems (the atomic bomb, radar, computers, jet engines, rockets, etc.), sanctioning the shift to "big science" as well as "big technology." The links between science and technology became so close that their advance became increasingly interdependent.

The characteristic feature of this stage is that science became increasingly capital-intensive, dependent on huge investments in manpower and specialized equipment. This was partly because research programmes were far more expensive than before and partly because the research programmes were also far more ambitious in terms of both scale and expectations of quick results. "This change is as radical as that which occurred in the productive economy when independent artisan producers were displaced by capital-intensive factory production employing hired labour" [46, p. 44]. Science became indispensable to industry, while industry imposed itself on science, forcing science to adopt its concerns, making science dependent on its contracts, influencing the moral code even to the extent of sometimes

37

preventing the publication of certain results or, conversely, insisting on patenting things that previously had remained in the public domain (e.g. computer software or biological cloning). The industrialization of science also altered and extended the scientist's role so as to become simultaneously: in the university a teacher, administrator, and research scientist; with various state agencies, a contractor for research, an assessor for research proposals, an official adviser on existing projects, a military or diplomatic adviser, a specialist in strategic problems such as the management of advanced weapons systems or the negotiations on arms control; with commercial industry, a private consultant to firms, and a businessman manufacturing equipment of his own invention. These transformations did not occur without causing problems, challenging traditional values, and exposing researchers to conflicts of interest and forcing them to make political, ideological, or commercial commitments that their predecessors had been sheltered from (or alleged they were) thanks to the "neutrality" of science.

Habits change in time: the "detached" academic researcher came to be replaced by the scientific entrepreneur struggling for recognition and maximum profit. Henceforth, many more scientist-researchers worked in industrial laboratories, public or private, and for the military than in the universities. In the era of industrialized science, businesses organize themselves with a view to science-based production and technical innovation. The distinction between science and technology has become blurred: as technologies have become increasingly sophisticated and complex, the innovation process has become increasingly dependent on the findings and methodology of science. From now on, the practice and advance of science are far more dependent on technology than vice versa. Important discoveries are as likely to be made in industrial laboratories as in universities (e.g. nylon by Du Pont, the transistor by Bell Laboratories, enzyme synthesis by Merck, superconductors by IBM). And the system of management, control, and evaluation typical of industry is increasingly applied to research activities, including those in universities.

The expansion of modern science and technology

As we have seen, the link with political power was present from the beginning of modern science, but that link was all the less effective, institutionalized, and systematic because science had little influence on economic, military, and technical development, and at the same time, because the state intervened little in its affairs. The age of institutionalized science policy really started only when scientific activities began to have a direct effect on the course of world affairs, thereby causing the state to become aware of a field of responsibility that it could not neglect [20, 52, 4]. To give an idea of the change of scale that occurred as a result, we need only point out that the entire Federal R&D budget of the United States was less than $1 billion in 1939 (agriculture and health accounted for the lion's share); the Manhattan Project alone, which was responsible for the first three atomic bombs produced by 1945, cost $2 billion over three years, while the Apollo Program to put a man on the moon cost $5 billion per year over 10 years. In 1989, the total American gross domestic expenditure on R&D went up to $135,150 million, of which a little more than 50 per cent was financed from public sources. Even the countries that are the most vociferous upholders of free-market principles and abhor state intervention, from the United States to Germany, have seen public support for R&D, both direct and indirect, considerably increase and expand.

By science policy we mean the collective measures taken by a government in order, on the one hand, to encourage the development of scientific and technical research and, on the other, to exploit the results of this research for general political objectives. Today these two aspects are complementary: policy *for* science (the provision of an environment fostering research activities) and policy *through* science (the exploitation of discoveries and innovations in various sectors of governmental concern) are on a par in the sense that scientific and technological factors affect political decisions and at the same time condition the development of various fields (defence, the economy, social life, etc.). The historian of science will find it easy to show that neither the idea nor the thing itself was really absent from the development of science as an institution before the Second World War. However, if these two aspects did exist beforehand, they rarely did so simultaneously, and in any case only for short periods marked by the interest of the state in military exploitation of the results of scientific research, for instance during the French Revolution, the American Civil War, or the First World War [53].

The rise of science policy
In the West, the examples of a closer link between science and the state provided by the First World War and the post-war period were only a rough sketch of a process that was to be accelerated and firmly

established by the time of the Second World War. In particular, even though the Depression of the 1930s caused some people to become aware of the role that science policy might play in economic and social development, this awareness did not go so far as to provide the state with the means to guide the direction of scientific research, or even to organize it in a more coherent manner [9]. France alone among the market economies endeavoured to recognize the jurisdiction of politics over scientific affairs by setting up, under the Popular Front, the post of Under-Secretary of State, which was given first to Irène Joliot-Curie, then to Jean Perrin. The fact that the two Nobel prizewinners occupied in 1936 a ministerial position and the establishment of the Centre National de la Recherche Scientifique (an institution mainly concerned with the promotion of basic research) are the first signs in the West of the recognition on the part of the state of both the role played by science in economic and social affairs and the political concern that it should be integrated into the general fabric of government decisions [40].

This case, unique in the West, was inspired in part by the Soviet experience. For it was indeed in Russia that the closest link ever to be forged between science and politics was established by the triumph of the Revolution. The progress from ideology to action provides a model of organization inasmuch as it attempted to integrate science into the social system as a "productive factor" among other productive forces. Certainly, scientific activities enjoyed a status and a support at that time that had no equivalent in other countries before the Second World War; research was considered inseparable from the political system of which it was both the means and the end. Nevertheless, as heavily as political factors may have weighed on the development of science as an institution, the model presented by the Soviet regime did not give rise then to a real science policy [14, 62].

It was at any rate that model that served as reference to Bernal when, just before the war, he wrote his book *The Social Function of Science*, a pioneer work heralding the enormous changes that were soon to affect the relations between science and the state [3]. No other work has done more to ensure the recognition of scientific activities as a social institution that both affects and is affected by the development of the social system as a whole. In many respects, Bernal's analysis still shows a utopian approach directly inspired by the hopes that the Enlightenment and nineteenth-century positivism had placed in the politically liberating and inevitably beneficial character of science. He is nevertheless the first to have perceived and analysed

40

(even though with the Marxist bias of that time) all the aspects that could make scientific and technological research activities themselves into objects of social research. As such, Bernal appears as the founding father of the new field that is, in relation to development issues as well as to the industrialized countries, the subject of this whole volume: science policy, or science, technology, and society "studies." Bernal deplored the lack of public interest in science at that time and the scarcity of resources, but he had no doubt as to the immense progress science would accomplish and the great service that, associated with technology, it would render to society. Two conditions at least needed to be met in his view if these promises were to be fulfilled: far greater resources allocated for research activities, and the implementation of deliberate science policies.

It is now commonplace to point out that the Manhattan District Project, the name given to the programme that developed the first atomic bombs, marked an irreversible turning-point in the relations between science and the state: the establishment of science as a "national asset," the direct intervention of governments in the direction and range of research activities, the recruiting of researchers for large-scale programmes [21]. The change in scale of research activities goes hand in hand with the major technological developments that had a direct effect on the relations between countries: there were 100,000 researchers (scientists, engineers, and technicians) in the world in 1940, and 10 times this number 20 years later [10]. In the OECD area alone, the total R&D personnel was estimated at 1,754,430 in 1983, of which the United States accounted for a little more than 700,000 [38].

Indeed, the nature and the scale of the scientific research undertaken during the Second World War and, above all, the strategic importance of its results, have had consequences beyond anything Bernal had foreseen. According to his own words in the preface to the new edition of his book, "the scientific revolution entered a new phase – it became aware of itself" [3]. During and after the Second World War, scientific and technical research, conceived with military strategic ends in mind, became the source of newly discovered forms of technology that were to be applied on a vast scale in civil life: nuclear energy, radar, jet planes, DDT, computers, missiles, etc. From then on it became impossible for political power to leave science to its own devices, and at the end of the war, the demobilization of researchers, far from signalling the end of "mobilized" science as such, gave rise to systematic efforts to take

advantage of research activities in the context of "national and international" objectives [18].

The perfecting of nuclear weapons, missiles, and computers
altered the most traditional law of the balance of power: it was no
longer enough to avoid being at the mercy of the enemy, one had
now to forestall him. In this new kind of international competition,
between the "balance of terror," the arms race, and the fear of "technological gaps," scientific and technical research constituted a
powerful strategic, diplomatic, and economic resource. Science policy
developed in this context of strategic competition as a consequence of
the impossibility of establishing real peace at the end of the Second
World War. In this sense it is obviously one feature of an overall
policy determined by rivalry, struggles, and clashes between nations,
ideologies, and will for power. But in another sense the growing influence exerted by technological and scientific affairs on politics in
general could be regarded as a cause as well as an effect of the international climate of insecurity. No doubt, the "tyranny" of the arms
race and escalation operated through a "scientific-military-industrial
complex" that is very real and the irony (or wisdom) of history is that
it was a senior army officer and president of the United States who
uttered the first and gravest warning against this complex. In his
farewell speech as president, Eisenhower referred to the risks of a
public policy becoming the captive of a scientific and technological
élite and of the military-industrial complex to which this élite owes its
existence (*New York Times*, 22 January 1961).

Actually, it was only from 1957 – the date of the first sputnik – that
institutions really concerned with science policy were set up. Even in
1963, when the first Ministerial Meeting on Science took place at the
OECD, the ministers specifically in charge of scientific affairs could
be counted on the fingers of one hand [28]. In the space of only three
years, they made up the majority. As a field of government competence, science and technology were no longer intended merely to follow in the wake of educational or cultural policies. Whatever the institutional arrangements, the organizations concerned with science
policy, wherever they were, all fulfilled at least three functions: information, consultation, and coordination. Science policy of any kind
had to be prepared by administrative services, clarified by the advice
of experts, coordinated between the various ministries and agencies
concerned with research activities, and finally, of course, decided
upon and implemented in conjunction with the private industrial sectors. National traditions and structures provided a framework for

these functions and, within that framework, specific bodies (e.g. the Office of Science and Technology in the United States, the Délégation générale à la recherche scientifique et technique in France). According to whether the political system was centralized, decentralized, or pluralistic, science policy was developed in different institutions, linked more or less closely with bodies concerned with economic and strategic planning. Everywhere these bodies started their functions by collecting statistics on R&D activities, drawing up an inventory of researchers and laboratories, and allocating resources to sectors considered to have priority [5].

From the 1950s to the 1970s, science policy in the industrialized countries went from an age of pragmatism to the general awareness of the role played by scientific and technological research in the "wealth of nations" and in the struggles for international competition. However, there were important changes not only in the aims but in the political and cultural contexts. The first period, which corresponded to a climate of high tension, the Cold War, strategic competition and economic development impervious to the social and environmental costs it engendered, came to an end in 1968–1969. In the aftermath of *détente*, the campus revolts, the growing awareness of the limits to economic growth, and the American fiasco in Vietnam, the positivism induced by the methods and achievements of science was questioned not only by movements outside the scientific community but also by scientists themselves [49]. An American walked on the moon, but the very success of the Apollo Program marked a turning-point: the great options that had fed science policy during two decades ceased to be taken as articles of faith. The previous priorities were being re-examined critically, and reordered in a manner that, it was felt, would be more concerned with social well-being than with technological progress as such.

It is instructive to underline some of the conceptual changes that have taken place in the field of science and technology policy research and that show how this area of policy-making, although defined and nurtured by science, is heavily dependent on social structures and pressures. The OECD has been one of the leading institutions in highlighting the importance of science and technology policy; the first report prepared by the Secretariat in 1963, *Science, Economic Growth and Government Policy*, was quite optimistic and focused on the formulation of government policies, the building of scientific and technological infrastructures, and on the need to expand science and technology education as a lever for increasing economic growth. Nearly a decade later, in 1971, another report on the subject, *Science, Growth and Society: A New Perspective*, stressed the social impact of scientific and technological advances, paid attention to the American challenge in technology, and focused on both the role of innovation as an engine of growth and the need to anticipate and assess the negative aspects of technical change. The OECD reports published in 1980, *Technical Change and Economic Policy*, and in 1981, *Science and Technology Policy for the 1980s*, put greater emphasis on the economic and social changes that characterized the industrialized nations during this period and acknowledged that after three decades of unprecedented growth in the world economy, the situation was likely to be different. The oil crises led to focusing research priorities on possible energy alternatives, but issues such as the interaction between technology and employment, the dominant role played by micro-electronics and informatics, the growing importance of biotechnology and new materials, the restructuring of world industry and international competitiveness became central concerns of science and technology policy makers.

Thus in less than 20 years, a new perception of the interactions between science, technology, and society has emerged in the industrialized countries, one in which the optimistic views have been replaced by increased concern regarding the impact of advances in science and technology on society. The scientific crisis simply reflected the crisis taking place in society. As the Brooks Report pointed out, "science policy is in disarray because society itself is in disarray, partly because the power of modern science has enabled society to reach goals that formerly were only vague aspirations, but whose achievements had revealed their shallowness or has created expectations that outrun even the possibilities of modern technology or the economic resources available from growth" [33]. The problems posed by the deterioration in living standards, the chaotic state of urban development, the difficulties of transportation, pollution, the threat to the environment, and the growing inequalities within most of the industrialized countries and between them and the developing countries – all of this called for some control over the course of technical progress and the building of new paths that would reconcile technical progress to a more harmonious type of development. The notion emerged that the solution to these problems does not lie solely in the technocratic application of instruments that would reduce history to its physical constraints. Even in the case of strategic weapons and arms control, some scientists became aware that the "dilemma of steadily increasing military power and steadily decreasing national security has no technical solution" [61].

It is in this context of challenge and disenchantment that technology assessment was launched: a new function that would enable possible undesirable effects to be foreseen or the costs of the introduction of new technologies to be considered in relation to obvious or disregarded social needs. Subsequently, following the example of the United States, most of the industrialized countries created special bodies, within or outside their parliaments, whose function was not only to anticipate and regulate the effects of technological change but also to involve the public more closely, if not make it participate in the decision-making process relating to science and technology activities. However, this period of questioning and reappraisal did not see any reduction (rather the contrary) in the predominant strategic and prestige objectives concentrated in the most important industrialized countries on defence, nuclear, space, and computer research. And the malaise felt in relation to social issues was soon to be superseded by the economic difficulties precipitated by the oil crisis of 1973. The barely attempted efforts to redirect research activities toward the solution of social problems were limited, if not stopped, by the economic crisis, growing unemployment, and more intense international economic competition in relation to the "new technologies."

The defence-related R&D endeavour
Science policies were the consequence of the Second World War and the absence of peace that followed it. For the most industrialized countries, and in particular those with nuclear weapons, the Cold War was a period of full-scale mobilization of scientific resources, with huge investments in R&D in three key sectors: nuclear, space, and information and communications technologies. For the United States, Britain, and France, these investments accounted for two-thirds of their total R&D expenditure, public plus private. For the USSR, the defence budget was an even greater drain on resources, with the statistics for the 1980s indicating that military expenditures varied between 20 and 28 per cent of GDP – an enormous proportion when compared to that of the United States, where military spending equalled 6.5 per cent of GDP in the same period, even if the American GDP was much higher [62].

The arms race was one of the most spectacular features of the Cold War, but there was also fierce competition for world renown, ranging from the first sputnik to the first men on the moon. These

45

struggles forced the state to intervene in research and innovation, even in countries claiming to be unshakeable upholders of free-market capitalism. Questions may indeed be raised about the cost of the exaggerated level of armaments and the links between economic and strategic reasoning; it may be argued that the arms race diverted scarce resources (capital and skills) that could have been used for more socially and economically constructive purposes. The debate about the cost-benefit analysis of the "spin-offs" from military R&D for the civilian economy is not over, but it is impossible to under-estimate the importance of the innumerable innovations generated by military R&D during this period, and especially the role they played in the conception and development of the new technologies that characterize the "new technical system" just now beginning to flourish [58, 26, 55].

On the Soviet side, it is clear that the priority given to the military-industrial complex in R&D expenditure and production made a de-cisive contribution to the collapse of the economic system. It cannot be ruled out that Reagan's challenge via the Strategic Defense Initiative (Star Wars) helped Gorbachev to realize that the centrally planned Soviet system had reached its limits, with a civilian economy in a des-perate state and a military sector unable to keep up with the rapid progress of American technology. For the capitalist democracies, the costs in terms of economic growth were far smaller, but still not zero. One has only to compare the rates of productivity growth in countries with high levels of defence-related R&D to those with low levels. Germany and Japan, forbidden to invest in military activities after 1945, have had far higher productivity growth and much greater tech-nological success in commercial terms than the United States, Brit-ain, and France. Furthermore, in the 1970s, the innovations gener-ated by the defence sector seemed increasingly remote from the needs of ordinary consumers. The military demands for technical ex-cellence in terms of reliability, miniaturization, resistance to extreme conditions, etc., have created products that are harder and harder to adapt for civilian purposes. At the same time, in certain high techno-logy areas (especially "chips," components), commercial users have tended to overtake military orders in stimulating innovation. It is likely that the spin-offs from military R&D will be far less useful for the civilian economy in future, so that the economic growth rates of the countries most committed to such programmes will suffer accordingly.

Military R&D efforts have not been monopolized by the most advanced, industrialized countries. Among the developing countries, nations such as Brazil, China, and India have strengthened their manufacturing potential at the same time as their ambitions to build up an independent armaments industry, and even their own nuclear and space facilities. The growth in the arms trade in developing countries and the appearance of new producing countries are a sign of·both the relative success of some industrialization policies and the feelings of insecurity that rightly or wrongly beset the purchaser nations. Military ambitions have been able to stimulate industrial modernization in a context of policies of economic nationalism; yet, it is obvious that this choice of manufacturing and exporting weapons has diverted scarce resources that could have contributed to a more balanced economic and social development.

The Cold War justified everywhere the growth of a vast public sector and increasing state intervention in the private sector. Business interests were able to cash in on the arms race precisely because both sides felt insecure. "A war with no fighting neatly avoids the risk of fighting coming to an end. Obsolescence in a technological competition is a nearly perfect substitute for battlefield attrition" [12]. As long as the Cold War lasted, stopping the race was deemed more dangerous than the race itself. The post-war period has ended with the collapse of the communist system, the abolition of the Warsaw Pact, and the fragmentation of the Soviet empire. The signing of the START agreements means a 30 per cent reduction in long-range nuclear weapons. The end of the confrontation between the two systems and the collapse of the communist economies lead to the end of the arms race, and hence mean facing the problem of how to convert some (if not most) of the arms industries to civil purposes – a very difficult issue, which will take many years to resolve and which will quickly generate large-scale redundancies to add to the economic crisis in the republics of the new Commonwealth of Independent States.

There are already signs of a new race beginning, this time either to attract the best scientists from these countries to work in the West or else to "anchor" them in their laboratories, helping them to destroy the existing weapons systems or to redirect their research towards peaceful ends. Either way, the aim is to hold onto them and discourage them from selling their services to developing countries that would like to build up their own nuclear weapons and space capability. The OECD ministerial conference on science and technology in March 1992, attended for the first time by representatives of Russia, Hungary, Poland, and Czechoslovakia, was almost entirely devoted to this problem. And the sole purpose of the International Centre for Science and Technology established in Moscow with funding from the European Community and the United States is to prevent the growth of "mercenary science," where nuclear scientists rather than hired soldiers offer themselves to the highest bidder.

The reduction in nuclear weapons is not the same thing as disarmament, and the scaling down of the arms race by cutting the number of weapons does not necessarily mean scaling down military R&D programmes – even if there is now less urgency to perfect some of them. For one thing, the agreements deliberately leave open the possibility of increasing the numbers of cruise missiles, and the removal of some intercontinental missiles will in fact lead to even greater R&D efforts to improve the "quality" of conventional arms. For another, although the end of the Cold War undermines the traditional basis for the legitimacy of the military-industrial complex, the subsequent upheavals that are likely in central Europe and above all in the former Soviet republics will encourage the West to "lower its guard." It is clear, after the experience of the Gulf War, that the research into electronic warfare, in particular the anti-missile systems, is likely to expand rather than diminish, because of the threats of nuclear proliferation from peripheral countries.

Although the spectre of global nuclear war is fading for the first time, local conflicts are far from over. Military R&D efforts will continue to concentrate on miniaturization and on improving the precision of conventional weapons, as well as perfecting the systems of surveillance, monitoring, and response peculiar to electronic warfare. As General Poirier [42] has stressed, nuclear weapons, paradoxically, restrained the level of violence, because potential enemies knew that they must act and stop each other from acting in a haze of shared uncertainties, which led to political moderation and strategic prudence. In the "balance of terror," uncertainty brought a degree of order to relations between the superpowers, as deterrence only works when the enemy acknowledges the same rules. Nuclear proliferation may lead to an "imbalance of terror," where uncertainty generates disorder and where disorder on the periphery in fact adds to general uncertainty. The death of communism and the collapse of the Soviet system have removed the basis for the whole post-war strategic confrontation, and it is hard to imagine the biggest nations relying upon their nuclear deterrence in the event of hostilities initiated by "non-rational" smaller countries without atomic weapons. However, given that the sources of conflict throughout the world have not been eliminated, the "watch" will continue to mobilize substantial scientific resources. The heyday of the military-industrial complex is not yet over; that of defence-related R&D even less so.

The era of innovation policy

Whichever country – and no matter its political ambitions or strategic commitments – the primary objective of the industrialized nations now is to achieve and if possible improve economic growth, without which nothing else is possible, in economic as well as in all other spheres. Economic growth depends more than ever on firms' competitiveness, which in turn is very closely linked to the capacity for innovation, not only of firms but also of the entire system of social and economic organization (especially in relation to education and technical training). The research effort of these countries can be defined today as more and more oriented towards this goal, and it is complemented by a set of measures aimed at increasing the diffusion and application of technology in a large array of traditional industries and activities, as much as in the industries with a high R&D intensity.

This is the most important and revealing change: innovation policy appears as an extension of (or an alternative to) what was previously called science and technology policy. The concept emerged in the course of the 1970s as a result of three developments: first, economic and sociological analysis of the factors responsible for the performances of firms and especially of the roles played therein by technical innovation; second, the economic problems starting with the oil crisis that stopped the post-war period of rapid growth and full employment; and third, the upsurge of the "new technologies," particularly the information technologies, which brought about great changes in products and services throughout the economy. During the 1980s, the "structural policies" followed by the industrialized countries reshaped the continuum of their research systems to adjust to and overcome the consequences of the crisis (industrial restructuring, competition from the "newly industrializing" countries, unemployment, etc.) and the changes in the system of production and consumption introduced by the "Information Revolution." To these should be added the recent concerns about the environment, which are generating more and more public and private R&D efforts to bring products, processes, and industrial waste into line with new regulations. These changes in standards reflect changes in attitudes and values that oblige industry to innovate so as to satisfy the new consumer demands as well as the new legislative requirements regarding safety and pollution.

In brief, while state intervention in R&D activities has evolved in a context of privatization and deregulation, the American model has been replaced by the Japanese model, involving a package of long-term measures with a common target covering education, research, industry, foreign trade, and environment aimed at ensuring and sustaining the dynamism of firms in a global context [34, 35, 37]. The idea that innovation and entrepreneurship were among the basic factors underlying industrial expansion was certainly not new, since it dates back to the writings of Schumpeter. But the period of expansion after the war caused it to be overlooked. Although many studies were undertaken, notably those of the OECD on the "technology gap," the "Charpie Report" in the United States, and the research of economists like Edwin Mansfield, Richard Nelson, and Christopher Freeman, governments did not pursue them beyond affirming the importance of a well-thought-out policy for scientific and technological research activities: their gaze fixed on the input, they barely concerned themselves with the ways of ensuring a better diffusion of the output [11, 8].

All these efforts nevertheless arrived at the same conclusion: the problems of innovation depend less on the size of the investments in R&D than on basing the management of university and industrial resources on the entrepreneurial model. By emphasizing the importance for the innovation process of these factors, which are not properly scientific or even technical, all these studies recommended concentrating on policies that at first sight appear to have little in common with science policy as such. They stressed that it is not enough for a country to have excellent universities and research teams, to turn out increasing numbers of Ph.D.'s, to devote vast resources to R&D activities, or even to pile up Nobel Prizes in order to be one of the leading innovators. Winning the productivity battle, capturing and keeping new markets, and developing the full potential for innovation does indeed require a well-run research system, but that is just one prerequisite among many others. For innovation to be successful, the diffusion process is much more critical than that of either discovery or invention.

This period of introspection and research led to a better understanding of the sources, determinants, and nature of innovation [22]. In particular, it came to be realized that commercial viability depends

50

as much, if not more, on the social and institutional factors that provide the environment for the management of innovation as it does on the technical sophistication of the new products or services that it generates. To a large extent, the success of the "American model" could be attributed to the combination of two factors: the capacity of the universities to adapt very rapidly to the new needs generated by advances in knowledge, and the ability of industry to exploit the results of research more efficiently. And yet most of the European policy makers paid less attention to these factors and their combination than to the magnitude of the United States' expenditures for R&D (the "magic" target of 3 per cent of GNP) and the role exerted by the Federal government in stimulating the national research endeavour in the name of strategic and defence-related challenges.

In fact, even before the crisis of the 1970s, the example of the United States itself, where a few people had begun to be concerned with the falling rate of productivity growth, gave food for thought. Clearly, there was no direct link between the amount invested in R&D and the performance of the economy: champions in most categories of science and technology, the United States still had a productivity growth rate below that of Europe and, most important, of Japan. The question has been debated for more than a decade, and the Americans are still pondering the answer [30]. The fascination with the success of the "American model" made observers overlook the take-off conditions of a very different model, which more than ever confirmed that innovation should not be confused with scientific research: the model adopted by Japan, soon followed by the "little dragons" of South-East Asia. This raises at least the question of how much basic research does really contribute to growth and development at large. The modernization of Japan and its most recent success story in industrialization, like that of the newly industrialized countries, was not until recently accompanied by major contributions to scientific progress as such. The situation started to change in Japan because the very nature of its industrial development now requires a greater input of theoretical research. But this change is connected as much with the greater economic prosperity of the country as with the new prerequisites for producing technical innovations that are increasingly "sophisticated" and linked to laboratory research [56].

In Europe, it was not until the crisis of the 1970s that the significance of these limits to science policy began to be appreciated. By shifting from science in the strict sense to the broader field of innovation, governmental concern demonstrated an awareness of the fact that economic development was increasingly dependent upon constraints affecting industrial competitiveness and international trade. In the preceding period, the main concern had been to make basic research an integral part of the research system and to rely for technological innovation on "major programmes" supported, if not directly managed, by the state. Henceforth, there was debate about the extent to which the state should provide support for basic research and these "major programmes" that were financed (or subsidized) by public resources. Now, in the new context of privatization and deregulation, the question is how far the state should go, and under what institutional conditions, in intervening in the market in order to stimulate technological innovation.

Thus the criteria, as well as the instruments, involved in science policy have been profoundly altered. Science policy as such concerns individuals, institutions, and issues involved in measures related to scientific training, higher education, and academic research. As illustrated by the recent OECD report, *Technology and the Economy: The Key Relationship* [39], which is entirely devoted to an analysis of technological innovation in the context of increasing international competitiveness, innovation depends on a much wider range of actors, institutions, and issues – from industry, the banking system, and the overall economic environment to vocational training and even the general level of technical and scientific literacy. What is at stake is the need to "integrate" science and technology policies with all other government efforts, especially economic, industrial, energy, and social policies, as well as policies on education and employment. This was all the more obvious because of the need to cope not only with the consequences of the economic crisis but also with the changes introduced by the "new technologies." The products and processes created by these new technologies led to new modes of production and consumption that spread through all sectors of economic and social life; these products and processes are developed mainly by flexible, decentralized firms that are able to adapt quickly to market changes and are highly aware of consumer needs and preferences. In this context of market economies, if the role of the state cannot be limited to merely supporting scientific and technological activities, how far should it intervene, under what circumstances and on what criteria?

In some areas, state intervention is traditionally unquestioned (or, in some countries challenged less than in others): defence, basic research, the environment, health, large-scale technological systems such as those involving large infrastructures and networks (energy, transport, telecommunications). These areas concern society as a whole and require strategic action; in short, they are outside the market framework, and the private sector cannot be expected to take on the risks involved, or to safeguard and respect the public interest. The decisive competitive battle is now being waged among the small and medium-sized firms rather than among the major public programmes. Here, innovation involves entrepreneurial initiative, for which the management structures of public enterprises are badly (or rarely well) prepared. If the state has to intervene directly, it can be in the preliminary stages, where an "infant" technology or an "infant" industry threatens to be stifled before it reaches maturity by pressures from competitors. Yet the state cannot forever stand in for firms, or at least not without allowing its programmes to be guided by non-economic considerations, and unthinkingly subsidize their products in order to protect them from foreign competition; there is no lack of examples of these risks and failures, from the Brazilian "reserved market" for information technologies to the Anglo-French supersonic Concorde and the French "Plans Calcul" [32].

In the past, the state could start from scratch or could promote an industry (e.g. metals, shipbuilding, railways, oil) where the aim was to satisfy national needs without having to face the pressure of international competition. If need be, it could nationalize existing firms, even if they were foreign. But when the new technologies are involved, which deal mainly with intangibles (i.e. information, from hardware to software), the state has far less room for manoeuvre. Nationalizing firms in this sector would mean buying only the factories without having any control over the flows of intangible data that are the real source of technical and commercial success. In this context, the trend towards deregulation appears to be the result not only of economic (if not ideological) considerations, but also of institutional and technical factors: on the one hand, the organizational and social setting, which reveals the limits of the management and control of the monopoly hitherto enjoyed by publicly owned firms (e.g. the post office), and on the other, the new technical system, which imposes strategies and even an entrepreneurial approach closely linked to consumer demand and international markets. Once outside the programmes that are its concern for strategic reasons, it is through indirect measures (especially fiscal, but also educational in general) and above all a macroeconomic policy favouring investment that the state is best placed to stimulate technological innovation

53

efficiently – and more economically [50]. Most of these changes will continue to affect this new "strategic posture" of the industrially advanced countries, a posture that is basically defined by the growing economic competition, more concern for the regional and global environment, and the possibility – still to be confirmed – of an effective levelling off not only in the military budgets at large but also more specifically in the defence-related R&D endeavour.

Cultures and coexistence of rationalities

The radical change accomplished by modern science generates a major debate and many questions. For instance, what is it in the rationality of this science – European in origin and destined, as Needham says, to become "ecumenical" – that distinguishes it from other types of knowledge and culture? Or again, why did this version of science make its rapid rise in western Europe at the time of Galileo? The paradox involved in these questions is that much is said about the universality of modern science, while stressing the peculiar nature of its Western origins. The debate is all the more difficult in that it leads one to ask why, at a given moment in their histories, one civilization was so far ahead of others, for example the Chinese or the Muslim cultures, which turned in on themselves and missed the boat of "progress"?

Needham's life's work shows us that certain societies, certain cultures, at various periods of history, reveal themselves as far more efficient than others in the mastery of scientific knowledge and the exploitation of technical progress. But it is not only the past that tells us this. At this very moment, even as there is talk of a new stage in the history of the Industrial Revolution, it is clear that considerable disparities exist in the ability of different societies to take advantage of the possibilities opening up and, a fortiori, in their capacity to contribute to the conception, development, and production of the "new technologies." Needham's conclusion has the merit to exclude from the very outset any "physical-anthropological" or "racial-spiritual" factor involved in what may explain the advance or the lateness of societies in relation to each other: "The answer to such questions lies, I now believe, primarily in the social, intellectual and economic structures of the different civilizations" [31, pp. 127–128]. Moreover, it rightly suggests that catching up – as much as fading away and decline – is possible as a function of the efforts made to adjust and modernize these structures.

Scientific and other knowledge

Indeed, there is a common postulate subsumed in the approach of modern "hard" sciences: the constancy of the laws of the universe. This postulate went almost unchanged from Lucretius, who spoke of the laws of nature as contracts (*foedera*), to Einstein, who proclaimed that "God is subtle, but does not have a malicious nature." Or as Norbert Wiener wrote: "Nature plays fair and if, after climbing one range of mountains, the physicist sees another on the horizon before him, it has not been deliberately put there to frustrate the effort he has already made. The devil whom the scientists are fighting is the devil of confusion, not of wilful malice" [60]. The postulate of this rationality is that the universe functions according to commands that are like decrees. In fact these would seem to be the decrees of a supra-rational legislator, decrees that the founders of modern science – Galileo, Descartes, Kepler, Newton – thought to be "revealed" to the human spirit.

This postulate is what led Needham to highlight the essential difference between the conception of the order of the world in traditional China and that in Europe of the Renaissance. In the latter, the laws of nature are valid for the earth and heaven according to "orders" given by a rational legislator; in the former, there is no superior authority instituting a system of causal relations but an organic cooperation defining a cosmic reality: the law has no clear representation outside human affairs so that the intelligibility of the world is never guaranteed. Needham cited the example of medieval Europe, struggling against sorcery, where trials were held in which charges were brought against roosters that laid eggs. These roosters were condemned to be burned alive because they had betrayed the divine order. Needham used every opportunity to show that Taoist China would never have dreamed of conducting similar trials. Such phenomena were considered to be "rebukes of heaven," "celestial misfortunes," and not a perversion of the order of the world guaranteed by God.

Western science was finally developed and imposed itself by doing without the guarantee of a supreme legislator; nevertheless, statistical regularities and their mathematical expressions guarantee somewhat the hypothesis of an "honoured contract," of an order removed from the whims and arbitrary moods of either a magical or a malicious intervention: it is by definition impossible to hold the rational functioning of natural phenomena in default (which does not mean that there is neither deep complexity nor even disorder and chaos in the functioning of some of these phenomena, as shown by the most recent developments in theoretical physics). Hence the remark by Needham, which marvellously locates the boundary between the cultures ready to adopt a Western rationality and those that are closed to it: "Perhaps the kind of spirit which could make of an egg-laying rooster a being to be persecuted by the law was necessary in a culture so that this same culture would later be capable of producing a Kepler?" [31].

Until the seventeenth or eighteenth century, China and the West shared the same capital of knowledge, and China was in many aspects more advanced technologically. The compass, gunpowder, and printing were all transfers of technology from China to the West, and the end of the seventeenth century marked, thanks to the Jesuits' "technical assistance," reciprocal exchanges between the two civilizations in the common area of mathematics. "The Europeans at my court have presided over mathematics for a long time already. During the civil wars they rendered an essential service to me with the cannon which they have cast," states the Edict of Tolerance of K'ang-hsi in 1692. And the Chinese "model" defined a good part of European literature during the entire eighteenth century. But it was from the seventeenth century onwards that the parting of the ways occurred, with rivers that no longer flowed into the same ocean up until the nineteenth. Economic and social structures in Europe prepared the way for the scientific and technical revolution, while in China the "celestial bureaucracy" refused entrepreneurship, innovation, and change. Along with economic and social structures, there came – some would say that they are dependent on them – moral attitudes and new values.

Modern science is not content merely to substitute one model of knowledge for another (mathematics and experimentation for perception by the senses), but sets up a conception of the world in which the capacity for action is directly linked to speculative knowledge. It is from this angle that the rationality of Western science is the opposite of that of traditional science, whose influence is still present in most developing countries, especially in Asia. For instance, it has been shown why the world view found in traditional India could not have produced natural sciences in the sense understood in the West since Galileo [65]. The principle underpinning ayurvedic science in fact is that of law, and deals with rites and legends; its action depends on doing things in the ways stipulated in the traditional texts and not at all on research into causes that then leads to changes in the way things are done and to technical progress. Scholarly medical treatments in Asia seem to be outside history, ignoring the idea of change over time; they have links with the divine world and divination that can be traced back to the earliest sacred texts, providing complete responses from the outset. The principle behind the application of these treatments cannot be extended to have universal applicability, whereas for Western science, the constant search for and identification of causes lead to discoveries and innovations whose effects can be universally reproduced.

The complementarity of rationalities

The universality and the universalization of science are postulates of scientific thinking as it was formed in the classical period and developed in the course of industrialization. Indeed, these postulates were adopted by non-universalist cultures for reasons that have less to do with the definition of scientific research than with the power of economic-military-industrial complexes [57, 41]. Yet, although the operational power of modern science provided European imperialism with a means of unprecedented efficacy, the universality postulated by modern science did not (and could not) thereby make Western civilization universal. The desire for knowledge is truly universal no matter what form the knowledge may take. However, the universality of scientific knowledge in the Western sense affects only the network formed and developed by the adoption of the model of scientific institutions – from structures of education, training, and research to social and political institutions – that was created in Europe.

It is therefore easy to appreciate the limits and too often the failures of certain experiments in modernization conducted at headlong speed without regard for the economic, social, or cultural realities of the societies in which they were being conducted; the utilization of science and technology cannot be reduced to the insertion of knowledge or know-how, techniques, and methods into a social fabric that is unprepared. This fact underlies the equivocal nature (for some the illusion) of the notion of "technology transfer," a transfer that involves much more than the movement of a physical object from one place to another. Transfers of technology require the preparation of education, management, and production structures appropriate to the mastery of the production of knowledge and know-how themselves.

Do such structures have to be identical to those that produced modern science in Western countries? Not necessarily, given the example of Japan, where the Meiji "Restoration" led to the political decision to import the European scientific and technical model. The initiation into, and the rapid mastery of, Western scientific thinking came about not in terms of a rejection of a Japanese approach, but rather as its fulfilment. What distinguishes Japan from the European speculative heritage that dates back to ancient Greece is an attitude to science defined more in terms of its ability to produce practical applications rather than in terms of its purely scientific creative power. It is obvious that Japan never tried to follow the West blindly; instead, it tried to incorporate into its own system only those elements that would be of advantage in its task of modernization. This prudent and selective process of learning is often referred to as *wakon yosai*, meaning "Japanese spirit and Western learning" [15, 16].

The international network of scientists trained in the same institutions of higher learning and research, speaking the same language and publishing in the same journals, meeting one another periodically in the same places for colloquia and conferences, is indeed based upon the shared language, methods, and results of a universal scientific community in the Western sense. For a researcher, the notion of belonging to the extended community of science is highly significant and supportive. But this international network of science, in much the same way as the airline routes, is not universal in the sense that absolutely everyone can join in: belonging to the network is not the same as sharing in the conceptual framework that gave rise to that network. From this viewpoint, the "universality" of modern science is illusory.

It is not enough to rely upon the universal methods of science and technology in order to reproduce a model of development based on a tradition, history, and reality alien to that of most developing countries. What has been written about India in the aftermath of Independence is equally applicable to many other cases: "Science has grown as an oasis in an environment which, if not antagonistic, is also not sympathetic to it, with the majority of people steeped in superstitions and traditionalism of which many of the leading scientists are also victims" [45, p. 94]. It may appear obvious that one of the major aims of any development process must be to acknowledge science and technology as crucial elements in social and cultural life. But this is much easier said than done, and it is not surprising that Nehru and subsequent Indian leaders have constantly fought to spread the "scientific temper" among the vast population of the subcontinent.

At the same time, this does not mean dismissing sciences based on a different rationality from that of Western science, particularly since these have ceased to be strange and exotic in the West. They are thriving even in the midst of the scientific establishment, as is clear from the way that the teaching of acupuncture has spread in Western medical schools, or from the return to herbal remedies and "soft" technologies. The criticism of Western medicine for offering "aggressive" treatments and drugs, which do not respect the "harmony" of the balance between the psyche and the soma, is another example of a cultural transfer from East to West. The range of rationalities needs to be recognized by stressing the way they complement one another, rather than setting them against each other. Nor is their coexistence neutral: it leads to positive interactions, and it is well known that non-Western medicine can have beneficial effects on cases of chronic and functional disorders.

For the health services of developing countries, this complementarity in fact accords with social necessity. In Asia, the popular medicine provided by herbalists, soothsayers, spirit mediums, Taoist or Buddhist priests carries on alongside scholarly traditional medicine practised by people trained in recognized schools and hospitals [17, 29]. This scholarly medicine is supported by the World Health Organization, because its usefulness is all the greater in that Western medicine is costly, beyond the reach of the majority, and impossible to provide in country areas. Moreover, the two styles of medicine do not merely complement each other in providing treatments, but also in research, with studies combining traditional remedies and modern chemotherapy techniques. Since the 1970s, in Japan, Hong Kong, and Taiwan, publications on traditional Chinese medicine (*kanpo*) have enjoyed a tremendous boom, and it is not uncommon for a Japanese doctor trained in Western methods to practise *kanpo* at the same time, just as there are acupuncturists who increase the effectiveness of the needles by passing an electric current through them.

Yet, one must recognize that these transfers of practices from East to West are examples closer to what might be called "soft technologies" than to the technologies represented by the giant scientific complexes that are the mainstay of advanced physics and biology. The coexistence of different systems of rationality refers to institutions and practices from different levels, and what is valid for medicine, still more an art than a science, and even more so for the social sciences, may not be applicable to the "hard" part of scientific research. At the same time, within industrialized countries, the growing awareness of the social costs of the process of industrialization and the related threats faced by the environment leads some to question the foundations of Western rationality. Thus the coexistence of rationalities demands reflection not only on the limits to knowledge that does not meet the criteria of modern science, but also on the limits encountered by the very application of this kind of rationality. Even modern science, which based its claims to universality on the association of knowledge and power, is rediscovering that it is necessary to pay heed to the gap between knowledge and wisdom.

The search for new paths that would provide a legitimate and more viable framework for the pursuit of alternative development strategies requires a change in the perspective from which the concepts of "development" and "progress" are viewed. Despite its unquestionable achievements, the Western scientific-technological culture cannot be considered as the universal model to be imitated by the developing countries. A more ecumenical perception of the processes of development and progress is required, in which the potentialities of the many cultures that are part of the developing countries have to be revalued and appreciated, particularly if one tries to visualize what could be achieved through a harmonious integration of their cultural heritage with modern science. There have been many discussions on the question of whether it is possible to evolve a Latin American, Islamic, Asian, or African science, in contrast with the universal character of modern Western science that would not admit local variants. In a sense, this debate is an outgrowth of the much wider (and long-standing) debate between the "internalist" and "externalist" schools of thought in the history of science, which respectively attribute the main driving force of science to causes internal to the scientific enterprise and to the social context of science [51].

It is clear that the rate and direction of scientific progress is affected by considerations both external and internal to the conduct of scientific activities. If science is to be integrated with the cultures of the developing countries, so as to lead to the growth of science and technology capabilities, it is necessary to pay more attention to the factors that confer on science a local flavour and condition the necessity of its being combined with the cultural heritage of the developing countries. For instance, the process of identifying, selecting, and formulating problems so they would be amenable to attack through scientific research is clearly influenced by economic, social, political, and cultural factors. And while the choice of an individual research project may be more affected by considerations closely linked to the conduct of scientific research, the overall thrust of the scientific effort of a given nation is clearly conditioned by the general context in which science is inserted. The postulation of hypotheses and the building of theories to be tested are also influenced by broader considerations of a cultural character. This is a process where creativity finds room for expression, and where there is room for the modes and habits of thought that characterize different cultures to manifest themselves. Finally, the process of testing and verifying hypotheses must allow for the possibility of independent corroboration, and should comprehend rigorous comparison of the hypotheses – and the predictions derived from them – with the actual behaviour of the phenomena under scrutiny. This aspect of the scientific process is obviously the least amenable to the introduction of local considerations, and verification methods should, at least as an ideal, be truly universal.

This shows that a "local flavour" can be imparted to the conduct of science through the first stages of problem identification and formulation of hypotheses, and that in the stage of verification it becomes necessary to acknowledge the universal character of the scientific enterprise. And thus it is possible to orient the growth of science or at least an important part of a "national" scientific enterprise in the developing countries in directions that would respond more to the local conditions and problems and take into account their cultural heritage, while at the same time maintaining the crucial aspects of methodology and subsequent universality that are essential for the conduct of modern science. Indeed, furthering scientific knowledge and (all the more) mastering technological change make up a social process in which individuals and groups make choices about the allocation of extremely scarce resources. There is the saying: "Tell me who you know, and I will tell you who you are." When it comes to development and the uses of, as well as the support for, scientific and technological resources, this can be rephrased: "Tell me what you are researching and which innovations appeal to you, and I'll tell you what you really care about."

References

1   Bachelard, Gaston. *La formation de l'esprit scientifique*. Paris: P.U.F., 1938. 13th ed. Paris: Vrin, 1986.
2   Ben-David, Joseph. *The Scientist's Role in Society: A Comparative Study*. Englewood Cliffs, N.J.: Prentice-Hall, 1971.

3   Bernal, John D. *The Social Function of Science*. London: Routledge & Kegan Paul, 1939. Reprint. Cambridge, Mass.: MIT Press, 1967.

4   Blume, Stuart S. *Toward a Political Sociology of Science*. New York: The Free Press, 1974.

5   Brooks, Harvey, and Chester L. Cooper, eds. *Science for Public Policy*. Oxford: Pergamon, 1987.

6   Cardwell, D.S.L. *The Organisation of Science in England: Retrospect*. London: Heinemann, 1957.

7   Cohen, I. Bernard. *Revolution in Science*. Cambridge: Harvard University Press, 1985.

8   Dosi, Giovanni et al., eds. *Technical Change and Economic Theory*. London: Frances Pinter, 1988.

9   Dupree, A. Hunter. *Science in the Federal Government*. New York: Harper Torchbook, 1964.

10   Freeman, C., and A. Young. *The Research and Development Effort in Western Europe, North America and the Soviet Union*. Paris: OECD, 1965.

11   Freeman, Christopher, and Luc Soete, eds. *Technical Change and Full Employment*. Oxford: Basil Blackwell, 1987.

12   Galbraith, J. K. *The New Industrial State*. Boston: Houghton Mifflin, 1967.

13   Goldsmith, Maurice, and Arnold Mackay. *The Science of Science – Tribute to J. D. Bernal*. London: Souvenir Press, 1964.

14   Graham, Loren A. *The Soviet Academy of Sciences and the Communist Party*. Princeton, N.J.: Princeton University Press, 1967.

15   Hayashi, Takeshi. *Transformation and Development: The Experience of Japan*. Tokyo: United Nations University, 1980.

16   ———. *Historical Background of Technology Transfer – Final Report*. Tokyo: United Nations University, 1984.

17   Huard, P. et al. *Les médecines de l'Asie*. Paris: Seuil, 1978.

18   Kuehn, Thomas J., and Alan Porter, eds. *Science, Technology and National Science Policy*. Ithaca, N.Y.: Cornell University Press, 1981.

19   Kuhn, Thomas S. *The Structure of Scientific Revolutions*. Rev. ed. Chicago: University of Chicago Press, 1970.

20   Lakoff, Sanford A., ed. *Knowledge and Power: Essays on Science and Government*. New York: The Free Press, 1966.

21   ———. "Scientists, Technologists and Political Power." In: Price and Spiegel-Rösing, eds., pp. 355–391. *See* ref. 44.

22   Landau, Ralph, and Nathan Rosenberg, eds. *The Positive Sum Strategy: Harnessing Technology for Economic Growth*. Washington, D.C.: National Academy Press, 1986.

23   Lenoble, Robert. "La révolution scientifique du 17è siècle." In: René Taton, ed. *Histoire générale des sciences*, vol. 2: *La science moderne*. Paris: P.U.F., 1958.

24   Lindberg, D.C., and R.S. Westman. *Reappraisals of the Scientific Revolution*. Cambridge: Cambridge University Press, 1990.

25   Mendelsohn, Everett. "Science, Technology and the Military: Patterns of Interaction." In: Salomon, ed., pp. 49–70. *See* ref. 55.

26   Mendelsohn, Everett et al., eds. *Science, Technology and the Military: Sociology of Sciences Yearbook*. Vol. 12 (2 vols.). Dordrecht: Kluwer Publishers, 1988.

27   Merton, Robert K. *Science, Technology and Society in Seventeenth-Century England*. 1938. Reprint. New York: Harper Torchbooks, 1970.

28   Mesthene, Emmanuel. *Ministers Talk About Science*. Paris: OECD, 1965.

29   Meyer, Fernand. *Gso-Ba Rig-Pa: Le système médical thibétain*. Paris: CNRS, 1983.

30   NAE (National Academy of Engineering). *National Interests in an Age of Global Technology*. Washington, D.C.: NAE, 1991.

31   Needham, Joseph. *The Grand Titration: Science and Society in East and West*. London: Allen and Unwin, 1969.

32   Nelson, Richard. *High Technology Policies: A Five Nation Comparison*. Washington, D.C./London: American Enterprise Institute, 1984.

33   OECD (Organisation for Economic Co-operation and Development). *Science, Growth and Society*. Paris: OECD, 1971.

34   ———. *Technical Change and Economic Policy*. Paris: OECD, 1980.

35   ———. *Science and Technology Policy for the 1980s*. Paris: OECD, 1981.

36   ———. *Science and Technology Indicators*. Paris: OECD, 1984.

37   ———. *New Technologies in the 1990s: A Socio-Economic Strategy*. Paris: OECD, 1988.

38   ———. *Science and Technology Indicators Report: R&D Production and Diffusion*. No. 3. Paris: OECD, 1989.

39   ———. *Technology and the Economy: The Key Relationship*. Paris: OECD, 1992.

40   Papon, Pierre. *Le pouvoir et la science en France*. Paris: Centurion, 1978.

41   Petitjean, P., C. Jami, and A.M. Moulin, eds. *Science and Empires: Historical Studies about Scientific Development and European Expansion*. Dordrecht: Kluwer, 1992.

42   Poirier, Lucien. *Stratégie théorique*. Vol. 2. Paris: Economica, 1987.

43   Price, Derek de Solla. *Little Science, Big Science*. New York: Columbia University Press, 1963.

44   Price, Derek de Solla, and Ina Spiegel-Rösing, eds. *Science, Technology and Society: A Cross-Disciplinary Perspective*. London/Beverley Hills: Sage, 1977.

45   Rahman, Abdul. *Triveni: Science, Democracy and Socialism*. Simla: Indian Institute of Advanced Study, 1977.

46   Ravetz, Jerome R. *Scientific Knowledge and Its Social Problems*. Oxford: Clarendon Press, 1971.

47   Roland, Alex. "Science and War." *Osiris* 1 (2nd series) (1985).

48   ———. "Technology and War: A Bibliographic Essay." In: Merritt Roe Smith, ed. *Military Enterprise and Technological Change*. Cambridge, Mass.: MIT Press, 1985.

49   Rose, Hilary, and Steven Rose. *Science and Society*. London: Penguin Books, 1970.

50   Rothwell, Roy, and Walter Zegfeld. *Industrial Innovation and Public Policy: Preparing for the 1980s and the 1990s*. London: Frances Pinter, 1981.

51   Sagasti, Francisco. "The Two Civilizations and the Process of Development." *Prospects* 10 (1980).

52   Salomon, Jean-Jacques. *Science and Politics*. Cambridge, Mass./London: MIT Press and Macmillan, 1973.

53   ———. "Science Policy Studies and the Development of Science Policy." In: Price and Spiegel-Rösing, eds., pp. 43–70. See ref. 44.

54 ———. *Le destin technologique*. Paris: Balland, 1992.

55 ———, ed. *Science, War and Peace*. New York: St. Martin's Press; Paris: Economica, 1989.

56 Salomon, J.-J., and A. Lebeau. *Mirages of Development*. Boulder, Colo.: Lynne Rienner, 1993. Originally published in French as *L'écrivain public et l'ordinateur*. Paris: Hachette, 1988.

57 Salomon-Bayet, Claire. "Modern Science and the Coexistence of Rationalities." *Diogenes* 126 (1984).

58 Sapolsky, H.M. "Science, Technology and Military Policy." In: Price and Spiegel-Rösing, eds., pp. 443–471. *See* ref. 44.

59 Watson, James. *The Double Helix*. 1968. Rev. ed. London: Weidenfeld and Nicolson, 1970.

60 Wiener, Norbert. *The Human Use of Human Beings: Cybernetics and Society*. 1950. Rev. ed. New York: Avon Books, 1967.

61 York, Herbert F. *Making Weapons, Talking Peace*. New York: Basic Books, 1987.

62 Zalevski, Eugene. "Les dépenses militaires en URSS." *Futuribles* 158 (1991).

63 Zalevski, Eugene et al. *Science Policy in the USSR*. Paris: OECD, 1969.

64 Ziman, John. *Public Knowledge: An Essay Concerning the Social Dimension of Science*. Cambridge: Cambridge University Press, 1968.

65 Zimmerman, Francis. *La jungle et le fumet des viandes: un thème écologique dans la médecine hindoue*. Paris: Hautes Etudes Gallimard-Seuil, 1982.

2

The story of development thinking

Nasser Pakdaman

After the Second World War, academic economics began to tackle the problem of how to deal effectively with the poverty and destitution that weighed upon two-thirds of the human race. Development economics is "a comparatively young area of inquiry . . . born about a generation ago" [27, p. 372] "[It] did not arise as a formal theoretical discipline, but was fashioned as a practical subject in response to the needs of policymakers to advise governments on what could and should be done to allow their countries to emerge from chronic poverty" [47].

Its birth occurred in an unusual historical context and under the decisive influence of a range of political and cultural factors. The historical background was the aftermath of the Second World War, whose end led to optimism that new forms of international cooperation and solidarity would be effective in resolving the problems of "backward" countries and regions and would create new opportunities in this regard. The disintegration of the colonial empires as a result of movements for national independence brought to prominence a new factor that hitherto – like the Third Estate in pre-Revolutionary France – had been "nothing" but now wanted to become, if not "everything," at least "something." "To become something" expressed a desire or an intention to bring about change that can be found underlying all the plans for development conceived by those parts of the planet later labelled by Alfred Sauvy as "the Third World" [61]. The development of this "neglected, exploited and despised" world, to quote Sauvy, was a very important international cause for concern, "a major problem that should fill the next half-

65

century, and perhaps the one after that as well, provided that no serious accident occurs to give a new twist to the conflict between the two power blocs" [60].

It was against this background that the problems of development acquired far greater urgency than ever before and attracted the attention of economists. This is also the impression given by reading the accounts of the "pioneers in development," who were drawn from all sorts of backgrounds and for all sorts of reasons to study the problems of underdevelopment at that time [47].

Pioneers in development

In the post-war period, orthodox economics still had no interest in the problems of growth and in what occurred in the long term, so that W.A. Lewis could write in 1955 that "the last great book covering this wide range was John Stuart Mill's *Principles of Political Economy*, published in 1848," adding, "after this economists grew wiser; they were too sensible to try to cover such an enormous field in a single volume, and they even abandoned parts of the subject altogether, as being beyond their competence" [40].

There has been a tendency to think of economics as a discipline founded by the classical school, and rounded out and perfected by the neoclassical school. The contribution of other economists to the shaping of the discipline tends to be presented as marginal, secondary, if not actually insignificant. Nevertheless, the success of Keynesian analysis meant that orthodox economics was forced to acknowledge the existence of Keynes's followers and the "new economics" they proposed. "Mainstream" economics thus split into two: orthodoxy and its concomitant heresy, with everything supposedly belonging to either the neoclassical or the Keynesian school.

As in any polarized situation, the two protagonists had a common interest in defending the validity of this dichotomy and denying that other views had any great significance or even existed. However, matters are not so clear-cut in practice, and if ever there was a branch of economics that managed to develop quite independently of the two main schools, it is development economics. Indeed, the problems of development must be analysed over the long run, i.e. in the time span where, as Alfred Marshall said, "real life begins" – or in which we shall all be dead, to quote Keynes's famous remark.

It is true that "development economics took advantage of the unprecedented discredit orthodox economics had fallen into as a result of the depression of the 1930s" and the victory of the Keynesian re-volution [27, p. 375]. Nevertheless, development economics did not grow out of "new economics." The problems of development relate to the problems of change, i.e. they arise only in the long term and moreover require an interdisciplinary approach – but neither the neoclassical nor the Keynesian school provided appropriate concep-tual tools for this purpose, as is clear from the writings of the "pioneers" [47]. Some of the early development economists were familiar with Keynes's ideas and those of his circle, but they did not consider themselves Keynesians. Several writers tried to adopt a Keynesian approach to the problems of development, one of the first and most famous being Kurt Mandelbaum [44]. But the relevance of Keynesian concepts for underdeveloped economies was already being questioned in the years immediately after the Second World War [53].

Among the "non-Keynesians," W.W. Rostow explained that a study of economic history made him aware of the narrowness of the neoclassical approach and led him to develop a "Marshallian long period," taking account of the contribution of social, political, and technological factors in real life [47]. Paul Rosenstein-Rodan, having parted company with the marginalist analysis, was forced to abandon the Marshallian theory of static equilibrium and to acknowledge the virtues of interventionism in order to devise a strategy for tackling poverty in the less advanced countries of southern and south-eastern Europe. He himself described the starting-point of his thinking about development in terms of a motto: in economics, "Nature does make a jump," which is the opposite of Marshall's belief that "Nature does not make a jump" (*Natura non facit saltum*).This led to the formula-tion of the well-known theory of the "big push," whereby "back-ward" economies needed a development strategy based on a kick-start to set in motion the "disequilibrium growth process."

The insignificance of the contribution of the neoclassical school to the emergence of development economics has been acknowledged by one of its most prominent representatives, Gottfried Haberler. He explained this in terms of "the decline of liberalism": "a sharp de-cline. . . started with the onset of the Great Depression of the 1930s (or possibly earlier – the precise date does not matter)" and reached its low point after the Second World War, when "faith in liberalism, in free markets and in free enterprise was probably at its lowest point since the early 19th century." He therefore argued that it was be-cause economic liberalism had become discredited that the neoclassical school failed to make any real contribution to the creation of development economics (in Meier [46]).

In order to identify the sources of development economics, we must therefore look instead to economists who worked outside mainstream economics. The problems of development have been a central concern for several branches of the subject. W.A. Lewis notes that "the theory of economic development established itself in Britain in the century and a half running from about 1650 to Adam Smith's *The Wealth of Nations* (1776)." Lewis defines development theory as "those parts of economics that play crucial roles when one tries to analyze the growth of the economy as a whole," and he demonstrates

how much of modern development theory was already available in the year 1776. . . . This was quite a good beginning, that gave us the constraints imposed on growth by the agricultural surplus, or foreign exchange, or saving. Also we had Say's Law, the "Quantity Theory of Money", inflation, continual unemployment, entrepreneurship as a separate factor of production, the theory of bank credit, human capital and the incidence of taxes. Just ahead of us, in the first half of the 19th century, would come the law of diminishing returns, the law of comparative cost, the theories of population and of land tenure. After that, interest in development theory would almost die out until the theoretical explosion of the 1950s and after. (in Chenery and Srinivasan [12])

Amartya Sen also stresses the importance of development problems to seventeenth- and eighteenth-century writers:

Indeed, in the early contributions to economics, development economics can hardly be separated from the rest of economics, since so much of economics was, in fact, concerned with problems of economic development. This applies not only to Petty's writings, but also to those of the other pioneers of modern economics, including Gregory King, François Quesnay, Antoine Lavoisier, Joseph Louis Lagrange and even Adam Smith. *An Inquiry into the Nature and Causes of the Wealth of Nations* was, in fact, also an inquiry into the basic issues of development economics. (in Chenery and Srinivasan [12])

The quality and importance of the contributions of pre-classical economists to the problems of development should not, however, make us neglect those of the German school or of Marx. The study of actual economic change in order to identify the mechanisms and the types, "stages," "periods," and "phases" was one of the main preoccupations of the German historical school, which consequently introduced into its analysis a notion of relativity in the "laws" of evolution, and adopted a multidisciplinary approach [28].

As for Marx's contribution, Schumpeter maintains that "development" is "the central theme" in the general schema of Marx's thinking [62]. Indeed, one of the first instances of the term "development" occurs in Marx, in a passage in the preface to the first German edition of *Das Kapital*, dated 25 July 1867, that suggests a special view of historical evolution: in order to forestall the criticisms of German readers who might question why he used England "as the chief illustration in the development of [his] theoretical ideas," Marx stressed that "it is not a question. . . of the higher or lower degree of development of the social antagonisms that result from the natural laws of capitalist production. It is a question of these laws themselves, of these tendencies working with iron necessity towards inevitable results. The country that is more developed industrially only shows, to the less developed, the image of its own future" [45, p. 1718]. As regards his methods of investigation, in his afterword to the second German edition of *Das Kapital* (1873), Marx referred approvingly to one of his critics, who had described the way he applied these methods:

The one thing that is of moment to Marx, is to find the law of the phenomena with whose investigation he is concerned; and not only is that law of moment to him, which governs these phenomena, in so far as they have a definite form and mutual connexion within a given historical period. Of still greater moment to him is the law of their variation, of their development, i.e. of their transition from one form into another, from one series of connexions into a different one. This law once discovered, he investigates in detail the effects in which it manifests itself in social life. [45]

Among the first to be concerned with the problems of development were colonial authorities and those living under colonial rule. The former were mainly interested in "colonial development." It is not just coincidence that the first occurrence of the expression "economic development" is found in an essay written in Australia in 1861 on "the manufactures most immediately required for the economic development of the resources of the colony" [5]. Henceforth, investigation of the development/colonization/exploitation of colonial regions became the principal task of a new discipline, colonial economics, concerned above all with maintaining the status quo in "an essentially static world" [47], as well as with problems of foreign trade and overseas markets. The spirit and the concerns of colonial economics are well illustrated by British legislation, such as the Colonial Development Act (1928) and the Colonial Development and Welfare Act (1938).

Colonial economics could not avoid examining the reasons for the differences observed between the situation of the colonies and that of the mother countries, or saying something about the timeliness and the chances of success of measures (already taken or required) aimed at solving the problems of the "backward" countries. As a consequence, the unity of economics was challenged, and doubts were raised about the universal validity of the concepts and the analytical tools provided by "Western" economics. From early in this century, there are instances here and there of people stressing the insurmountable differences between two types of social and economic organization, and the uneasy coexistence of two distinct social and economic systems, one imported and imposed by the colonial power, the other belonging to the "native" population. A dualist theory was first formulated before the First World War, while starting in the 1930s there were references to differences in socio-economic "structures" as the main reason for the polarization of colonial societies and economies [9, 10].

The colonial approach was based on an ethnocentric viewpoint and a belief in Western supremacy, which in itself showed the "backward" countries the direction they should be going in in order to achieve Salvation: they must take the West as their model. At the same time, it was understood the West should take responsibility for and even actively implement this global scheme of social, economic, and cultural emulation. There thus arose a "development strategy" based on Westernization as a first version of what was to be thought of later as "modernization." The civilizing role of the developed world was even stressed in official documents: the League of Nations Pact of 28 June 1919 used the term "development" five times in its article 22 in talking about "peoples who are not yet able to run their own affairs themselves in the particularly difficult conditions of the modern world." "The welfare and the development of these peoples are a sacred mission of civilization." "The developed nations are entrusted with the supervision of these peoples." The conditions and the precise manner in which this supervision would operate depended on the *degree of development* of the people and the communities concerned [15]. It justified putting "under international mandate" countries that in fact were under the rule of a single nation.

It should be remembered that, already in the nineteenth century, there was persistent questioning in the "communities concerned" as to the reasons – political, economic, cultural, etc. – for the "lag" behind the "advanced" countries, and a variety of answers were given

70

to explain their state of political and economic subjugation, as well as to suggest swift, efficient, and lasting solutions that would get them out of poverty and decline [8]. "How to achieve economic development?" and "What should be done in order to catch up?" were the main preoccupations of the colonial world. The responses were diverse, but all of them made industrialization the key element in any development strategy, since that had been the critical factor in revolutionizing the West and generating its economic growth. For proof, one has only to read the passionate debates stimulated by plans to set up a bank, build a railway, to exploit mineral resources in countries such as Iran, Egypt, or in the Ottoman Empire. It was no coincidence that Sun Yat-sen published a book in 1922 on the international development of China, in which he set out an impressive programme for the country's economic development [5]. It would be easy to find other examples in other parts of the colonial world, indicating the same concerns with combating poverty and promoting progress.

By the interwar period, everyone believed that industry was more important than agriculture: you had to have begun to industrialize in order to have an industrial revolution. This craze for industrialization, explained by some observers as ultimately derived from the theories of Saint-Simon, was apparent in the discussions at conferences, from Baku in 1920 to Bandung in 1956, gathering together representatives of countries rebelling against the colonial status quo. One example must suffice here. Among the resolutions following the Asian Relations Conference in New Delhi (23 March–2 April 1947) attended by the representatives of about 30 countries, points 4 and 5 dealt with the transition from a colonial to a national economy, the problems arising from "the development of a national economy" and "agricultural reform and industrial development." Point 5 included the statement that "the real criterion for Asian independence will . . . depend on the capacity of Asia to achieve a substantial level of industrialization" (in Queuilles [55]).

The new discipline of development economics was thus created where several points of view came together, all of which had some impact upon it: those wishing to identify the laws of economic evolution, others seeking to build a new and better world, others trying to maintain colonial regimes, and yet others trying to throw off colonial rule. Although development economics was not entirely a product of the post-war period, it was none the less strongly influenced by the atmosphere of the Cold War [22] and decolonization, Western ethnocentrism and the emergence of new sovereign states in the third world seeking "good advice."

The discipline develops

Since the years following the Second World War, development economics has continued to evolve in a climate of optimism and confidence, sometimes arrogantly and aggressively, and often with doubt and depression. We therefore find today a range of different and frequently contradictory arguments based on the shared concerns of a particular school of thinking or a particular body of problems. To gain an insight into the distance that has been covered since the war, it is interesting to observe the changing contents of successive editions of handbooks on development, such as Meier [46], or even better to compare a "textbook" written in the early 1950s with a more recent one. Alternatively, one might look at the accounts by development economists of their experiences in recent decades, such as the two volumes produced at the instigation of the World Bank on "the pioneers on development" [47, 46]. Fifteen such "pioneers" were asked to make a critical examination of their own working hypotheses, concepts, analytical tools, advice, and policy recommendations. Contributions came from P.T. Bauer, C. Clark, C. Furtado, G. Haberler, A.C. Harberger, A.O. Hirschman, W.A. Lewis, H. Myint, G. Myrdal, R. Prebisch, P.N. Rosenstein-Rodan, W.W. Rostow, T.W. Schultz, H.W. Singer, and J. Tinbergen. Each essay is followed by comments from one or more younger economists, so that in all, 23 currently active economists offer their critical assessment of the work of the "pioneers." It was hoped that these studies would provide an exceptional opportunity for a review of what had happened to development economics since its early days.

This type of study invites the usual remarks about the selection and the representativeness of the sample and the reasons for notable absences. For example, one may reasonably wonder why no French-speaking economists were included (such as C. Bettelheim, R. Dumont, F. Perroux, or A. Sauvy), nor any of the many African and Asian specialists in the field, such as those who took part in the planning efforts in India after Independence. The result is a somewhat incomplete picture of the pioneer age, with very few "local" representatives. Were they not concerned about their own development, or is this another example of (Anglo-Saxon) ethnocentricity? Or are those third world economists right who consider studying development within the social sciences as yet another "product of the West," "an outsider's view of our development, in particular from the countries that once ruled us" (Goonatilake quoted in [8])?

The literature on development economics has been so rich and various that numerous attempts have naturally been made to classify it (e.g. Hirschman [27], Dockès and Rosier [19]) or to present it in historical and analytical terms (Roxborough [58], Kitching [36], Harris [25], Stern [72], Oman and Wignaraja [49]), but these still leave the reader pondering on the absences and oversights, or the reasons why a given author has been classed under one heading rather than another. This situation is partly a product of the way that development economics has evolved.

The subject has grown up in different continents simultaneously, across many cultures and at different levels related to the problems encountered, the differing schools of economic thought, and the models of society created by or for the developing countries. These levels were clearly not independent of one another; on the contrary, the many links that were forged among them helped strengthen the multifaceted character of development economics. As regards the problems encountered, in both development theories and discussions about policy choices, there was a gradual shift away from a purely economic approach in favour of a more interdisciplinary one.

The first formulations of the "problématique" of development focused on capital formation, seeing that as the engine of economic growth. Lack of capital was the distinctive feature of low-income economies that relied heavily on low-productivity agriculture. How could they be transformed into industrialized economies with high incomes? The answer was simple and categoric: investment. But how and where was the capital to come from? According to W.A. Lewis,

the central problem in the theory of economic development is to understand the process by which a community which was previously saving and investing 4 or 5 per cent of its national income or less, converts itself into an economy where voluntary saving is running at about 12 to 15 per cent of national income or more. This is the central problem because the central fact of economic development is rapid capital accumulation (including knowledge and skills with capital). We cannot explain any "industrial" revolution (as the historians pretend to do) until we can explain why saving increased relatively to national income. [40]

The necessary capital would be found either through the free operation of the foreign exchange market, which would attract foreign capital, especially public or private aid, to the underdeveloped countries; and/or through the interventionist policies of the state in planning the national economy and mobilizing "hidden" resources in order to achieve an increase in national income.

The first argument was proposed by the neoclassicists, but in the post-war years it was the second argument that tended to influence the design of development policies. Industrialization offered the key to growth and was presented as the main hope of most poor countries hoping to raise their incomes. This policy of industrialization – whether aimed at achieving "balanced" [48] or "unbalanced" [26] growth, whether conducted via a "big push" [57] or via "growth poles" [52] or via the choice of "industrializing industries" [18] – concerned above all the domestic market, where it was expected to satisfy existing demand for "modern" products, previously met by imported goods manufactured abroad. Import substitute industrialization was thus meant to bring about growth in developing countries by creating a modern industrial sector that would replace imports with locally made goods. It was argued that the benefits would "trickle down" to reach all parts of society, and in consequence the implementation of such development policies did not require any political or social transformations to the status quo. As regards strands of economic thinking, these first attempts at formulating the problématique of development economics were supported by the structuralists, the institutionalists, and the proponents of the dualist approach – in short, groups outside the orthodox camp.

The neoclassicists – the orthodox camp – argued that market forces unfailingly provided the engine of economic growth: the interplay of supply and demand in both domestic and international markets would ensure economic success. The market was seen as a tool of social and economic management, and as such was thought to be the most efficient way of making decisions about the optimal allocation of available resources. The free marketeers, who were extremely critical of the interventionist and protectionist positions of the structuralists and institutionalists, thought that opening up to world markets could only bring benefits to third world countries, as suggested by Ricardo's theory of comparative advantage or the improved versions of it proposed by Heckscher and Ohlin. Theories like that of Jacob Viner [78] argued that, through trade, the growth occurring in the advanced countries would be transmitted to the developing ones. Full integration into world markets therefore became the key aim of every development strategy, and nothing would be spared to achieve it.

This choice had enormous consequences: economic development would be promoted by free enterprise and not the state; *laissez-faire* would replace all attempts at planning, and the main policy emphasis, instead of import substitution, would be on encouraging exports. Third world countries therefore ought to stick to exporting raw materials and should do everything to expand production of these commodities, while waiting patiently for growth to be transmitted to them from outside.

This idyllic vision of the world economy was vigorously challenged by all those who argued that international economic relations were shaped by mechanisms of domination, submission, and dependency. As early as 1948, François Perroux offered an analysis in terms of domination of the world economy, which he argued was divided into dominant (firms, countries, or regions) and dominated elements, with the former having an extremely uneven impact on the latter [52]. The notion of general and mutual interdependence offered by the neoclassical theory of general equilibrium was therefore replaced by a notion of "the dynamics of inequality" arising from and maintained by the dominant forces.

The centre and the periphery

At the same time, quite independently of one another, Raul Prebisch [54] and Hans Singer [70] highlighted the issue of worsening terms of trade for the developing countries. They argued that international trade worked against third world countries that relied on exporting primary products and importing manufactured goods. It was not a matter of mutual benefit, as the neoclassical theory maintained, but of an unfair transfer of economic gains. For Prebisch and his colleagues at the United Nations Economic Commission for Latin America (ECLA), the world economy was made up of two different and separate entities – the centre and the periphery – and the nature of their relations tended constantly to reproduce the conditions of underdevelopment and to widen the gap between developed and underdeveloped countries.

This was the first formulation, inspired by the structuralists, of a new paradigm of development: dependency. The underdeveloped countries were part of a network of international economic relations in which the industrialized countries, favoured by their position at the centre and by their early technical progress, organized the system as a whole to serve their own interests. The producers and exporters of raw materials were thus linked with the centre as a function of their natural resources, thereby forming a vast and heterogeneous periphery incorporated in the system in different ways and to different extents, depending largely on their resources and their economic and political capacity for mobilizing them. According to Prebisch,

this fact was of the greatest importance, since it conditioned the economic structure and dynamism of each country – that is the rate at which technical progress could penetrate and the economic activities such progress would engender. Similarly this system . . . exaggerated the degree to which income in the periphery was siphoned off by the centers. Moreover, the penetration and propagation of technical progress in the countries of the periphery was too slow to absorb the entire labor force in a productive manner. Thus the concentration of technical progress and its fruits in economic activities oriented towards exports became characteristic of a heterogeneous social structure in which a large part of the population remained on the sidelines of development. (in Meier and Seers [47])

Dependency theory marked a radical departure in development thinking: henceforth, underdevelopment was thought to be an inescapable consequence of the world economic system, and to analyse it required that all the links of dependency between the periphery and the centre be taken into account. Any development strategy that hoped to be efficient should therefore make the restructuring of the world economic order its principal goal.

More radical versions of dependency theory were proposed by Marxist economists and sociologists. In his analysis of the political economy of development, Paul Baran [6] uses the concept of economic surplus, defined as the difference between production and consumption. In every society, two main types of economic surplus are found: actual, which is the difference between current production and consumption; and potential, which is the difference between the potential production of a given economy and what is considered its "basic consumption." According to Baran, much of the potential surplus remains unexploited in the capitalist developing countries, while much of the actual surplus is transferred to the industrialized countries. The capitalist world is made up of two organically interlinked parts, the development of the one being the reason for the underdevelopment of the other. The relations between the developed and underdeveloped parts (i.e. Prebisch's centre and periphery) prevent any chance of normal capitalist development in the underdeveloped countries.

In its radical versions, dependency theory is an extension of Marx, taking further the Marxist analysis of imperialism, of the dynamics of advanced capitalism or the characteristics of different types of development in social structures that have a "backward" sector. Underdevelopment is thus taken to be an inescapable concomitant of the laws of unequal development inherent in the capitalist system, and it arises from the way the capitalist mode of production in the dominant countries interacts with pre-capitalist or semi-capitalist modes in the dominated economies. The links between the centre and the periphery create and maintain underdevelopment and at the same time constantly exacerbate the disparities between the two parts of the system – which in turn fosters underdevelopment. Arghiri Emmanuel [20] argues that the economic relations between the centre and the periphery are based on principles of unequal exchange, and this thereby both overturns the theory of comparative advantage and provides decisive arguments in favour of dependency theory. Samir Amin [2] argues that, mainly because of the transfer of the surplus from the periphery to the centre, capital accumulation now occurs at the world and not the national level.

According to the radical exponents of dependency theory, the solutions to underdevelopment lie not in partial efforts to reform the system but in severing the bonds of dependency, then embarking upon various types of self-reliant development. A clean break with the capitalist world system is thus the main prerequisite in the struggle against underdevelopment, and countries must choose a completely different approach in order to put such problems behind them forever [3].

The virulence of the criticisms, the messianic tone, and the simplicity of the message expounded by the dependency school made their ideas very popular with some peripheral countries and were seen by the countries of the centre as an essential part of the prevailing third world ideology. In social science, there was also interest in other forms of dependency: e.g. political dependency [21] and dependent societies [74]. All in all, dependency theory generated considerable debate [64, 11, 8].

Another version of the strategy involving a clean break with capitalism must be mentioned so as to put all the schools of this period in a fair historical perspective: that produced by Soviet writers as part of the theory of non-capitalist development. This started life at the conference of 81 communist parties meeting in Moscow in November 1960 and became the main argument of the Soviet position on development in the 1970s [4]. The non-capitalist approach meant rejecting capitalism as a system and making a commitment to creating the material basis for a socialist society. This meant taking decisive steps against imperialism, capitalism, and feudalism, with an "attack" on major representatives of domestic and foreign capital, nationalization of the main means of production, the creation of a state sector, and the implementation of "radical" land reform. In fact, in order ultimately to achieve its economic goals, the strategy should start with certain political measures: the removal of "pro-imperialist forces" and the establishment of a policy of cooperation with the socialist bloc countries [71]. In the final analysis, implementation of this policy of cooperation constitutes the only valid indicator of success in carrying out such a strategy of non-capitalist development! After the implosion of the communist world, nobody knows what aftermath, if any, these notions of development may turn out to have.

Questioning and crises

It seemed a propitious moment, nearly 20 years after the first development programmes and plans were launched, to take preliminary stock of the results achieved thanks to the studies and advice provided by development economists. Three conclusions were drawn.

First, development did not always occur, and instead there were often disappointments and surprises. In the search for the miracle solution to the problems of underdevelopment, policies were changed too frequently (see, as a by no means exceptional example, the description of the successive development policies applied in Pakistan in an effort to keep up with the latest fashion: Haq [24]). The first disillusionments did not manage to dispel entirely the belief that miracle solutions existed, and the hunt for the developmental philosopher's stone was doggedly pursued. Nevertheless, every change of fashion drew attention to a new aspect of the complexity of the problems of underdevelopment.

Second, there is no miracle solution for generating development, which in any case cannot be achieved through the automatic workings of a set of economic variables and indices. Reality is not in the least like that, as is proved by the way that, where incomes did rise, the benefits did not spread evenly – instead, the rich became richer and the poor became poorer. The vaunted "trickle-down effect," so eagerly awaited, did not in fact occur.

Third, development is not accurately reflected in statistics in national accounts (e.g. gross national product or per capita national income), and it is not possible to monitor progress by such means. Other indicators besides growth in per capita income must be selected and examined.

This stocktaking led to a gradual broadening of the field studied by development economics, illustrated by a whole series of efforts to look critically at what was known and to extend it, as well as to take into account less purely economic aims in development strategies. This shift can be seen clearly in the various proposals made by international organizations and certain non-governmental organizations as they tried to define the goals of development policies. In 1969, the International Labour Office (ILO) launched its World Employment Programme, which gave priority to efforts to combat unemployment in all development strategies in the third world. In the early 1970s the World Bank announced its preference for "redistribution with growth" [13] as the only way to achieve equitable development. The second United Nations Development Decade, the 1970s, stressed "employment-oriented" development strategies that, with help from the prevailing populist tendencies, provided a good moment to revive the concept of "hidden unemployment," now renamed "the informal sector" and deemed likely to act as the engine of new types of development [69, 36, 51, 16]. The fight against poverty, which had been the prime mission of all development planners in countries like India in the early 1960s, also became a major concern of international organizations like the World Bank [80, 24]; the ILO, at its world conference on employment in 1976, adopted the strategy of "satisfying basic needs" [31]. According to the ILO's director-general, basic needs include not just material items like shelter, food, clothing and essential community services like supplies of drinking water, public health measures, public transport and education, but also non-material needs such as human rights, a job, and a share in decision-making.

Such attempts continue unabated, with each one highlighting new aspects of development problems that require attention: an alternative development strategy should be "need-oriented, endogenous, self-reliant, ecologically sound and based on the transformation of social structures" [17]. Unesco encouraged "endogenous development" that respected the cultural identity and lifestyles of each society [1]. Another, supported by the United Nations Conference on Trade and Development, sought to establish a "new world economic order," less hostile to the interests of developing countries and more in line with their development requirements. As we shall see in Ignacy Sachs's chapter in this volume, the theme of "sustainable development" was put forward in the Brundtland Report of the World Commission on Environment and Development [79]. The Commission defined sustainable development as being development that meets present needs without compromising the capacity of future generations to satisfy their own needs. The final declaration of the Earth Summit in Rio in June 1992, in affirming that human beings are central to the concerns of sustainable development and that they therefore are entitled to a healthy and productive life in harmony with Nature (Principle 1), was yet another instance of this new approach to development, looking to the future and being highly aware of the ecological and demographic problems of the Earth. This broadening of the field covered by development economics is also revealed in the "human development" reports of the United Nations Development Programme (UNDP), published every year from 1990 onwards.

These examples give some indication of the efforts to come to grips with a complex and multifaceted problem. Some features are unchanging (such as the deteriorating terms of trade for raw materials, poverty, or external domination, to mention only the economic ones), but there are also surprises, like the recent unexpected slowing of population growth in some countries in Asia, Latin America, North Africa, and the Middle East, the success of industrialization in the Far East with the emergence of the "little dragons," or the continually worsening position of the "least developed countries."

As the situation changes, things alter and disappear. Does "third world," associated as it is with the Cold War and the division of the world into two opposing blocs, have any meaning now? Some writers emphasize its diversity [82], or recognize its variety but are interested by what unites it [37], while others declare that it no longer exists [25]. The international organizations are now producing classifications based on new criteria and with more refined subgroups in an attempt to give a more accurate picture of the changed situation (see, for example, the World Bank's annual reports on world development from 1979 onwards, the UNCTAD reports on trade and development from 1981, or the ICSPS report published by Unesco in 1992).

The uneven pace of change, the resistance mechanisms, and the appalling problems of the developing world continue to stimulate and challenge development thinking. Each set-back is another clear refutation of the notion of miracle cures for underdevelopment, and each crisis offers a new occasion for finding and expressing serious doubts and reservations about elaborate claims for development. These criticisms, arising from disappointed hopes, poorly shouldered responsibilities, or disgust at instances of domination and repression, occur and recur according to a cyclical pattern. The amount of debate and its virulence are indeed one of the remarkable features of development economics. Each round of criticisms leads to new questioning, the formulation of new demands, and suggestions of new priorities.

During the 1980s major debates raged in many countries on the subject of development economics. In France, for example, criticism of "la vision tiers-mondiste" was accompanied by a declaration that development was finished [50]. Latouche wondered whether it would not be better simply to throw out development, on the grounds that it is the product of a technocratic attitude that considers only the economic aspects of the problem, whereas in fact underdevelopment results from the destruction of the cultural coherence of developing countries by the expansionist and imperialist forces of capitalism [38, 39]. In rejecting development that depends only on the solutions of economists and technocrats, that is based on copying other cultures and leads to the Westernization and acculturation of third world societies, Latouche is defending a historical and cultural approach that sees the solution to these countries' problems in a revival of their own cultural identity. Once their native cultural creativity is reestablished, they will be able to have an independent vision of their situation, identify their own problems, and find appropriate solutions for them (for a critical discussion of Latouche's views, see Kabou [34]).

At the end of his preface to the volume on the "pioneers in development" published by the World Bank, G.M. Meier acknowledges that readers of these accounts might well wonder whether the efforts of the pioneers in fact led to the creation of a new branch of economics, and if so, what proportion of their contribution is still valid and insightful, which questions have remained unanswered, and lastly where development economics is heading now. Many development economists try hard to provide answers to such questions.

Hirschman on the discipline in decline
Albert Hirschman, both an observer and a long-time participant, has examined the causes of the rise and decline of development economics [27]. In a text that Amartya Sen has called "an obituary of

development economics" [67], Hirschman suggests a typology of development theories based on two criteria: the monoeconomics claim (the belief that there is one form of economics that is valid everywhere for all time), and the mutual-benefit claim (the belief that reciprocal advantages are to be found in any bilateral relationship).

Using these two criteria, Hirschman distinguishes four types of development theory:

- orthodox neoclassical theories, which believe in the universality of economics and reciprocity of benefits;
- neo-Marxist and dependency theories, which reject these two postulates;
- those based on "Marx's scattered thoughts on development of 'backward' and colonial areas," which accept monoeconomics but reject notions of mutual benefit;
- development economics, which rejects the monoeconomics claim but accepts the mutual benefit one.

For Hirschman,

it is easy to see that the conjunction of the two propositions – (a) certain special features of the economic structure of the underdeveloped countries make an important portion of orthodox analysis inapplicable and misleading, and (b) there is a possibility for relations between the developed and underdeveloped countries to be mutually beneficial and for the former to contribute to the development of the latter – was essential for our subdiscipline to arise when and where it did: namely, in the advanced industrial countries of the West . . . at the end of the Second World War. [27, p. 375]

According to Hirschman, the first proposition was the prerequisite for creating a separate theoretical structure, and the second was needed "if Western economists were to take a strong interest in the matter."

Today, however, this "fledgling and far-from-unified subdiscipline" is in a state of crisis thanks to the impact of two factors: on the one hand, a double attack from both the Right and the Left; on the other, the succession of development disasters that have occurred in several third world countries. The right-wing attack, from the neoclassicists, criticized development policies for denying the "universal validity of economic laws" and consequently alleged they were the main reason for the misallocation of resources in underdeveloped countries. The left-wing attack (from neo-Marxists and dependency theorists) argued, among other things, that these "so-called development policies only created new forms of exploitation and 'dependency.'" As for the disasters, these were "clearly *somehow* connected with the stresses and strains accompanying development and 'moderniza-tion,'" and they include everything "from civil war to the establish-ment of murderous authoritarian regimes."

In the climate created by these criticisms and political disasters, development economics switched from optimism to deep pessimism. There then followed "a Freudian act of displacement" in which, so as to make up for their anguish at the political situation, some special-ists in development theory and practice attacked the weak points in the economic results. Both the political and economic balance sheets were disappointingly bad, so that Hirschman could write:

development economics started out as the spearhead of an effort that was to bring all-round emancipation from backwardness. If that effort is to fulfill its promise, the challenge posed by dismal politics must be met rather than avoided or evaded. By now it has become quite clear that this cannot be done by economics alone. It is for this reason that the decline of develop-ment economics cannot be fully reversed: our subdiscipline had achieved its considerable luster and excitement through the implicit idea that it could slay the dragon of backwardness virtually by itself, or at least, that its con-tribution to this task was central. We now know that this is not so. [27, p. 387]

Seers on the death of the discipline
In his article on "the birth, life and death of development economics" [63], Dudley Seers goes even further. He too considers that develop-ment economics started in the 1950s, and traces its ancestry in part to colonial economics. The other part, according to Seers, was political opportunism with regard to the development of "backward" coun-tries, on the part of both their own governments and the major cap-italist countries, who saw in development an efficient means of fighting the communist threat.

Based on simplistic arguments, development "became increasingly identified with economic growth, as measured by the national income (defined according to Keynesian conventions)." The "developed" countries, seen as the social and political models, had high per capita incomes, so that high per capita income became both a necessary and sufficient condition for creating a welfare state with low unemploy-ment. In order to raise incomes, what was needed was capital. There was general agreement, shared even by the Marxists, that the goal was higher incomes and that capital investment was the means of achieving it. This vision of development was strengthened by a range of innovations in economics and statistics (national accounts, growth models, development plans, etc.), and their elaboration and use became a prerequisite for success in implementing development strategies.

The force of circumstances and the complexity of the situation in practice soon led to disillusionment. Already in 1964, at a conference at Manchester University on "the teaching of development economics: its position in the present state of knowledge," serious doubts were expressed not only about the efficacity of approaches to development based on concepts such as economic growth, but also about the usefulness and appropriateness of neoclassical economics when applied to underdeveloped economies. Seers reminds us how these doubts, expressed more and more clearly, developed and ended up by discrediting the discipline as a whole: "development economics in the conventional sense has therefore proved much less useful than was expected in the vigorous optimism of its youth. In some circumstances, it may well have aggravated social problems if only by diverting attention from their real causes" [63, p. 712].

Seers's low opinion of what development economics had achieved is matched by unambiguous scepticism as to the discipline's chances of survival, which he considers slim for two reasons. For one thing, it has become clear that the economic aspects of development cannot be studied in isolation from the social, political, and cultural factors. A macroeconomic analysis of changes in consumption patterns cannot pretend to be exhaustive unless it is accompanied by a study of foreign cultural influences or the way they are transmitted. Secondly, contrary to the understanding of the term development economics in the 1950s and 1960s, the problems of development are no longer confined to the developing countries. According to Seers, recent changes in the "developed" world, especially since the oil shock of the early 1970s, show that there is no longer a distinct frontier between North and South. Consequently, development economics should be disposed of, and greater emphasis should be placed on the similarities rather than the differences between countries.

Thus development economics died young, after much suffering. "The history of economic thought shows that, in the end, irrelevant theoretical frameworks are discarded." Henceforth, "the logical future . . . is the study and teaching of development in a social and political as well as economic sense, with a wider geographical coverage and special emphasis on European development needs" [63, p. 717].

Streeten: The dichotomies overcome
The reactions to these arguments were many and various. Paul Streeten believes that development theories suffer from not just one but several dichotomies, and if these are taken into account, it is possible to make other classifications that are closer to reality. In an article on "development dichotomies," reprinted as the conclusion of the Meier and Seers volume on the pioneers of development [47], Streeten discusses the typology proposed by Hirschman and argues that the theories of development, which at the outset were devised on the basis of broad generalizations and abstractions, as they are based increasingly on concrete examples, is becoming more precise and realistic. It is true that there is a wide range of development theories, but with time this diversity appears both relative and highly complex as it is realized that many of the South's problems are shared by the North, and that few problems are common to all the countries of the South. This diversity is not always an expression of a split in the discipline, dividing it into opposing camps, but may be rather the result of the fact that there are many possible solutions to every social problem. Seen from this angle, the dichotomies are not, as Hirschman believed, the reason for the inevitable decline of development economics, but are on the contrary a sign of its great richness and intellectual vitality.

It should also be stressed that these different theories of development are not always separated from one another by insurmountable barriers and that in many cases the diversity of theories hides the beginnings of convergence, in so far as the tools created by development economists prove useful and efficient in analysing developed economies, as has been the case, for example, with the application of structuralist theories to the study of inflation in the industrialized countries. According to Streeten, therefore, we are not witnessing the death of development economics but its transition "from the 'economics of a special case', viz. Third World economies, to a new global economics of shared problems, but with greater differentiation of approaches and analyses" [73, p. 886].

Thanks to these analytical methods, development economists should concentrate on building up three hitherto neglected aspects of their research:
– the historical dimension, so as to aid understanding of how things came to be what they are now;
– the global dimension, which would entail a study of international

relations transcending national frontiers, of the interactions among the various national policies and the international system, and finally of "the alliances of interests across national boundaries";
– the "micro-micro" dimension, so called because it deals not only with what happens within a country, but also within the firm, the household, and "possibly within one individual, with conflicting desires." Of the three institutions – the public sector, the market, and the household – it is the last that has been the most neglected by economists.

Based on these areas of research and with attitudes favouring synthesis, development economics could yet usefully provide "imaginative but carefully worked out visions of alternative social possibilities" [73, p. 875].

Sen and the economics of "entitlement"

In a speech to the Development Studies Association in Dublin on 23 September 1982, Amartya Sen, too, discussed the questions raised by Hirschman and offered an assessment of development economics. In his view, the evaluation is far from negative: traditional development economics "has not been particularly unsuccessful in identifying the factors that lead to economic growth in developing countries" [67]. To achieve this growth, development economists prescribed a policy based on several major strategic themes: industrialization, rapid capital accumulation, mobilization of underemployed manpower, planning, and an economically active state.

A swift examination of the results of efforts at economic development in underdeveloped countries between 1960 and 1980 leads Sen to conclude that there is "still much relevance in the broad policy themes which traditional development economics has emphasized. The strategies have to be adapted to the particular conditions and to national and international circumstances," but these themes are not "rejectable" and "the time to bury traditional development economics has not yet arrived" [67, p. 753].

This position does not prevent Sen from recognizing the "real limitations" of this new discipline: in fact, "it has been less successful in characterising economic development." Here, the limits of development economics appear much more clearly; they arise "not from choice of means to the end of economic growth, but in the insufficient recognition that economic growth was no more than a means to some other objectives." In order to fill these gaps, Sen proposes a new definition of economic development:

86

Perhaps the most important thematic deficiency of traditional development economics is its concentration on national product, aggregate income and total supply of particular goods rather than on "entitlements" of people and the "capabilities" these entitlements generate. Ultimately, the process of economic development has to be concerned with what people can and cannot do, e.g. whether they can live long, escape avoidable morbidity, be well nourished, be able to read and write and communicate, take part in literary and cultural pursuits, and so forth. It has to do, in Marx's word, with "replacing the domination of circumstances and chance over individuals by the domination of individuals over chance and circumstances." [67, p. 754]

For Sen, "entitlement" refers to "the set of commodity bundles that a person can command in a society using the totality of rights and opportunities that he or she faces" [67], which allow that person to acquire some capabilities and not others. The process of development can be seen as a process of expanding people's capabilities and entitlements. This study of entitlements should not cover just purely economic factors but should take into account the political arrangements "that affect people's actual abilities to command commodities, including food" [67]. Thus, here again, an examination of development economics reveals the need for a new definition, stressing the non-economic and above all the political aspects of development phenomena.

Lewis: A discipline in good health
Defining the aim and the contents of development economics was also the main concern of W.A. Lewis in his speech to the ninety-third congress of the American Economic Association in San Francisco on 29 October 1983 [41]. Examining the "state of development theory," he proposes a new definition of development economics as that branch of economics dealing with "the structure and behavior of economies where output per head is less than 1980 US$2000" [41]. The justification for making this a separate discipline lies in the need for analytical concepts and tools appropriate to the special problems of these economies.

These problems fall into two main categories: problems of resource allocation in the short term and of growth in the long term. With regard to the first category, the difference between the developed and underdeveloped economies is one of degree and not kind, since the same phenomena are found everywhere that the market, influenced by a variety of factors, no longer brings about the famous state of equilibrium. In these circumstances, prices no longer reflect accurately the trends in supply and demand, thus preventing a sound allocation of economic resources. Among the examples of market failure are where prices no longer reflect real social cost, where "the unregulated market constrains productive capacity," or where low elasticities of supply and demand coupled with low levels of stocks mean that the economy moves too slowly towards equilibrium. In addition to these cases where the price mechanism cannot function properly, there are others where, for "non-economic considerations," production and trade are not governed by the desire to maximize profits. To analyse these problems, recourse must be had to economic anthropology, the only discipline to have studied them.

Lastly, another non-economic factor must be taken into account in allocating resources in developing countries: the role played by the government. In poor countries, the government is far more active in the modern sector of the economy. Moreover, where there is market failure, it is up to the government to correct the errors and omissions of the price mechanism. Furthermore, it should not be forgotten that in these countries, it cannot always be assumed that the government represents the people, and that there are many models of government: "military (with generals), military (with sergeants), technocratic, aristocratic, popular front, peasant, kleptocratic – which react differently to similar stimuli" [41, p. 4]. Here, recourse must be had to the expertise of sociologists and political scientists.

Analysis of long-term growth involves quite distinctive problems peculiar to developing countries: the search for the motor driving development and the models of development. "The economist's dream would be to have a single theory of growth that took an economy from the lowest level of, say, $100 per capita . . . up to the level of Western Europe and beyond" [41, p. 4]. The problem is that such a theory does not exist. We have plenty of models for the final state of economic maturity and for those at the bottom of the ladder. As to the countries halfway up, our knowledge is too fragmentary and inadequate for us to answer a crucial question: "how output would be affected by policies that gave, say, an extra five percentage points of the national income to the bottom 80 percent of the population, assuming a peaceful transfer over, say, ten years. Would output per head rise faster, more slowly, or at the same rate? One must also ask whether it matters, or should the change be made in any case" [41, p. 6].

The pattern and size of these changes are poorly understood, and there is minimal knowledge about the "engine" of economic growth.

What is it? Investment (whether in plant or in people) is not the only factor in growth, yet there is a strong correlation between the two, which allows us to consider it "as a proxy for the forces propelling the economy." How therefore does investment occur and what are the driving forces behind it? They may be economic (credit institutions, the tax system, etc.) or, more likely, non-economic, and the relationship between them and social institutions has always been one of the concerns of development economics. "Given the importance of incentives and institutions, are there particular circumstances that favor growth?" Each school of economists has chosen its own engine of growth: agriculture for the physiocrats, foreign trade surplus for the mercantilists, the free market for the classical school, capital for the Marxists, entrepreneurship for the neoclassical school, and so on.

According to Lewis, there is no single engine. "Growth occurs wherever there is a gap between capability and opportunity. Capability covers skills (domestic and foreign), government, savings and technology. Opportunity can be of any kind, including markets, rainfall, access to licenses, infrastructure. The engine may be at home or abroad, an innovation, a good site for a transportation center, or much else" [41]. Hence the problem of growth is extremely complex, demanding a whole package of complementary theories rather than a single, universally applicable one. However, there is no general agreement about the theories to be included in the package.

For Lewis, the central one must be the theory of distribution, because this will provide incentives and savings. Among the others are the theories of government, of training, and the class struggle, of the firm and entrepreneurship. "Thus, a theory of the growth of the economy as a whole brings together what we know of its parts." But it is not enough merely to explain growth, or to have a model that produces growth. Countries may grow strongly for a short while, then growth slackens or decline may even occur, so that "one must also be able to explain why some countries fall out of the line while others keep up the pace" [41].

Attention must also be paid to the problems arising from "self-sustaining" growth, which should be analysed with regard to both resources and leadership (public and private). As regards the first, a country can be said to enjoy self-sustaining growth when it is "more or less self-sufficient in savings, in its managerial cadre, in skilled workers, and in other infrastructure. The physical part we can quantify, even if rather arbitrarily" [41]. As regards the second, there is no way of predicting or prescribing the qualifications required for leadership; all that can be done is to observe that for the moment the developing countries lack leaders of sufficient quality to deal with the tasks of a self-sustaining economy.

After this survey, Lewis maintains that his subject is still just as much alive as any other branch of economics. "If conflict and dispute are indices of intellectual activity, our subject seems adequately contentious. Development economics is not at its most spectacular, but it is alive and well" [41].

Prospects

What conclusions can be drawn from this long debate? Let us first acknowledge with W.A. Lewis that the life of the discipline is not in danger. Nobody can deny that there have been set-backs in some of the efforts to promote development, causing bitterness among the various partners involved. But the existence and survival of an academic discipline do not depend entirely on the success of policies recommended or inspired by its specialists, let alone their unfailing success in dealing with problems. In social sciences, failure may be a source of vitality, since it shows that the problem still has to be solved and therefore the business of analysis and research must be pursued further.

Development remains one of the crucial topics in economics: the functioning of any economy means that it generates and experiences change. The main aim of development economics is to identify the laws, causes, forms, and manner of these changes. Another aim is to study the direction and nature of these changes: are we witnessing "the natural movement towards opulence and improvement" (Adam Smith), a decrease in poverty (W.A. Lewis), or a growth of entitlements (Sen)? What should be done to trigger, speed up, or control this trend?

All economic systems have to cope with these problems, but development economics is concerned above all with studying them in the context of the former third world. Now, however, there is greater awareness of the diversity within different regions and countries, each with its own culture and its own experience of change generated by development in particular conditions and at varying speeds. The recognition of this variety has disturbed some development economists, as it means that universal remedies are misleading and useless. The critical analysis of these set-backs should be conducted with perseverance and clear-sightedness, and should make us less ambitious: let there be no more talk of "miracles," since they threaten to turn quickly into "mirages" [42]. Development is a long and slow process of change. The various stages cannot be rushed without suffering the consequences of trusting in the logic of formal models. And in fact, do these "stages" really exist? We must therefore start work all over again, taking account of the diversity of circumstances and the many dimensions of change, which can be economic, but also cultural, social, and political. Hence the need for constant recourse to multi-disciplinary methods of analysis.

We should also not forget the prime importance of global, multi-national, not to say imperial, economies in the process of development. This aspect has been highlighted above all by the work of the dependency school, and to neglect it would be to leave gaps in the analysis and in our understanding of the phenomena being studied.

Lastly, we must also stress the importance of the political dimension. In practice, any experience of development is simply the illustration of a development strategy, drawn up and carried out by a political authority. It cannot therefore be assessed without reference to the political context of its conception and implementation. Those development economists who like to think of themselves as development strategists and thus become "the Prince's advisers" should not be surprised to find their names on the list of those responsible for any given development catastrophe, figuring among those with "dirty hands," suspected of having helped the creation and growth of bloody regimes.

How can one be political without working for the "unenlightened despots" of the developing world? Development economists find themselves faced with Max Weber's dilemma of the scientist and the politician. In order not to collapse under the weight of "disillusionment" after rude awakenings, which also recall the well-known "white man's burden" caustically described by those with now-fashionable anti-third world views, development economists must ensure that their analysis is always a critical one. If they do, development economics will not become, as some fear, a branch of knowledge dedicated to the maintenance of the status quo, but rather one that helps "individuals to dominate chance and circumstances" [33]. The story of development thinking is no longer circumscribed by the intellectual weight of the most industrialized countries. No longer is it captured by economic references alone. It will continue as long as disparities between nations delineate an unbearable frontier of poverty and injustice. And thus, it will continue as long as it is a matter of both scientific research and political struggle that transcends the understanding of economics *per se*, as well as the heritage of the "Northern/Western" paradigm.

References

1   Abdelmalek, Anvar et al. *Clés pour une stratégie nouvelle du développement.* Paris: Presses de l'Unesco/Éditions ouvrières, 1984.

2   Amin, Samir. *L'accumulation à l'échelle mondiale.* Paris: Anthropos, 1970.

3   ———. *La déconnexion.* Paris: La Découverte, 1986.

4   Andriev, I. *The Non-capitalist Way: Socialism and Developing Countries.* Moscow: Progress Publishers, 1977.

5   Arndt, H.W. "Economic Development: A Semantic History." *Economic Development and Cultural Change* 29 (April 1981), no. 3: 462–472.

6   Baran, Paul. *The Political Economy of Growth.* New York: Monthly Review Press, 1957.

7   Bhagwati, Jagdish N. *Wealth and Poverty: Essays in Development Economics.* Vol. 1. Edited by G. Grossman. Oxford: Blackwell, 1985.

8   Blomstrom, Magnus, and Björn Hettne. *Development Theory in Transition: The Dependency Debate and Beyond: Third World Responses.* London: Zed Books, 1984.

9   Boeke, Julius Herman. *The Evolution of the Netherlands Indies Economy.* New York: Institute of Pacific Relations, 1946.

10   ———. *Economics and Economic Policy of Dual Societies as Exemplified by Indonesia.* New York: Institute of Pacific Relations, 1953.

11   Cardoso, Fernando H. *Les idées à leur place: le concept de développement en Amérique latine.* Translated by D. Ardaillon and C. Tricoire. Paris: A.-M. Métailié, 1984.

12   Chenery, Hollis, and T.N. Srinivasan. *Handbook of Development Economics.* 2 vols. Amsterdam: North Holland, 1948.

13   Chenery, Hollis et al. *Redistribution with Growth.* Oxford: Oxford University Press, 1974.

14   ———. *Structural Change and Development Policy.* Oxford: Oxford University Press, 1979.

15   Coquery-Vidrovitch, Catherine, Daniel Hemery, and Jean Piel, eds. *Pour une histoire du développement Etats, sociétés, développement.* Paris: L'Harmattan, 1988.

16   Coquery-Vidrovitch, Catherine, and S. Nedelec, eds. *L'informel en question? La recomposition des sociétés dépendantes.* Paris: L'Harmattan, 1991.

17   Dag Hammarskjöld Foundation. "What Now: Another Development?" *Development Dialogue* Nos. 1, 2 (1975).

18   Destanne de Bernis, G. "Industries industrialisantes et contenu d'une politique d'intégration générale." *Economie appliquée* 19 (July–December 1966): 415–473.

19   Dockès, Pierre, and Bernard Rosier. *L'histoire ambigue: croissance et développement en question.* Paris: Presses Universitaires de France, 1988.

20   Emmanuel, Arghiri. *L'échange inégal: essai sur les antagonismes dans les rapports économiques internationaux*. Paris: Maspero, 1969.

21   Evans, P. *Dependent Development: The Alliance of Multinational State and Local Capital in Brazil*. Princeton, N.J.: Princeton University Press, 1979.

22   Gendzier, Irene L. *Managing Development Change: Social Scientists and the Third World*. Boulder, Colo.: Westview Press, 1985.

23   Griffin, Keith. *Alternative Strategies of Economic Development*. London: Macmillan, 1989.

24   Haq, Mahbub ul. *The Poverty Curtain: Choices for the Third World*. New York: Columbia University Press, 1976.

25   Harris, Nigel. *The End of the Third World: Newly Industrializing Countries and the Decline of an Ideology*. London: I.B. Tauris, 1986.

26   Hirschman, A.O. *The Strategy of Economic Development*. New Haven: Yale University Press, 1958.

27   ———. "The Rise and Decline of Development Economics." In: *Essays in Trespassing: Economics to Politics and Beyond*. Cambridge: Cambridge University Press, 1981.

28   Hoselitz, B.F., ed. *Theories of Economic Growth*. Glencoe, Ill.: The Free Press, 1960.

29   ICSPS (International Council for Science Policy Studies). *Science and Technology in Developing Countries: Strategies for the 90s, A Report to UNESCO*. Paris: Unesco, 1952.

30   ILO. *Strategies for Employment Promotion*. Geneva: ILO, 1973.

31   ———. *Employment, Growth and Basic Needs: A One World Problem: Report of the Director General of the International Labour Office to the Tripartite World Conference on Employment, Income Distribution and Social Progress, and the International Division of Labour*. Geneva: ILO, 1976.

32   ———. *Meeting Basic Needs: Strategies for Eradicating Mass Poverty and Unemployment*. Geneva: ILO, 1976.

33   ISS (Institute of Social Studies). *Rethinking Development*. The Hague: ISS, 1982. Pp. 1–19.

34   Kabou, Axelle. *Et si l'Afrique refusait le développement?* Paris: L'Harmattan, 1991.

35   Kemp, Tom. *Industrialization in the Non-Western World*. London: Longman, 1983.

36   Kitching, Gavin. *Development and Underdevelopment in Historical Perspective*. London: Methuen, 1982.

37   Lacoste, Yves. *Unité et diversité du Tiers Monde*. 3 vols. Paris: Maspéro, 1980.

38   Latouche, Serge. *Faut-il refuser le développement?* Paris: Presses Universitaires de France, 1986.

39   ———. *L'occidentalisation du monde: essai sur la signification, la portée et les limites de l'uniformisation planétaire*. Paris: La Découverte, 1989.

40   Lewis, W. A. *The Theory of Economic Growth*. London: Allen and Unwin, 1955.

41   ———. "The State of Development Theory." *American Economic Review* 74 (1984), no. 1: 1–10.

42   Lipietz, Alain. *Mirages et miracles: problèmes de l'industrialisation dans le tiers monde*. Paris: La Découverte, 1985.

43   ———. *Toward a New International Order*. Cambridge, Mass.: Polity, 1992.

44   Mandelbaum, Kurt. *The Industrialization of Backward Areas*. Oxford: Blackwell, 1945.

45   Marx, Karl. *Capital*. Ed. F. Engels; trans. S. Moore and E. Aveling. Vol. 1. New York: International Publishers, 1963.

46   Meier, Gerald M. *Pioneers in Development*. 1964. 2nd series. Oxford: Oxford University Press, 1987.

47   Meier, Gerald M., and Dudley Seers, eds. *Pioneers in Development*. Oxford: Oxford University Press, 1984.

48   Nurkse, R. *Problems of Capital Formation in Underdeveloped Countries*. Oxford: Blackwell, 1953.

49   Oman, Charles P., and Ganeshan Wignaraja. *The Postwar Evolution of Development Thinking*. London: Macmillan, 1991.

50   Partant, François. *La fin du développement: naissance d'une alternative*. Paris: La Découverte, 1982.

51   Penouil, Marc, and Jean-Pierre Lachaud, eds. *Le développement spontané: les activités informelles en Afrique*. Paris: Pedone, 1986.

52   Perroux, François. *L'économie du XXᵉ Siècle*. Paris: P.U.F., 1961.

53   ———. *Pour une philosophie du nouveau développement*. Paris: Aubier, 1981.

54   Prebisch, Raul. *The Economic Development of Latin America and its Principal Problems*. New York: United Nations, 1950.

55   Queuilles, Pierre. *Histoire de l'afro-asiatisme jusqu'à Bandoung: la naissance du Tiers monde*. Paris: Payot, 1965.

56   Ranis, Gustav, and T. Paul Schultz, eds. *The State of Development Economics: Progress and Perspectives*. Oxford: Blackwell, 1987.

57   Rosenstein-Rodan, P. "Problems of Industrialization in Eastern and South-Eastern Europe." *Economic Journal* 53 (1943).

58   Roxborough, Ian. *Theories of Underdevelopment*. London: Macmillan, 1979.

59   Sachs, Ignacy. *Stratégie de l'éco-développement*. Paris: Editions ouvrières, 1980.

60   Sauvy, Alfred. "Introduction à l'étude des pays sous-développés." *Population* 6 (October–December 1951): 601–608.

61   ———. "Trois mondes, une planète." *France Observateur*, 14 August 1952.

62   Schumpeter, Joseph. *History of Economic Analysis*. London: Allen and Unwin, 1954.

63   Seers, Dudley. "The Birth, Life and Death of Development Economics." *Development and Change* 10 (1979): 707–719.

64   ———, ed. *Dependency Theory: A Critical Reassessment*. London: Frances Pinter, 1981.

65   Sen, Amartya. *Employment, Technology and Development*. Oxford: Oxford University Press, 1975.

66   ———. *Poverty and Famines: An Essay on Entitlement and Deprivation*. Oxford: Oxford University Press, 1981.

67   ———. "Development: Which Way Now?" *Economic Journal* 93 (1983): 745–762.

68   ———. *Resources, Values and Development*. Oxford: Blackwell, 1984.
69   Sethuraman, S. V., ed. *The Urban Informal Sector in Developing Countries: Employment, Poverty and Environment*. Geneva: ILO, 1981.
70   Singer, H. "The Distribution of Trade between Investing and Borrowing Countries." *American Economic Review*, no. 40 (1950): 473–485.
71   Solodovnikov, V., and V. Bogoslovsky. *Non-capitalist Development: An Historical Outline*. Moscow: Progress Publishers, 1975.
72   Stern, Nicholas. "The Economics of Development: A Survey." *Economic Journal* 99 (September 1989), no. 4.
73   Streeten, P. "Development Dichotomies." *World Development* 11 (1983), no. 10: 875–889.
74   Touraine, Alain. *Les sociétés dépendantes*. Paris: Duculot, 1976.
75   UNCTAD. *Trade and Development Report*. Geneva: UNCTAD, 1981.
76   ———. *Trade and Development Report*. Geneva: UNCTAD, 1991.
77   ———. *The Least Developed Countries: 1991 Report*. Geneva: UNCTAD, 1992.
78   Viner, J. *International Trade and Economic Development*. Oxford: Clarendon Press, 1953.
79   WCED (World Commission on Environment and Development). *Our Common Future* (The Brundtland Report). London: Oxford University Press, 1987.
80   World Bank. *The Assault on World Poverty: Problems of Rural Development, Education and Health*. Baltimore: Johns Hopkins University Press, 1975.
81   ———. *World Development Report*. Washington, D.C.: World Bank, 1977.
82   Worsley, Peter. *The Three Worlds: Culture and World Development*. Chicago: University of Chicago Press, 1984.

3

Measuring science, technology, and innovation

Jan Annerstedt

The Earth is round, but it does not appear perfectly spherical if we examine the worldwide distribution of resources devoted to research and experimental development (R&D) – defined to include fundamental and applied research. Far from being evenly distributed throughout the world, these resources are concentrated in a small number of countries. In the early part of the 1970s less than 3 per cent of the world's R&D expenditures were made by the developing countries, and just a little more than 11 per cent of its researchers – its R&D scientists and engineers – were employed there [1].

According to more recent, though less complete, data to be presented later in this chapter, changes influencing this North-South relationship have occurred, but the overall pattern has remained much the same through the 1970s and 1980s until today. The highly industrialized countries have kept their dominant position while strengthening their R&D capabilities. But notable shifts of positions have taken place within the developing world as countries like Brazil, India, and the Republic of Korea have increased spending on R&D, while, in the same period, a number of other developing countries have been forced to reduce their science and technology base.

On the basis of this crude statistical picture, it is no exaggeration to claim that not all countries are able to undertake the scientific and technological activities that they desire. As regards resources, the majority of the nation-states in the world are a research desert, and the remaining countries can still be looked upon as a small number of R&D oases, some of which are very large [3].

To government policy makers and corporate managers in the

highly industrialized countries, the global distribution of R&D resources may appear quite attractive. In their countries resources seem to be abundant. The social production and diffusion of knowledge have turned into specialized professional activities, financed and performed by many different firms, institutions, and other organizations throughout society.

In principle, science and technology could play similar roles in the economies of all other countries as well. Lack of resources, low levels of skills, few training opportunities, inappropriate curricula in higher education, weak technology-supporting institutions, etc., may prevent some of them from exploiting all of the nation's innovative capabilities. "Science and technology can only take root in a given society if their structures and goals are well matched to prevailing modes of thinking and of doing, in particular to local traditional technologies" [20, p. 52].

At the end of the twentieth century only a small number of developing countries have been able to create and maintain comparatively strong national R&D capabilities. Most developing countries do not possess the scientific instruments and other highly specialized equipment needed for advanced research in many fields of study. Instead of scientific research, they have to rely heavily upon other varieties of organized knowledge to better utilize and further develop their productive potential.

Typically, in Africa, Asia, and Latin America, the industrial firm wanting to innovate has no choice but to copy or simply accept incremental technical change for the renewal of its products and manufacturing capabilities. Indigenous technological capabilities at the level of the firm do exist – and local engineering and advanced consulting services are actually expanding in most developing countries – but technologically significant inventions are generally generated outside the company or industry. Science-based technical change in these countries is rarely indigenous.

At a lower technical level than in the industrial part of the world, economically significant inventions in the developing countries are more frequent. As these countries innovate, and they certainly change technologically, the sources of innovation seldom include endogenous R&D. There are examples of industrial firms and even branches of industry that have proved their capacities to renew and innovate without access to a specialized and multifaceted scientific and technological base.

Strong R&D capabilities are in fact not the same as strong innovative capabilities at the level of the firm. There may be a correlation, but – apart from scientific results and laboratory practices – it should be emphasized that there are many sources of invention. Nor is the effective diffusion and application of new inventions caused by strong capabilities in R&D. However, since new innovations in industry are increasingly science-based or high tech, the relatively scarce and scattered R&D activities among the developing countries have to be complemented by a steady stream of information and ideas, new goods and services, production methods and "best practices," patents and licences created elsewhere.

The growing need for R&D and innovation indicators

Since the early 1960s, basic data on resources devoted to research and experimental development have been collected by an increasing number of countries. R&D is by far the best measured category of innovation and may even be integrated as such into the UN System of National Accounts (SNA).

Some 140 countries have succeeded in producing statistical maps of their national R&D landscape (Unesco 1991 [55] includes only 128). By no means are all of the maps up to date. A few are just listings of government budgetary allocations for R&D with estimates of other R&D expenditures. We must remember, however, that the majority of today's developing countries have reached independence only since the Second World War; well over 100 nation-states have emerged on the political scene since the late 1940s. When they achieved political independence, most of them had only rudimentary R&D capabilities.

By aggregating national statistics, it becomes possible to make regional and even global aggregates of resources devoted to R&D. But these summaries cannot be better than the often fragmentary data available nationally. In order to produce statistically sound descriptions of the global R&D effort, more detailed and standardized data are needed.

Regardless of their position or general outlook, planners and decision makers in both the industrial and developing countries have a common need of R&D data that are more suitable for fully fledged descriptions and appropriate analyses of the international or global changes in the economy. Less and less are R&D statistics regarded as an independent category of data: since they measure a crucial part of society's innovative activities, they are seen as only one component of several significant sets of data on innovation and economic growth.

At present, and at least for some sectors of society, R&D statistics are being re-examined in a wider context of innovation and adjusted to fit better into national and international surveys of innovation under different socio-economic circumstances. Especially among the industrialized countries, further details are being asked for, e.g. on the flow of R&D funds between countries – in general and between companies located in different countries but within the same economic zone or region. Given the predominant role in overall R&D activities of large industrial firms operating in several countries, figures describing national R&D resources alone lose some of their value. They are gradually being supplemented by internationally comparable data at the company or industry sector level.

Such general pictures cannot be painted without an extensive use of reliable indicators. It is not enough just to order categories of basic figures and draw simple conclusions from unrefined tables. True, it may be of value to highlight important, though elementary, comparisons between countries and regions, but to become analytically useful, the statistics will have to be shaped into indicators that are defined within – or at least closely related to – a specific conceptual framework or analytical model. R&D and innovation statistics may serve several such models; and the models could change over time and still exploit the same series of data. The models may also link data on R&D and innovation with existing statistics on other economic and social activities, thereby creating new, more sophisticated indicators.

Step by step, over the past 10 years, there has been a move from input indicators towards output and impact indicators. Examples from the latter category are combined data on high-technology investments and trade; patents taken out at home and abroad; co-operative agreements on the transfer of know-how; strategic "technological alliances" between firms; and imports or exports of components and services with a significant technology content. For a developing country such output indicators could serve important purposes in the assessment of innovative capabilities and of technology gaps between countries or between branches of industry. Although significant, the move towards output and impact indicators has been slow and the statistics produced are still fragmentary, even among the industrially most advanced nations.

From macro-phenomena to innovation processes

Among policy makers as well as economists and other social scientists there is a widespread consensus that current R&D statistics should be further extended and developed by way of broader "innovation surveys." This widening of the statistical realm should improve the understanding of the role that R&D plays in innovation and help explain differences in performance between firms, sectors of industry, and (even) national economies.

However, differences in the level of development between countries may easily cause measurement problems. The same set of R&D and innovation indicators could give rise to different interpretations in different economic contexts (see Madeuf [21]). Probably, as among the industrialized countries, statistical analyses within a specific developing country grouping – with common economic characteristics – may prove to be more analytically fruitful. For instance, a developing country government that promotes export-oriented industrial strategies may understand technology-transfer data very differently from a government that supports inward import-substitution strategies. Likewise, countries operating similar economic policies are easier to compare.

Until recently, internationally comparable R&D statistics have been collected and processed only for macro-phenomena in the economy. Except for an increasing number of case-studies in industry, relatively little is known about innovation processes at the level of the firm or in subsectors of industry. Now, however, policy interests have stimulated R&D statistical studies of the linkages between the macro- and micro-levels with a view to assessing the flow of resources and evaluating the relative economic impact of investments in R&D and related activities. Ongoing international statistical efforts may help to overcome the current lack of transparency and compensate for the imperfect knowledge of the processes of innovation.

In both government and industry, policy makers and analysts have expressed a growing need for more sophisticated and usable R&D and innovation indicators. Such indicators should reduce uncertainties and help advance plans and decisions regarding national and sectoral science and technology efforts. By way of international comparisons, the specific conditions for innovation in areas such as industry and trade, education and training, public health and social security, could be further elaborated. But we are still far from viable international comparisons even among the highly industrialized countries, e.g. between Germany, Japan, and the USA.

For the developing countries, the relevance of available R&D output indicators varies. Output data commonly used in industrialized countries – such as rates of publication and the number of citations in internationally available journals, as well as statistics on patents and licences – are not easy to interpret in a third world setting. This is due to the lack of uniform and non-biased data in relation to publication and other communication practices. More importantly, the structural features of each developing economy demand a different framework for the analysis. The diversity among developing countries in organizing a national R&D system, in linking endogenous research to international (or Western) science, in improving the techno-scientific infrastructure, in furthering manpower development, etc., make output indicators complicated and even controversial, particularly for comparisons between industrial and developing countries [20, p. 53]. "There are no adequate, comprehensive indicators of development, which can reflect the complex cultural, social, economic, and political factors at play when the concept of 'development' is considered with all of its multidimensional implications. At best, there are some indicators of the penetration of western patterns into different societies" [20, p. 52].

As regards the particular needs of developing countries, R&D and innovation indicators should not only permit systematic international comparisons, but also provide information in order to assess the efficiency of science and technology capabilities, measure the flow of technology through various channels, and help analyse the contribution of both foreign and domestic sourcing of science and technology. Ideally, these and similar indicators should further the analysis of R&D and innovation policies aiming at balancing foreign and domestic sourcing of technology and enhancing the local science and technology base [21].

There is a general need to develop more sophisticated methods of surveying the diffusion of technology and other kinds of innovative activities, particularly methods to be used for advanced international comparisons.

Towards a worldwide standard for R&D surveys

Since the 1930s, and particularly during and after the Second World War, a dominant attitude among policy makers in the industrialized countries has been that of a necessary mobilization of science and technology for economic purposes as well as for national security and related strategic objectives. In the larger industrialized countries, the building of strong sectoral R&D capabilities responded to the needs of the military and, later, also to the reconstruction and economic recovery during the first 15–20 years of the post-war period [17]. Accordingly, the emphasis by statisticians was very much on the "supply side" of the national R&D system.

The first national surveys were based on approximated expenditure data for science and technology and on crude numbers of scientists and technologists in government and industry. "Looking through the various national statistical yearbooks, one is impressed by how many countries have felt the need to count their donkeys and how few their scientists," Stevan Dedijer wrote in a summary of the 1950s [11–13]. He and other pioneers of R&D statistics had to draw upon all kinds of primary data to quantify the resources devoted to R&D while attempting to make international comparisons. There were, in those early years, no serious attempts by intergovernmental agencies to provide quantifications of the global R&D effort. Instead, examples were set by individual scholars like John D. Bernal, who calculated national "budgets of science (and technology)" for several countries as early as the 1930s [8]. (Dedijer mentions Soviet studies of the mid-1920s with similar ambitions.)

In its study of science policy for the 1960s, the OECD (Organisation for Economic Co-operation and Development) found existing R&D statistics "grossly inadequate. Most countries have more reliable data on their numbers of poultry and their egg production than on their numbers of research scientists and engineers and their output of discoveries and inventions" [24, p. 21]. During the second half of the 1960s and in the early 1970s, the situation improved significantly. This was a period when statistical resources were activated all over the world in the quantitative study of R&D. All tables and charts that quantified the resources of the national R&D systems were dominated by rather simple data on given inputs into science and technology, only rarely supplemented by easily available output data of the system such as scientific papers, patents, licences for technology, etc. But advances were on their way.

Among the industrial countries the interest by the main users of R&D statistics had shifted to the "demand side" as opposed to the "supply side" of the earlier period. The market pull of technology, know-how, and other specialized knowledge was coming into focus after the long period of reconstruction of the national economies. Industrial innovation had become a competitive advantage. Accordingly, several of the national efforts by R&D statisticians were initiated by the drive towards better international comparisons (see, for example, Freeman and Young [18]). The relative economic performance of the different R&D systems had come into focus along with the growing interest in the role of science and technology for industrial innovation.

Still, while R&D statistics improved in certain highly industrialized countries, other countries approached the tricky problems of sources and methods with "quick-and-dirty" solutions in order to be able to present national R&D statistics with at least some of the required international comparability.

Among the first regionally based organizations to advance R&D statistical methodology and to promote comparative studies of R&D efforts was the OECD. Already in 1963, at Frascati in Italy, the OECD had convened a group of experts that soon developed a standard practice for surveys of research and experimental development, officially termed the Frascati Manual, which has been revised and updated ever since [25, 30].

In line with the Frascati Manual, statisticians of other regional organizations, such as the European Community and the CMEA (Council for Mutual Economic Assistance, formally dissolved in early 1991), have developed separate survey techniques and other analytical tools for national surveys and for cross-country comparisons. With early assistance by OECD experts, the Organization of American States (OAS) specified a standard for Latin America. Over the years, several such statistical endeavours have converged towards an international norm or standard for R&D data. But there is still no detailed, worldwide guide for R&D statisticians. Only a limited number of countries, most of them highly industrialized, have fully harmonized their statistics in this field.

For many years, Unesco (the United Nations Educational, Scientific and Cultural Organization) was a prime mover in the attempts to create a worldwide standard for R&D surveys (a comprehensive version is given in [51]; see also [52] and its later versions). Many developing countries have followed the suggestions by the organization not to design their own statistical methodology, but to accept that of Unesco. However, the problem of harmonizing already existing country standards and relating them to internationally accepted statistical methodologies has not been easily resolved. Moreover, the focus by

Unesco on developing countries has fostered survey techniques that are not always suited to the specialized policy needs of the highly industrialized countries. These latter countries were discouraged from using the Unesco R&D statistical methodology simply because it produced statistics that were too crude.

Subsequently, during the 1970s and 1980s, the industrialized countries settled with their own standards. In fact there were two: a western one for the OECD member governments (cf. [25]) and an eastern one for the CMEA members [10]. Nevertheless, without adopting a common standard, the two country groupings came close to matching their R&D statistical methodologies, although some basic statistical categories remained unrelated. Following the changes toward a market economy in eastern Europe and in the former Soviet Union and its successors, it is likely that the Frascati Manual will be adopted by all industrialized countries.

Despite continuous efforts, neither Unesco nor any other international agency has yet been able to implement, through the many national statistical units, a world standard on how to collect R&D data and further specify the kinds of innovative activities that should be measured as well. What has been agreed through Unesco is a general recommendation concerning the statistical categories by which data should be collected, processed, and presented. Agreeing on a general methodology is one thing; implementing it has proved to be quite another.

With or without a worldwide standard for R&D surveys, the majority of countries have regularly provided the Unesco statistical office with basic data on their R&D manpower and related expenditures. Most of this material has been published in the Unesco *Statistical Yearbook*. Other national data have been further processed for regional summaries and even global estimates of resources devoted to R&D (for the most recent global survey, see [55]; regional surveys have been conducted for, e.g., Latin America, Africa, and the Arab countries).

Following several revisions of the Frascati Manual over the last 10 years, the OECD secretariat has become a clearing-house for both national and international advances in R&D statistical methodology. Most importantly, the OECD has provided a permanent forum for expert consultations and responded actively to new statistical requirements. Its large unit of professionals engaged in the development of indicators have spent lengthy periods of exploratory work, involving the collaboration of national agencies and international organizations such as Unesco. Consultations and week-long seminars for R&D statistical staff of non-member countries, i.e. from eastern Europe, the former Soviet Union, and selected developing countries, are a relatively new feature among its activities.

For the OECD member countries, the benchmark source of R&D data is the biennial survey, conducted by national statistical agencies using the Frascati Manual's detailed questionnaires. These nationally collected data are fed into a "main science and technology indicators" database containing the variables most widely used over the past 20 years correlated with other data such as that of industry and trade. A data exchange system is operated in collaboration with national agencies and with the "Eurostat" of the European Community. Close relations are maintained with other international agencies as well. To meet specific policy needs, this exchange system should permit the design of special data segments.

Quantitative descriptions and qualitative assessments

Among the industrialized countries, it was not until the second half of the 1970s that the methodological work resulted in a deliberate push towards comparable "science and technology (S&T) indicators." Policy deliberations on industrial competitiveness in a new economic context and conflicts around the place of organized knowledge in society created a strong demand for this kind of internationally comparable data.

More importantly, the specific needs of actual and potential national users were better articulated. For the first time, R&D statisticians were placed in the centre of economic policy-making and forced to produce much more timely and appropriate indicators. In the OECD member countries, comprehensive sets of S&T indicators were generally available already by the end of the 1970s, following national attempts by, for instance, the United States National Science Foundation (NSF). The OECD indicators included inputs to the R&D system and outputs such as detailed patent statistics and the technological balance of payments, as well as impact indicators, which quantify trade in R&D-intensive products and give productivity indices, etc. (Representative indicators can be found in refs. 26, 27 and the *STI Indicators Newsletter*.) Many more innovative activities than before were brought into the realm of quantitative analysis.

This new type of more comprehensive indicators, produced by national agencies as well as by regional organizations such as the

OECD, became more widely used during the 1980s. Nevertheless, the new indicators only pointed out the salient similarities and differences among countries and economic sectors. They made possible a more thorough analysis of patterns and trends in both overall and specific innovative activities. They did not, however, bring about what was later to be called innovation indicators.

According to Christopher Freeman, a participating observer, the first stage in the development of today's variety of R&D indicators emphasized the efforts to expand the national R&D system "without too many questions about output and efficiency" [17, p. 115]. In the second stage "accountants, economists and managers began to ask more awkward questions about performance and responsiveness to the needs of the market," but mainly in terms of controlling expenditure and preventing waste. Now, in the third stage, the focus is put on more direct ways of stimulating economic growth and competition in world markets while combining technical change, industrial modernization, and trade strategies.

Differences between the three stages should not be exaggerated. Elements of both "supply-side" and "demand-side" economics have been present in the policy communities over the last 50 years. "Nevertheless," Freeman claims, "anyone who goes through the various reports of national science and technology advisory bodies – or of parliamentary debates on science and technology or of economic policy documents – cannot fail to be struck by this change in emphasis and focus over the post-war period" [17, p. 115]. The three stages in the production and use of R&D indicators can also be described as a move from quantitative descriptions by way of broad categories of data towards more qualitative assessments of R&D capabilities for industrial innovation and competitiveness.

Lately, and this is a new feature, government authorities in several OECD countries have reduced or even terminated a number of these surveys, while other statistical services have gradually been farmed out [34, p. 25]. The reasons are several. The rising costs of comprehensive statistical analyses have been increasingly difficult to reconcile with the need for budgetary restrictions. And the policy needs for general – or very particular – surveys of science, technology, and industrial innovation to be carried out by public agencies are not always clear to top government decision makers.

Although the field of R&D statistics is young, and innovation studies even more recent, routine procedures by government statistical agencies make it difficult to initiate new surveys, implement them, and then rapidly analyse the findings in order to answer urgent questions posed by planners and policy makers. As a result, quite a few surveys and other studies are now being carried out under flexible, short-term contracts by academic institutions or, more often, by companies that operate commercially.

As firms start producing politically and otherwise important analyses based on R&D and innovation indicators, the information may become a private rather than a public good. Availability is restricted or delayed; some statistical studies are made secret to all but those who finance them. This new situation causes problems for the quality and international comparability of R&D and innovation indicators. If the local customers find it more convenient to design surveys for their own particular purposes, the chances are limited that collected data will be processed in a way that would serve other potential users or the international statistical community.

As R&D and innovation indicators develop locally, while the internationally standardized survey techniques change only slowly, some countries have already expanded their range of indicators and adopted concepts and definitions without waiting for improvements at the international level.

The overall scope of R&D statistics among developing countries

The major problem in international R&D comparisons that include the developing countries is not the intricacies of statistical methodology but the simple question of reliable sources.

The evolution of R&D and innovation indicators among the developing countries does not follow the same trajectory as that of the industrialized part of the world. Several developing countries were among the pioneers in national R&D statistics and, since the late 1960s, many more of them have begun producing. Until now, however, only a few developing countries have followed the more ambitious R&D statistical path set by the OECD countries. Useful summaries of S&T indicators for a whole developing region have been provided by GRADE, a centre for development studies in Peru (see ref. 42, with the revised figures for the early 1980s in ref. 41).

For a number of national statistical agencies, R&D and innovation are not among the top priorities. In dozens of developing countries the status and quality of R&D statistics have even deteriorated. During the 1980s some governments simply stopped publishing internationally comparable R&D data, while others provide only fragmented R&D statistics on an ad hoc basis. In fact, Dedijer's characterization earlier of national statistical priorities is not outdated, although many more statisticians have been trained to produce R&D data. These professionals have had few chances to improve their skills.

Today, the universe of R&D statistics in Africa, Asia, and Latin America ranges from crude summaries of the total number of formally trained scientists in a given country to very detailed data, describing in full-time equivalents the number of scientists engaged in R&D activities and the current expenditures available. Most developing countries have low statistical ambitions in the field of R&D and innovation.

Given the relatively poor state of R&D statistics in developing countries, no general diagnosis can be made of the more than 130 "national R&D systems" in that part of the world, neither of their efficiency nor of their economic and other effects. And no demarcation lines can be drawn between scientific research, technological development, and other innovative work. Anyone interested in detailed international comparisons ought to be concerned about the deplorable fact that "science and technology," "applied research," "experimental development," and similar notions refer to slightly different activities in different countries and are usually performed by different organizations with different objectives. Using currently available R&D statistics, little can be said about the strength and relative performance of R&D activities in what was once called the third world.

For a developing country it is not enough to count the stock of resources currently available for innovative activities. It has become increasingly important to measure the R&D potential and also – by way of statistics – to reveal immediately available R&D resources such as highly qualified manpower not involved in R&D activities. In order to lay a basis for change, the flow of resources over time should be included.

Even if statisticians of all countries can agree on the feasibility and usefulness of a detailed world handbook of R&D and innovation statistics such as Dedijer proposed [14], many elementary problems of a common methodology remain to be solved. Unesco experts in the developing countries do not lack work opportunities. Due to the differences in the availability as well as the quality of data, one cannot be careful enough in drawing empirical conclusions from international

R&D statistics, especially when comparing industrialized and developing countries.

My intention has not been to criticize R&D statistics in general, but to underline the differences in quality and, hence, the problems of reliability and validity in international comparisons. Everyone should be aware of the elementary state of the art. Later in this chapter I shall discuss the opportunities. With limited resources and relatively simple means, much more professional statistical work could be done in the developing countries – and useful comparative data on both R&D and other innovative activities would become generally available.

In the following pages, I concentrate on basic input data such as resources devoted to research and experimental development and tackle a much-debated issue: Is the relative position of the developing countries in science and technology improving? Are the developing countries actually spending more on R&D than they did 20 or 30 years ago? As a key indicator, I use their expenditures on R&D, comparing them with the industrialized countries.

Has R&D spending by developing countries increased?

During the first two "development decades" proclaimed by the United Nations, one set of figures was frequently used to describe the international division of labour within world science and technology. The numbers 70 – 28 – 2 were quoted in most documents dealing with global science and technology policy. The figures showed that 70 per cent of total R&D funding was spent by the USA, 28 per cent by the other market economics, and, so it was claimed, only 2 per cent by the developing countries. This set of figures was widely quoted by the United Nations and others, simply because there were no other data available.

This uneven relationship in R&D between the three major groups of countries was a fact in the first half of the 1960s. But the figures of those years did not project a global picture of all R&D resources. The centrally planned economies – what were then the socialist countries at different levels of development – were not included. Moreover, R&D statistics did not even exist for a number of developing countries and were of poor quality in some developed market economies. But still, for many years, these figures for R&D funding were the best available.

Nearly US$29 billion was spent on R&D by the countries included

Table 1 **Distribution of R&D expenditures (estimated in billion US dollars [current prices] and in percentage of annual totals) among selected industrialized market economies and developing countries in 1963/64, 1977, and 1988**

	1963/64	1977	1988
United States of America	70	46	40
Other industrial market economies			
(except Australia and New Zealand)	28	51	56
Developing countries (sample)	2	3	4
Total (%)	100	100	100
Total (billion US$)	29	97.5	340

Sources: The statistics for this chapter, including the data presented in this table, are drawn from four major sources: (1) An OECD Development Centre study of the world distribution of R&D resources (refs. 1,2), which was subsequently up-dated (ref. 3); (2) R&D data collected and processed by regional organizations such as the OECD (refs. 26–28, 31, 47, 6, 42, 41, 37); (3) Unesco's *Statistical Yearbook* and R&D data processed for comparative purposes (refs. 49, 50, 54, 55); (4) National R&D statistical publications and selected independent studies (refs. 8, 11, 12, 18, 23, 22).

in the 1963/64 UN comparison (for further details see ref. 2). Thirteen years later, in 1977, the same country grouping spent about US$97.5 billion. In 1988, nearly a quarter of a century after the first estimate, their R&D spending was about US$340 billion (all dollar values in current prices). Not only has the magnitude of the R&D efforts changed, but so have the relations between the country groupings and within the group of developed market economies.

During the past 25 years, as shown in table 1, the share of global R&D spending by the USA has decreased from 70 per cent to 40 per cent, while the other Western industrialized countries, which were included in the original sample, have doubled their common share to 56 per cent of the grand total. Since 1980 the redistribution of the financial R&D inputs among the industrial nations has continued. In particular, Japan, France, Germany, Italy, and some of the small, highly industrialized countries like Sweden and the Netherlands have expanded their gross domestic expenditures on R&D more than the average OECD member country.

During the late 1970s and in the 1980s some of the western European countries did not expand their gross expenditures on R&D as much as the largest industrial countries. In relative terms this was a decline in European R&D activities.

During the same 25 years, the R&D position of the developing countries evolved a little differently. (Here the "developing countries" are those in the original sample from 1963/64 only.) In the 1960s their share of total R&D expenditures grew from 2 per cent to 3 per cent. In current US dollars, R&D spending went up more than three times over a 10-year period, to US$1.7 billion in 1973. Towards the end of the 1970s, it is estimated that the same grouping of countries spent a little more than US$3 billion on their R&D activities. This was 3.1 per cent of the grand total. Mainly because of a decline in R&D spending among the OECD countries, the relative share of R&D funds available among the countries in Africa, Asia, and Latin America grew to about 5 per cent of the grand total during the first half of the 1980s.

None the less, improvements in the relative position of the developing countries in the late 1970s and early 1980s were not the result of a general increase in R&D spending. First and foremost the countries with the largest R&D systems in Asia and, to some extent, in Latin America increased spending on R&D. So did a few, but not all, of the Arab states. These enhancements were big enough to affect the position of the developing countries as a group. Then came a relative decline. By 1988–1989, the developing countries had a little less than 4 per cent of total R&D funds, and preliminary estimates for the early 1990s show a further decline.

Yet, before concluding that the North-South division of labour in R&D proved to be rather stable up until the late 1970s, and that changes then appeared that implied a better, but far from enduring, position of the developing countries, we should delve a little deeper into available statistical data. A more representative global picture, based on both national and regional statistics, is presented in table 2.

In which regions are the world's R&D resources concentrated?

If all countries – and not just the previous sample of countries – that have recently produced R&D statistics are included in a world total, they devoted about US$435 billion to R&D in 1988. More than 96 per cent was spent by the industrialized countries, while the develop-. ing countries accounted for the remaining 3.9 per cent of global R&D finance. In current prices, the developing countries – still taken as a group – had a 1988 R&D budget that was nearly six times that of 1973. Their R&D growth rate was significant in the late 1970s and continued to grow for a few years in the early 1980s, but it has since declined.

In a global perspective, until the early 1980s, the R&D position of

Table 2 **Distribution of world R&D expenditures (estimated in billion US dollars [current prices] and in percentage of annual totals) among major groupings of countries in 1973, 1980, and 1988**

	1973	1980	1988
Developing countries	2.8	6.5	3.9
Latin America and the Caribbean	0.8	1.7	0.7
Africa (except the Arab states			.
and South Africa)	0.1	0.3	0.1
Arab states	0.2	0.4	0.1
Asia (except the Arab states,			
Japan, and South Korea)	1.7	4.0	3.0
Industrial countries	97.2	93.5	96.1
Japan and South Korea	7.9	10.2	19.3
Australia and New Zealand	1.2	0.9	0.9
USSR and Eastern Europe	33.0	27.2	17.3
Western Europe	21.6	24.2	25.8
North America	33.7	31.1	32.8
World total (%)	100	100	100
(in billion US$)	97	218	435

Sources: Table 1; percentages are calculated and rounded within each grouping of countries.

the developing countries greatly improved. Using only 2.8 per cent of all R&D money in 1973, and even a somewhat smaller share during the rest of the 1970s, the share came to 6.5 per cent in 1980 and higher in the next few years. However, during the rest of the 1980s most developing countries did not expand their R&D budgets relative to the industrialized countries; they contracted. The industrialized countries, particularly those with large market economies, regained some of their lost positions from the developing countries.

It must be remembered that the developing world is a heterogeneous entity. In 1980, nearly two-thirds of their R&D dollars were spent by countries in Asia, particularly those with relatively large R&D systems such as China and India, but also by Indonesia, Taiwan, and Thailand. Other countries with small or medium-size R&D systems, e.g., Pakistan and Malaysia, have also expanded their R&D finance, although not as much as the largest of the emerging industrial countries of Asia. For 1988, it was estimated that 3 out of 4 R&D dollars in the developing world were spent by East and South-East Asian countries. Today, more than six out of ten developing country researchers are Asians. Latin America and particularly Africa lost their previous strengths in R&D finance during the 1980s.

As a region with more than a quarter of R&D funds among the developing countries in the early 1980s, Latin America with the Caribbean has lost ground to other regions. But the countries with the largest R&D systems, e.g. Brazil, Argentina, and Mexico, seem to have kept relatively high rates of expansion even in the years with severe fiscal problems. In the early part of the 1980s, two of the three countries mentioned were spending more on R&D relative to other economic activities than the average developing country. In the last few years, however, the situation seems to have changed. By the early 1990s, the average Latin American country was not following the pace set by leading Asian countries; the continent as a whole is now lagging behind in R&D spending.

If the Republic of South Africa is excluded from comparisons, Africa south of the Sahara is still very much part of the old third world R&D desert. In all but a few African countries, R&D resources are comparatively scarce. Still, as a region, there are signs of change. Measured in percentages, sub-Saharan Africa's share of global R&D expenditures more than doubled between 1973 and 1980, then dropped to the earlier level by the end of the decade. There was no growth of expenditures; in fact, in the average African country there was a decline during most of the 1980s.

Among the industrialized regions of the world, the country grouping with the highest growth rate is Japan and South Korea. During the last seven years of the 1970s their annual R&D budgets nearly trebled (in current US dollars). By 1980, these two countries in East Asia together accounted for considerably more than the whole third world R&D spending. According to more recent statistics, their yearly R&D inputs have grown even further. As shown in table 2, their gross domestic spending on R&D in 1980 represented a tenth of total R&D finance in the world; eight years later it was more than 19 per cent.

In the 1970s, the western European countries were nearly as fast-growing spenders on R&D as Japan. Their investments in R&D grew by nearly 2.5 times in current US dollars and their international position went up from 21.6 per cent in 1973 to 24.2 per cent in 1980. More recent statistics show a steadily high average growth rate for many of the European countries. But now, the growth is not as significant in relative terms. Western Europe kept its strong position during the 1980s, but only with some difficulty.

On the other side, the eastern part of Europe and the USSR – here still presented as a single bloc of countries – have weakened their position as big spenders on R&D. In 1973 their gross national R&D expenditures were estimated to represent a third of the global total, while they spent only 27 per cent in 1980. With much less than a fifth of the world total by the end of the 1980s, the decline has continued.

In the aftermath of the 1989–1991 revolutions in eastern Europe and the former USSR, it has been revealed that R&D spending in the 1960s and 1970s may have been estimated and officially reported higher than it actually was. So far, however, there are few data available for better founded international comparisons [22].

In relative terms, the North American region, primarily the USA, also declined during the 1970s and into the 1980s. In seven years the share of world R&D fell from about 34 per cent to 31 per cent, while in absolute terms, in current US dollars, the R&D budgets doubled. The following seven years proved to be a period of stabilization, even growth. By the end of the 1980s the two countries accounted for nearly a third of the world's R&D expenditures.

In conclusion, financial resources devoted to R&D in the 1970s grew substantially faster among the developing countries than among the industrialized ones. But for the 1980s, it is fair to say that the highly industrial countries of the North regained nearly all of the lost ground. To the average industrialized country, R&D has become a much more significant element in the build-up of innovative capabilities.

Because of fluctuating exchange rates, variations in local purchasing power, and other problems of measurement, international comparisons of R&D expenditures do not always reflect the magnitude of available resources. Our North-South picture changes somewhat, though not dramatically, if we look at R&D manpower data.

By the end of the 1980s, the developing countries employed 18–19 per cent of the world's researchers (scientists and engineers engaged in R&D). This is a much larger share than for R&D expenditure, but differences between industrial and developing countries remain great. For instance, in Africa, Asia, and Latin America, average overhead costs for laboratory equipment are relatively small. The typical third world researcher in the 1980s did not have similar or the same working conditions as a scientist or engineer in an industrial country. The productivity of the R&D activity is affected.

Table 3 provides a global estimate for 1988–1989 of 4.1 million researchers in full-time equivalents [56]. The industrialized countries (including countries like South Korea) employed about 82 per cent of these. Leaving behind both the manpower indicators and the crude

Table 3 **Distribution of the world's researchers (scientists and engineers engaged in R&D; estimated in full-time equivalents) among major groupings of countries at the end of the 1980s**

	1988–1989
Developing countries	18.6
Latin America and the Caribbean	2.9
Africa (except the Arab states and South Africa)	0.6
Arab states	1.3
Asia (except the Arab states, Japan, and South Korea)	13.8
Industrial countries	81.4
Japan and South Korea	11.7
Australia and New Zealand	1.0
USSR and Eastern Europe	26.6
Western Europe	16.7
North America	25.4
World total (%)	100
(in full-time equivalents)	4,130,000

Sources: Calculated from Westholm [56] (which is based on Unesco, OECD, and national manpower statistics), supplemented by selected national data. Percentages are calculated and rounded within each grouping of countries.

and simple R&D expenditure data, a science and technology–related typology might be more useful in discriminating among the developing countries. Their strengths and weaknesses will be more visible if we look at their relative position in R&D-related technical change. Such a typology could reflect several criteria, such as the economically active population; the sectoral distribution of specialized manpower in relation to science and technology; and the size and structure of the domestic product (GDP), including the share of R&D (see, for example, ref. 20, pp. 55–77). Given the current state of international statistics, such a worldwide typology could not relate resources and capabilities in science and technology directly to the country's economic performance nor to the competitiveness among its main industries. Anyhow, by grouping all countries according to a set of available indicators, it will be easier to identify countries with (a) no science and technology base, (b) fundamental elements of a science and technology base, (c) a science and technology base well established, and (d) an economically effective science and technology base, notably in relation to industry. The last grouping is identical with the highly industrialized countries, while the three others belong to the developing world.

The first grouping of developing countries numbers about 55, in-

cluding most African countries. These are countries with no science
and technology base, still in the initial stage of development, with
low GDP per capita, low science and technology manpower potential, and a low share of manufacturing of total production.

The second grouping of countries, which have essential elements
of a science and technology base, are in the process of industrialization. With moderate GDP per capita, they have developed a limited
endogenous industrial production. Some of them may have a relatively high percentage of science and technology manpower that could
be activated in R&D, but the potential is low in absolute terms. This
second group represents nearly 40 developing countries and includes
Algeria, Ghana, Indonesia, Iraq, Malaysia, Paraguay, and Sri Lanka.

The third group of countries, with a high percentage of potential
science and technology manpower, have a solid science and technology base and a functioning industrial system. Their GDP per capita is
relatively high. This grouping covers about 40 developing countries,
including the "newly industrializing countries" (the NICs) in Asia and
some Latin American countries such as Argentina, Brazil, Mexico,
and Venezuela.

Two developing countries are difficult to fit into any of the above
categories or groupings of countries. China and India have to be
treated separately: they both have a low GDP per capita; at the
same time, due to their size, they have a huge science and technology manpower potential in absolute terms, but low as a percentage
of total population or in relation to the economy. However, manufacturing represents a large share of their total production.

This science and technology–related typology of countries does
not necessarily correlate with economic performance. As discussed
earlier, innovative capabilities may develop from many different
sources. And the linkages between the economic actors, including
government agencies, in a national, regional, and international setting might prove crucial for industrial competitiveness.

In the final sections of this chapter, I examine the intensity among
these linkages or couplings at the regional level of the world economy
and consider how they are being measured.

Science, technology, and new economic patterns

Following the emergence of not just one but of several major centres
of science and industrial technology, the world economy has become
much more integrated and interdependent. But there seem to be
limits even to the processes of internationalization.

Over the last two decades, data on R&D, innovation, and trade patterns in high-tech products make a distinction possible between three dominating regions of the world economy. Each of them forms a separate supply base for industrial development and production, although related to the other two and to other regions of the globe. There is an East Asian industrial space with Japan at the centre; a North American one with several industrial zones in the USA as its core; and a western European economic space with a handful of technologically important national economies. Based in these regions, about 1,000 major corporations control more than half of the world's manufacturing and almost two-thirds of international trade, much of which is in fact intra-regional trade. The three regions have been further consolidated in the last few years.

R&D and related economic statistics, which reflect both the diversity and integration of these regional supply bases, are available but not always beneficial for detailed comparisons. "The data collected systematically at the international level has until now primarily addressed international trade, patent applications and, to a lesser extent, capital movements (in a relatively aggregated form). In addition, private data banks have recently been developed for data concerning different categories of inter-firm cooperation agreements" [34, p. 10]. More is being done, especially through the OECD, to further develop data into sets of comprehensive indicators that could better describe the changing regional and national conditions for innovation.

From a statistical point of view, little is detectable of the "global reach" of large industrial corporations that operate from the three major industrial supply bases. There are no indicators at the firm or at the branch level of industry revealing the role of science, technology, and innovation in the expansion into other regional markets for products and services. Nor are there statistics of the transfer of technology and other knowledge in their quest for foreign supplies and new sources of production. The operations by corporations on patent protection, licence agreements, and royalty issues are not recorded in any databases.

In the polycentric economic context – with corporations based in three high-technology regions as the principal players – "globalization" refers to a set of emerging conditions in which value and wealth are produced and circulated by way of both regional and worldwide communication networks. The transnationally managed firms operate concentrated, even oligopolistic, supply structures by way of modern technology. For banks and other credit institutions as well as for large manufacturing firms and many service producers, modern communication technologies make it easier than before to manage intra-corporate information networks on a global level. Directly or indirectly, some of the small and medium-size firms in the three regions are linked to the same systems. As product and technology life cycles become shorter, this helps in the functional integration as well as in the economic fortification of the three dominating industrial regions.

Whether big firms or small, technological strengths and competitiveness are not determined solely at the level of the industrial firm, but also by the economic environment in which firms operate. Today, more than earlier, managers and entrepreneurs need to combine indicators in the economic environment that influence technical change and industrial innovation.

The developing country firms have similar needs, but face different challenges at home and, more importantly, in the international market-place. The pressure on them from abroad has increased tremendously. Technical change of their industry is very much needed. But the deterioration of the terms of trade, in particular the overall decline of prices of primary products over the past 10 years, the ups and downs of energy costs, the worsening balance of payments caused by the rise of interest rates on loans and credits, the repatriation of foreign investments, etc., have forced countries in the developing world to eliminate research projects, reduce experimental development, and downgrade or close R&D laboratories and related institutions.

To develop policies that could avoid further marginalization in foreign investment and technology transfer, the developing countries need much more detailed and statistically grounded analyses of the role of science and technology in the globalization process. We should not forget that most of the currently available R&D and innovation indicators were created in a specific national or regional context. Although they have been further developed out of broader policy needs, they do not take into account the current internationalization of the national economies. Available R&D statistics do not capture well the new forms of technology-based competition or contemporary economic interdependencies.

Innovation indicators in the making

Over the past three or four years, R&D statisticians in the industrialized countries have speeded up the improvement of indicators describing the role of technology in industrial innovation, human resource development, industrial performance, and international competitiveness. Similar work is done to reshape indicators of scientific research and of R&D performed within the public sector with government objectives. The reasons are obvious.

At present it is not possible to quantify such important tendencies as direct international investments by sector or product area, international flows of technology (licences, patents, know-how), inter-enterprise and inter-governmental technical cooperation agreements, and the international diffusion of high technology incorporated in goods. Data are available on various forms of localized R&D, but the level of aggregation is usually too high for detailed analyses. Typically, changes concerning different forms of relocated R&D are not very well described in current statistics. The same is true for detailed data on transborder flows of researchers or, more generally, of scientists and engineers.

For a developing country, these kinds of R&D and innovation statistics may become strategic for situating the country's economy and its industrial firms in the changing regional and global contexts. Data on technical standards and the protection of these, access to specific technologies, intellectual property rights, and competition policies in different regions and countries could be fundamental ingredients in government policies and corporate strategies.

The uneven performance of national economies sometimes leaves room for doubt as to the possibility of a balanced sharing of gains from trade among the industrialized and the developing countries. To shed light on specialization patterns and the changing international division of labour, more statistics in the form of economic and technical intelligence is needed. Subsequently – as in the highly industrialized countries – models for interpretation should first be constructed and, following this, the most relevant indicators be defined. Then the corresponding primary and secondary data should be processed with the most significant policy objective or corporate strategy in mind.

Data should be functionally organized to form indicators that could help describe dynamic situations involving a cluster of firms and other organizations involved in innovative activities. A "techno-economic network" – to take one illustration out of this context – may be defined as a coordinated set of heterogeneous actors such as public laboratories, technical research centres, firm laboratories, financial organizations involved in industrial investment, intermediate

119

and final users, as well as public authorities that participate actively in the design, development, and production/distribution of production processes, goods, and related services, some of which may entail a commercial transaction. The developing relationships between these actors in the innovation process may centre around the following three poles of attraction.

The first may be labelled the scientific pole. It consists of research centres, university laboratories, and company research units, where knowledge is generated. Here, activities are measured by bibliometric indicators, contracts between firms and research centres, training of personnel, migration of skilled labour, etc.

The second pole of attraction could be called the technical. Here new goods and related services are produced, i.e. prototypes, pilot projects, models, patent descriptions, etc. Activities are measured by surveys of major innovations, patent applications, the creation of new high-tech or science-based firms, licence agreements, and other forms of cooperation between technically advanced firms.

Thirdly, there is a market pole that focuses on the demand for goods and services by users and customers. Indicators here should measure and describe the main characteristics of the distribution system, provide information on user participation in the design of goods and services, particularly quality control and definition of standards.

In some developing countries, and definitely in most of the highly industrialized countries, the tendency now is to see R&D and innovation indicators as advanced tools for evaluations and assessments as well as for analysis and policy formulation. At the same time, there is widespread agreement that the traditional "factors of production" – measured in relation to science, technology, and innovation – do not help much in explaining the dynamic interplay between technical change, industrial competitiveness, and economic growth. Recent advances in economic theory have to be clearly manifested in new R&D and innovation indicators, particularly concerning the place of science and technology in both macroeconomic and microeconomic models, and in the interaction between tangible and intangible investments.

Among intangible investments, R&D is by far the best measured economic activity. But existing data on patents and licences, design and engineering, manpower training, information flows, and organizational structures are seldom defined in such quantitative terms that would allow for a more detailed, comparative understanding of the preconditions and driving forces of industrial innovation. A more detailed breakdown – by product rather than by branch or product group – is usually needed.

The complexity of such measuring tasks should not be underestimated. Already, rather simple quantitative analyses give rise to a variety of interpretations. And it is not enough to settle for existing indicators; the combinations of old and the creation of new innovation indicators must be placed at the top of the agenda, and substantial work is needed to reach a generally accepted quality in the statistical analysis [29].

The "second-generation" statistical manuals

At this point, the shopping list of R&D and innovation statisticians, recently itemized by OECD at a meeting on "Consequences of the Technology Economy Programme for the Development of Indicators," is becoming even more intricate and sophisticated. In addition to the ongoing fourth revision and subsequent expansion of the OECD standard practice for surveys of R&D (the Frascati Manual), three new statistical manuals are being launched by the OECD in close cooperation with national statistical agencies and professional user groups. Experts from the European Community and Unesco are participating as well.

First, at the international level, there is a proposed standard method of compiling and interpreting technology balance-of-payments data. Following the statistical deliberations on dynamic trade relationships, the "TBP Manual" [32] can be looked upon as a forerunner in a series of "second-generation" handbooks on the measurement of scientific, technical, and other innovative activity. These new handbooks link together input, output, and impact indicators in order to better describe and situate innovative activities, in this case the international transfer of technology. Various existing measures of output (e.g. patents and the technology balance of payments) and of impacts (trade in science-based or technology-intensive products and productivity indices) are combined with both economic statistics and new types of R&D data that are drawn from, e.g., bibliometric studies and innovation surveys. Experimental studies of this kind have been conducted by national statistical agencies, but the "TBP Manual" should make possible cross-country studies based also on long time series of data.

Secondly, a manual dealing with surveys of innovative activity in industry is in the making (the "Oslo Manual": [36]). This manual

121

goes far beyond the scope of the Frascati Manual by including a whole set of the indicators discussed earlier in this chapter.

Thirdly, with a more limited scope, a science and technology human resource manual is being developed to facilitate internationally comparable statistics on highly qualified manpower.

On top of this, as an extension of the Frascati Manual, a guideline for the interpretation of bibliometric data is being improved by OECD consultants. Bibliometric methods such as publication counts, citations, co-authorships, and co-work analysis are used for analysing the output of the R&D system. Hence, bibliometric techniques may be useful in evaluating the productivity of individuals, teams of researchers, laboratories, or even national R&D institutions, but they may also be relevant for the tracing of linkages between fields and/or researchers and, combined with other indicators, between science and technology.

For the further development of R&D and innovation indicators in a standard format at least four criteria must be used. The first criterion concerns the real demand or significant importance of the proposed indicators. Will it permit analysis and policy conclusion over and above what can already be done through existing indicators? The second criterion is quality, which is based on theoretical soundness, validity, and operational value. If the indicator is to serve as a basis for policy decisions it must also be reliable and, maybe also, internationally comparable. The third criterion is linked to appropriateness for the users and adaptability to the relevant socio-economic objective or the development stage in relation to the innovation process. The fourth criterion is availability, which links resource efficiency in the processing of data, timeliness, and realizability of the statistical task [34, table 2].

Taken together, the three international manuals of R&D and innovation statistics should help to improve a "second-level" analytical database to provide a comparative "scoreboard of indicators" of scientific, technological, and other innovative activities in relation to economic performance. This database, which is being created and maintained within the OECD "structural analysis programme," referred to as the "STAN Programme," is to become operational in the early 1990s. It should allow for internationally comparative measurements of links between science, technology, competitiveness, and structural change. Such analytical studies could examine determinants of international competitiveness, the contribution of technology to productivity, and growth patterns at the level of industrial branches or subsectors [33]. Nevertheless, much remains to be done to achieve international comparability.

References

1   Annerstedt, Jan. *A Survey of World Research and Development Efforts: The Distribution of Human and Financial Resources to Research and Experimental Development in 1973*. Paris: OECD Development Centre, and Roskilde: Roskilde University Center, Institute of Economics and Planning, 1979.

2   ———. "The World Research System: The Distribution of Human and Financial Resources Devoted to Research and Experimental Development in the 1960s and the 1970s." Department of Theme Research, Linköping University, Linköping, 1981. Unpublished report, 84 pp.

3   ———. "The Global R&D System: Where is the Third World?" In: Annerstedt and Jamison, eds., pp. 129–141. *See* ref. 4.

4   Annerstedt, Jan, and Andrew Jamison, eds. *From Research Policy to Social Intelligence*. London: Macmillan, 1988.

5   Archibugi, Daniele. "In Search of a Useful Measure of Technological Innovation (To Make Economists Happy without Discontenting Technologists)." *Technological Forecasting and Social Change* 34 (1988): 253–277.

6   Arregui, Patricia McLauchlan de. *Indicadores comparativos de los resultados de la investigación científica y tecnológica en América Latina*. Documentos de Trabajo de GRADE, no. 2. Lima, Peru: GRADE, 1988.

7   Basberg, Bjørn L. "Patents and the Measurement of Technological Change: A Survey of the Literature." *Research Policy* 16 (1987): 131–141.

8   Bernal, J. D. *The Social Function of Science*. London: Routledge & Kegan Paul, 1939. Reprint. Cambridge, Mass.: MIT Press, 1967.

9   Callon, Michel, John Law, and Arie Rip, eds. *Mapping the Dynamics of Science*. London: Macmillan, 1986.

10   CMEA. "Statistics on Science." In: *Basic Methodological Concepts on Statistics*, vol. 2, pp. 356–383. Moscow: CMEA, 1980.

11   Dedijer, Stevan. "Scientific Research and Development: A Comparative Study." *Nature* 187 (1960), no. 4736: 458–461.

12   ———. "International Comparisons of Science." *New Scientist* 21 (1964), no. 379: 461–464.

13   ———. "The Future of Research Policies." In: L.W. Bass and B.S. Old, eds. *Formulation of Research Policies*. American Association for the Advancement of Science, no. 87, pp. 141–162, Washington, D.C.

14   ———. "Wanted: A World Handbook of Research and Development Statistics." *Minerva* 7 (1968/1969), nos 1, 2: 88–91.

15   Dunn, William N. et al. "The Architecture of Knowledge Systems: Toward Policy-relevant Science Impact Indicators." *Knowledge: Creation, Diffusion, Utilization* 9 (1987), no. 2: 205–232.

16   Elkana, Y. et al., eds. *Toward a Metric of Science: The Advent of Science Indicators*. New York: John Wiley, 1978.

17   Freeman, Christopher. "Quantitative and Qualitative Factors in National Policies for Science and Technology." In: Annerstedt and Jamison, eds., pp. 114–128. *See* ref. 4.

18   Freeman, Christopher, and Alison Young. *The Research and Development Effort in Western Europe, North America and the Soviet Union*. Paris: OECD, 1965.

19   Gostkowski, Z. *Integrated Approach to Indicators for Science and Technology (CSR-S-21), Current Surveys and Research in Statistics*. Paris: Unesco, Division of Statistics on Science and Technology, 1986.

20   ICSPS (International Council for Science Policy Studies). *Science and Technology in Developing Countries: Strategies for the 90s, A Report to UNESCO*. Paris: Unesco, 1992.

21   Madeuf, Bernadette. *Technology Indicators and Developing Countries*. UNCTAD/ITP/TEC/19. Geneva: UNCTAD, 1991.

22   Mindeli, L., ed. *Research and Development in the USSR: Data Book 1990*. Moscow: Centre of Science Research and Statistics, 1992.

23   Norman, Colin. *The God That Limps: Science and Technology in the Eighties*. New York/London: W. W. Norton, for Worldwatch Institute, 1981.

24   OECD. *Science, Economic Growth and Government Policy*. Paris: OECD, 1963.

25   ———. *The Measurement of Scientific and Technical Activities: Proposed Standard Practice for Surveys of Research and Experimental Development (Frascati Manual)*. Paris: OECD, 1981.

26   ———. *OECD Science and Technology Indicators: Resources Devoted to R&D*. Paris: OECD, 1984.

27   ———. *Science and Technology Policy Outlook: 1985*. Paris: OECD, 1985.

28   ———. *Main Science and Technology Indicators, 1981–87*. Paris: OECD, 1988.

29   ———. *The Measure of High Technology: Existing Methods and Possible Improvements*. DSTI/IP/88.43. Paris: OECD Directorate for Science, Technology and Industry, 1988.

30   ———. *The Measurement of Scientific and Technical Activities: R&D Statistics and Output Measurement in the Higher Education Sector (Frascati Manual Supplement)*. Paris: OECD, 1989.

31   ———. *Main Science and Technology Indicators*. Nos. 90–91. Paris: OECD, 1990.

32   ———. *The Measurement of Scientific and Technological Activities: Proposed Standard Method of Compiling and Interpreting Technology Balance of Payments Data ("TBP Manual")*. Paris: OECD, Technological and Industrial Indicators Division, 1990.

33   ———. *Competitiveness and Structural Adjustment: Progress Report on the STAN Database and Analytical Studies*. DSTI/STIID. Paris: OECD, 1990.

34   ———. *Current Problems Relating to Science, Technology, and Industry Indicators*. Paris: OECD, 1990.

35   ———. *A User's Guide to the Status and Access of the Structural Analysis Databases*. Paris: OECD, 1992.

36   ———. *OECD Proposed Guidelines for Collecting and Interpreting Technological Innovation Data ("Oslo Manual")*. GD 92 (26). Paris: OECD, Directorate for Science, Technology and Industry. 1992.

37   PECC (Pacific Economic Cooperation Conference). *Pacific Science and Technology Profile, 1992*. Singapore: Pacific Economic Cooperation Council, 1992.

38   Raan, Anthony F. J. van, ed. *Handbook of Quantitative Studies of Science and Technology*. Amsterdam: Elsevier Science Publishers, 1988.

39   Raan, Anthony F. J. van, A. J. Nederhof, and H. F. Moed, eds. *Science and Technology Indicators: Their Use in Science Policy and Their Role in Science Studies*. Leiden: DSWO, n.d.

40   Rothschild, Lord. "Forty-five Varieties of Research (and Development)." *Nature* 239 (1972): 373–378.

41   Sagasti, F. R., and C. Cook. *Tiempos difíciles: ciencia y tecnología en América Latina durante el decenio de 1980*. Lima: GRADE, 1985.

42   Sagasti, F. R., F. Chaparro et al. *Un decenio de transición: ciencia y tecnología en América Latina y el Caribe durante el decenio de los setenta*. Lima: GRADE, 1983.

43   Sharif, M. Nawaz. "Measurement of Technology for National Development." *Technology Forecasting and Social Change* 29 (1986): 119–172

44   Sirilli, Giorgio. "Conceptual and Methodological Problems in the Development of Science and Technology Indicators." In: Hiroko Morita-Lou, ed. *Science and Technology Indicators for Development*. The United Nations Science and Technology for Development Series. Boulder, Colo.: Westview Press, 1985, pp. 188–197.

45   ———. *Models and Indicators for Science and Technology: A View to the 1990s*. Rome: National Research Council, Institute for Studies on Scientific Research and Documentation, 1990.

46   Soete, L. et al. *Recent Comparative Trends in Technology Indicators in the OECD Area*. DSTI/SPR/89-7. Paris: OECD, 1989.

47   *STI Indicators Newsletter*. Paris: OECD, Scientific, Technological and Industrial Indicators Division.

48   Torero, Maximo, and Patricia McLauchlan de Arregui. *Publicar y/o morir: la productividad e impacto de la investigación científica en América Latina*. Lima: GRADE, 1990.

49   Unesco. *Statistics on Research and Experimental Development in African Countries*. Paris: Unesco, 1974.

50   ———. *Statistics on Scientific and Technological Manpower and Expenditure for Research and Development in Arab Countries*. Paris: Unesco, 1976.

51   ———. *Guide to Statistics on Science and Technology*. ST.84/WS/19. Paris: Unesco, Division of Statistics on Science and Technology, 1984.

52   ———. *Statistical Questionnaire on Scientific Research and Experimental Development*. STCC/Q/882, STCC/Q/8901. Paris: Unesco, 1984.

53   ———. *Annotated Accessions List of Studies and Reports in the Field of Science Statistics*. ST.90/WS/7. Paris: Unesco, 1990.

54   ———. *Statistical Digest 1990: A Statistical Summary of Data on Education, Science and Technology, Culture and Communication, by Country*. Paris: Unesco, 1990.

55   ———. *Estimation of World Resources Devoted to Research and Experimental Development: 1980 and 1985, Current Surveys and Research in Statistics*. CSR-S-25. Paris: Unesco, Division of Statistics, 1991.

56   Westholm, Gunnar. "La science dans le monde." In: Georges Ferné, ed. *Science, pouvoir et argent*. Paris: Autrement, 1993.

Part 2: From history to current challenges

Part 2 marks the road from history to current challenges. Andrew Jamison puts modern science in perspective and discusses in what ways it bears the imprint of the civilization in which it emerged. Is there a Western bias built into the methods and uses of modern science and technology? There is a general lack of agreement about what Western science actually is, though clearly it has several dimensions: philosophical, including both cosmological issues and epistemological questions; sociological; and technological. Among the critiques, the author distinguishes the romantic critique, which has rediscovered the critical writings of poets and artists about the "single vision" of Western science, as well as reinterpreted the significance of mystical and occult science; the technological critique, taking its point of departure in the range of problems – from environmental destruction to structural unemployment and military escalation – that have been associated with science and technology; and the growing feminist critique of science that has emerged during the past 20 years, focusing on the gender biases at work in both the institutions and concepts of scientific research. The search for alternatives is marked by conflicts over the most appropriate way to develop "non-Western" ways of doing science: a traditionalist approach that has sought to revive the pre-colonial past in a more or less unadulterated form, and an integrative approach that has sought to combine elements of indigenous traditions (e.g. Islamic science). Finally, Jamison takes India as an example to illustrate how this tension between critical assimilation of Western science and a dogmatic reconstruction of non-Western tradition has played itself out.

Hebe Vessuri examines the historical process of institutionalization of Western science in developing countries, both as an instrument of the interests of the most advanced countries and as a result of active attempts by underdeveloped nations to master the knowledge that was the promise of modernity. Illustrating her argument with many examples, the author shows how at different times the major colonial powers and the new independent nations established science and technology institutions, giving rise to a variety of modes of institutionalization, with active government support and a wealth of cultural responses to Western learning. Nevertheless, it has been difficult for science to take root, particularly since it was expected to produce economic growth. Through all the differences of national contexts, Hebe Vessuri points to the tension present in developing countries between the assertion of national identity and autonomy and the socio-psychological feelings of peripherality, marginality, and invisibility.

Jacques Gaillard picks up and develops one specific dimension of the institutionalization process: the emergence of national scientific communities and styles of sciences in developing countries. The concept itself of "scientific community" has a variety of meanings, and most studies tend to conclude that none of the developing countries has a genuine scientific community and that there is a widening gap between the "least developed countries" and the "newly industrialized countries." The main conclusion of the survey carried out by the author on the origins, behaviour, and conditions of scientists in 78 countries is that scientists from developing countries find themselves faced with a dilemma: whether to participate in solving local problems or to follow the models and reference systems more or less imposed by the international scientific community. Third world scientists face several disadvantages compared with their colleagues in scientifically more advanced countries. Important handicaps relate to the visibility and the recognition of their scientific production in mainstream science. The challenge today for scientists is to gain in legitimacy, to find their place in a scientific community that has its own acknowledged place in society. Wherever scientific communities are emerging, the debate henceforth centres on the professionalization of their scientists, on the conditions under which scientific activities are performed, and on the capacity of the scientific communities to reproduce themselves and sustain their activities. Thus, Gaillard highlights a number of conditions that should be fulfilled for supporting the emergence and reproduction of endogenous scientific communities in developing countries.

128

To address the economics of technological change, innovation, and production organization in countries of "late industrialization," Jorge Katz starts by underlining the drawbacks of conventional neoclassical growth models and points to an emerging heterodox theoretical paradigm. With these analytical tools, he then studies the way in which firms, markets, and institutions behaved in relation to the generation, diffusion, and utilization of technological knowledge during the import substitution industrialization of the 1960s and 1970s – a successful process of economic development – and how their behaviour has changed in the 1980s as a consequence of macroeconomic stabilization policies, the de-regulation of markets, and the opening up of the economy to foreign competition. Structural unemployment, an extremely adverse impact upon income distribution, equity, and welfare standards, increasing and more complex forms of social conflict, etc., appear as consequences of the continuing socio-economic restructuring process. The impact this process is having upon the rate and nature of technological change and the functioning of the national systems of innovation calls for: (i) sound macro-policy management; (ii) explicit industrial and technological policies capable of dealing with market failure; and (iii) addressing questions of social equity and political governability.

Sanjaya Lall argues that the development of national technological capabilities is the outcome of a complex interaction of incentive structures with human resources, technological effort, and institutional factors, mediated by government interventions to overcome market failures. It is the interplay of all these factors in particular country settings that determines, at the firm level, how well producers learn the skills and master the information needed to cope with industrial technologies and, at the national level, how well countries employ their factor endowments, raise those endowments over time, and grow dynamically in the context of rapidly changing technologies. The experience of eight industrializing countries is described to assess the validity of the proposed framework, illustrating the multiple models of industrial development based on varying combinations of incentives, capabilities, and institutions, and that each carries its own set of concomitant interventions. For the author, a large role remains for government policies, carefully and selectively applied, in promoting each of the three determinants of technological industrial development.

To conclude part 2, Ignacy Sachs examines the current environmental challenge. At the root of the first debate on environment and development are two extreme and still influential views: the narrowly economic and the unconditionally ecological. The "middle path" – development, or ecodevelopment – attempts to harmonize the three concerns of social equity, ecological prudence, and economic efficiency. The variables of the harmonization game are situated at both the demand and supply levels, as well as in the location of productive activities. The 20 years that separate the UN Stockholm Conference in 1972 from the UN Conference on Environment and Development in Rio, though marked by slow progress towards ecologically and environmentally friendly development, reveal some progress conceptually, in particular with regard to the planners' and managers' toolbox, the debate on the ambiguity of sustainability, the emergence of a new paradigm in ecology and the global change. The author argues that at this stage little will be gained by pursuing the conceptual discussion of sustainable development; priority should be given instead to designing transition strategies towards the virtuous green path, taking into consideration the diverse configurations in the North and in the South in terms of wealth, technical capability, lifestyles, and compelling social problems. In this search for ecosystem, cultural, and site-specific responses to global problems, science and technology appear as a major, but by no means unique, variable capable of speeding up or delaying the transition. As the signposts for the future, Ignacy Sachs concentrates on four examples chosen because of their importance for a meaningful transition strategy, offering many opportunities for innovative use of resources: a one Kw per capita society, a modern plant (biomass) civilization for the tropical countries, the way this might be applied to the development of the Amazon region, and strategies for cities to be made livable in the twenty-first century.

4

Western science in perspective and the search for alternatives

Andrew Jamison

Modern science and technology have developed within Western civilization, and they are the results, or products, of particular historical events and cultural conditions. But what, if anything, does the fact that modern science developed in a Western context mean for the knowledge that is produced – and not produced – in non-Western developing countries? In recent years, it has become a matter of some importance and no little controversy to determine in what ways modern science bears the imprint of the civilization in which it emerged. Is there a Western bias built into the methods and uses of modern science and technology? And are there alternative, non-Western, traditions of knowledge production, long neglected and all but forgotten, that are in some sense more appropriate for developing countries?

What is Western science?

There is a general lack of agreement about what Western science actually is. For some critics, the "Westernness" of modern science lies in what is purported to be its characteristic world-view, its fundamental attitudes to Nature, reality, and knowledge; for others it is the social system and/or institutional framework within which knowledge production is embedded that is seen as being most Westernized; while for still others the problems lie in the technological applications and more general economic development strategies that are in some way seen to be derived from, or intertwined with, science. It may thus be useful at the outset to attempt to characterize the various dimensions of Western science before turning to the criticisms that have been levelled against it. Part of the problem with the critiques of Western science is that they have been partial critiques and have failed to provide what might be considered workable alternatives to the totality of Western science. The alternatives, like the critiques, have all too often been too narrowly focused to be effective.

The philosophical dimension

Let us begin with what might be called the philosophical dimension, which includes both cosmological issues (that is, discussions of dominant world-views and attitudes to Nature) and epistemological questions: methodology, truth criteria, etc. Indeed, it is sometimes considered to be characteristically Western to separate the two: to divide philosophers from scientists, to distinguish those who are concerned with the nature of reality from those who are concerned with discovery of true knowledge about reality [58]. From the early nineteenth century, when Auguste Comte saw in the rise of the "positive" sciences a new rational basis for society that supplanted religion and metaphysics, positivists have seen philosophy and moral issues in general as being irrelevant to the production of knowledge. The "spiritual crisis" of the West is, at least in part, to be seen as the ensuing separation between facts and values and the more deep-rooted secularization of society and knowledge that came with it: what the German sociologist Max Weber called "disenchantment of the world."

While not every scientist working in the Western world has shared the same philosophical assumptions, there has none the less been a characteristic Western approach to Nature, derived from Judeo-Christian traditions and applied to most areas of scientific investigation. The central components of this attitude are objectification and reductionism. Non-human nature is seen as existing for man, and Nature is viewed as a realm of objects for man's potential use and benefit without any inherent subjective interests of its own. Against all vitalist and animist teachings, Western science has come to represent an objectifying, mechanizing way of knowing and doing. Furthermore, it has sought to reduce an understanding of reality down to its basic elements, namely the atoms and subatomic particles – as well as genes – that are seen through scientific instruments to exist in the invisible world of microscopic reality. Epistemologically, Western science can be said to be a deconstructive way of knowing: knowledge of reality is derived through analytical deconstruction of Nature into its component parts. At the same time, the identity of Western science is of a kind of knowledge that has no higher metaphysical or religious justification. It is an intrinsically instrumental knowledge, neither moral nor immoral in its ulterior motivation, in that morality as such is irrelevant to its mode of operation. By objectifying Nature and reducing reality to its component parts, the defenders of Western science have claimed to be able to provide a knowledge that is superior to, and more useful than, knowledges based on more speculative or holistic philosophies [66]. Even though the "truths" of Western science are intrinsically limited to those processes that can be investigated in the form of experiments or experiment-like operations, the knowledge that is produced has a reliability – and, most crucially, a verifiability – that knowledge produced by other means does not possess [103, 48].

Reductionism – literally the reducing of Nature to experimental demonstration – has been the dominant methodological doctrine since the seventeenth century scientific revolution did away with holism and organicism in the name of objectivity. Since that time, the epistemological criteria by which Western science can be said to produce a distinct form of truth have been based on experimental, or "objective," methods of discovery and rational, or "logical," criteria of verification. In this respect, Western science is one possible way of ordering reality, with particular ideas of what is to be considered true and accurate.

From the seventeenth century onwards, science in the West has been largely defined in terms of its methods, although different philosophers have emphasized different aspects as being central. For some, following an inductive, or empiricist, tradition identified with British philosophers such as Francis Bacon and John Stuart Mill, science has been defined by its use of observational and experimental procedures, i.e. by the manner in which its practitioners go about discovering or constructing the empirical "facts" of reality. For others, following a deductive, or rationalist, tradition more associated with continental philosophers such as René Descartes and Immanuel Kant, science has been defined by its use of mathematically based logical reasoning. In this tradition, it is primarily through its rational methods of argumentation that science is seen to be able to produce true knowledge, procedures derived from mathematics and logic rather than from any necessarily observable external reality. Science, from this vantage point, is an adventure of the mind: the uniqueness of the modern, Western variety is due to the rigour of its logic rather

133

than the quality of its experimental techniques. Western science is thus most properly seen as not one but at least two different knowledge traditions, one associated with experimentation, the other with mathematical logic [37].

In the twentieth century, as the philosophy of science has itself become professionalized, a number of philosophers have attempted to combine the two epistemological traditions; one of the more influential efforts has been Karl Popper's theory of falsification, which seeks to depict a "logic of discovery" in the relationship between experimentation and theory building. Theories, according to Popper, are conjectures that are formulated in order to lead to refutations by experimental testing. Scientific knowledge is thus not the same thing as truth, but is better viewed as a process of growth toward ever closer approximations to truth. It is the process that is objective rather than any one particular result, a process that Popper has characterized as falsification [67].

A theory, for Popper, is always provisional; his view of science reflects a reaction to the dogmatic ideologies of his youth, the totalitarian Marxisms and fascist teachings with their absolute truth claims in both science and politics. Popper's philosophy of science depicts scientific research as an ongoing, living process, rather than a set of finished statements; science was a part of what he came to term the "open society" with a sceptical and critical attitude to truth [68]. His philosophy has the ambition, which is shared by many contemporary philosophers, to articulate the way in which scientific knowledge is actually produced, rather than an idealized vision of what science should be. For Popper and his followers, science progresses by continually subjecting its findings to criticism. And even though Popper's critical empiricism has come to be seen as ideological in its own right – for how many scientists really act the way Popper says they should? – it has helped to open the philosophy of science to a closer relation with sociology, history, and science itself.

In the 1960s, Imre Lakatos reformulated Popper's empiricism to take into account some of the background assumptions and "research programmes" that also affect the research process [38], and in recent years, philosophers have come to focus more on the process of experimentation itself rather than Popper's somewhat idealized portrayal of experimenting [26, 29]. Popper's empiricism, which seems to exclude a good deal of modern science from its exacting criteria (many theories simply cannot be experimentally falsified), has also come to be challenged by what might be called neorationalism, a kind

134

of common-sensical view of science that limits epistemology to the semantic reconstruction of scientific statements [70]. While some philosophers have moved closer to the actual research process, others have taken what has been termed a linguistic turn and have come to concern themselves with the way in which scientific theories are constructed, formulated, and expressed [23].

Whatever their differences, however, both modern day rationalists and empiricists usually consider themselves "realists" and tend to close ranks against the various relativist philosophies that have been developed in recent years and that form, as we shall see, part of the contemporary critique of Western science. Where relativists or constructivists see scientific methods as context bound and the resultant findings as limited in their applicability, realists stress the operational, even universal, nature of scientific truth. Because of the particular methods of science, especially their reliance on experimental investigation and thus repeatable interactions with reality that produce verifiable data, science provides the most objective and unbiased knowledge that humans are able to produce. The realist truth claim is thus limited but none the less universal in its range.

It seems safe to say that almost all philosophers of science – and even most scientists – have shared a common "scientistic" faith; whether inductivists or deductivists, empiricists or rationalists, they have taken more or less for granted the superiority of scientific methods over other systems of knowledge or belief. Scientism, in this sense, is an outgrowth of the positivism first systematized by Auguste Comte in the nineteenth century, who contended that the growth of science marked a decisive, historical break, a huge cognitive step forward beyond metaphysics and religion [33]. According to positivism, science is to be distinguished from religion, metaphysics, even philosophy itself, by its reliance on impersonal, rational, objective methods. Even more than any particular epistemology or attitude to Nature, it is the positivist legacy, which in our day has taken the form of a scientistic mentality or belief system, that most of the more philosophically minded critics of Western science are attempting to challenge.

The sociological dimension

In contrast to the philosophical discourse, which locates the Westernness of modern science in its epistemology and world-view, there can also be said to be a distinctly Western sociological or organizational dimension. What makes science Western at this level is the way it has come to be organized in society and the corresponding social ethos or norm systems that it has built up [85]. Modern science, now international and global, took on much of its present character in western Europe in the course of the sixteenth and seventeenth centuries, the period that has come to be known, among other things, as the time of scientific revolution. In the transition of European societies from feudalism to industrialism – or capitalism – the modern scientist emerged as a kind of synthesis of the medieval scholar and the traditional artisan, with precursors among the artists and engineers of the Renaissance and Reformation [77].

The scientific academies of the seventeenth century, such as the Accademia del Cimento in Italy, the Royal Society in England, the Académie des Sciences in France, provided some of the first organized social spaces anywhere in the world for carrying out scientific research and communicating scientific results. No longer was scientific experimentation confined to private or secret laboratories; instead, experiments were carried out in public, with new, often state-financed instruments and under the auspices of royal, state patronage and support. Already in 1928, Martha Ornstein wrote that "it cannot be sufficiently stressed that it was the experimental character of science which encouraged the creation of scientific societies" [63, p. 67]. Recently, their crucial importance in providing "experimental spaces" has been discussed by a number of social historians [86, 30].

The academies were the first institutions of modern science, although museums, schools, and observatories in classical Greece and Rome, as well as in China, Africa, and the Islamic Middle East, had earlier provided temporary homes for the development of systematic technical and natural knowledge [45]. The difference can be seen as one between collecting information and producing knowledge, or, more colourfully perhaps, between hunting-gathering (and speculation) and conscious cultivation (and accumulation). Science, in its Western guise, has been characterized by a particular institutional and organizational form, a distinct "social relations" of knowledge production [49].

With the seventeenth century scientific revolution, science in the West came to be identified with experimental practice, mediated by technical instruments; the conscious development of instruments and experimental apparatus to accumulate what Francis Bacon termed "useful knowledge" is an important part of Western scientific identity, as is the conscious combination of practical skill and speculative thought [76]. What remained separated in other parts of the world,

136

divided into the separate realms of scholarly endeavour on the one hand and practical learning on the other, was combined in Europe in an academic scientific praxis [102]. With the coming of the political and industrial revolutions of the late eighteenth century, science entered the universities and, in the process, what had until then been a relatively marginal societal activity came to be transformed into a profession.

The links with technology and industrial development were intensified during the nineteenth century, in new types of scientific universities, industrial research laboratories, and technological colleges, so that by the early twentieth century, science had become a legitimate and highly significant part of Western culture. It was this institutionalized science that was transferred to, or imposed upon, the rest of the world in the "age of imperialism," supplanting other, indigenously generated forms of knowledge production and dissemination [64]. By the time of the Second World War, modern science had been spread throughout the world, and it is as a global, international science, a shared possession of all mankind, that we know it today. But as a form of human praxis it bears the marks of a particularly Western mode of organization, with certain characteristic institutional imperatives or norms [2].

Modern science, it has been claimed, subscribes to a norm of universalism, by which its findings can be duplicated anywhere in the world by scientists of any race or nationality. In the words of the American sociologist Robert Merton, who formulated the norm in an influential essay in 1942, "The acceptance or rejection of claims entering the lists of science is not to depend on the personal or social attributes of their protagonist; his race, nationality, religion, class and personal qualities are as such irrelevant. Objectivity precludes particularism. . . . The imperative of universalism is rooted deeply in the impersonal character of science" [52, p. 553]. For Merton, writing in the midst of the Second World War, when Nazi Germany sought to impose a nationalist "Aryan" ideology on its science, the universalism of Western science was a progressive attribute, indeed a central condition of progress itself. Universalism was linked to objectivity, or what Merton called "organized skepticism" and "disinterestedness" to establish a set of values that could ensure a knowledge free from ideological bias and that was central to a Western democratic societal developmental process.

In the 1940s and 1950s, Merton's sociological approach complemented the neoempiricism that Karl Popper was developing within the philosophy of science. Throughout the international academic culture, science came to be identified as the type of knowledge that had emerged in western Europe in the seventeenth century, a combination of experimentation and logic, a "hypothetico-deductive" knowledge linking the worlds of the craftsman/inventor to those of the scholar/mathematician. This science emerged in a particular kind of institutional setting and it established particular roles and functions within the emerging industrial capitalist society [4]. Indeed, as an organizational form, Western science can be defined, since the seventeenth century, as that kind of knowledge production that has taken place in specifically designated scientific institutions: first academies, then research laboratories, and finally R&D establishments. It is thus an expert knowledge, a kind of understanding that is considered legitimate and professional within a certain kind of society. It was to be distinguished from religious knowledge and metaphysical knowledge not only through a more all-encompassing philosophical goal or ambition, but through its organizational structure and the roles it played in industrial society.

The technological dimension

It is particularly since the Second World War, with the rise of so-called Big Science, that the Westernness of science has come to be seen not merely in philosophical or sociological terms; as science has become ever more important in the industrial and "post-industrial" political economy, attention has come to be directed to the productive, economic uses of scientific knowledge. What is seen as characteristic of Western science is no longer merely the internal truth criteria and attitude to Nature nor the institutional norms and social roles: Western science has come to be seen as integral to industrialization itself [39, 75]. There has developed, among economists and engineers, the notion of the innovation chain, by which basic scientific results are transformed into industrial products. It is its place in the innovation process, the capacity of Western scientific ideas to be able to be turned into profitable products, that is now seen by many to be most characteristic of Western science. For those involved in the planning and administration of science, the particularly Western styles of management and application have come to be seen as most significant. Even more, it is the integration of science and technology, the very industrialization of science and the transformation of knowledge itself into a commodity, that is seen as most characteristic of the Western style of knowledge production [72].

138

The industrialization of science can be seen as having gone through three main stages since the Industrial Revolution of the late eighteenth and early nineteenth centuries [98]. In a first stage, scientific education came to be oriented toward industrial needs by the creation of new scientific universities and technical "high schools" – and the infusion of science and laboratory teaching into university curricula. The new technologies also led to new scientific discoveries and theories, in thermodynamics, electricity, organic chemistry, geology, etc. From the second half of the nineteenth century, industrial research laboratories started to be established, in both Europe and the United States, and, in this second stage, engineering grew closer to science in its organizational and conceptual identity. The final stage, which is more recent and still developing, involves a more systemic process of integration, connecting science, engineering, marketing, and management into a more all-encompassing techno-structure or techno-science. The industrialization of science is thus a pattern of interlinkages and mutual influencing, so that science in the late twentieth century is no longer the same thing that it was in the seventeenth century. It is now ever more difficult to separate science from its technical uses, or to extract it, as some kind of pure ideational essence, from the innovation chains and corporate strategies in which it has become enmeshed.

Of course, this particular dimension is no longer geographically confined to the "West"; indeed, the economic application of science is, if anything, more actively pursued in East Asia than anywhere else in the world, although it is possible in this age of relativity to view Japan and Korea as the – extremely – Far West, and thus their development of the economic dimension of Western science can be seen as an extension, rather than an alternative. The Asian countries are not challenging the underlying logic of science and technology; they are, on the contrary, following that logic with a dedication and commitment that seems to be weakening in many of the originating Western countries.

However we are to view the Japanese assimilation of science and capitalism, the economic dimension involves the ways in which scientific knowledge is linked to the commodity form characteristic of the historical development of Western industrial capitalism. It was in the age of what Karl Marx called modern industry, from the mid-nineteenth century onwards, that economic development has come to be based on the results of systematic scientific investigations into the properties of natural phenomena and, increasingly, the functions of man-made artefacts. Western science is thus that form of knowledge that is "oriented" to technological use and application [36]. It is also, and perhaps most centrally for many of those who have criticized it, that form of knowledge production that has lent itself to technocratic visions and developmental strategies. It is, as such, indistinguishable from Western technology, which in its "neo-imperialist" pattern of transfer to non-Western societies is often identified as one of the main contributors to underdevelopment itself [74].

The critiques

There is nothing particularly new about criticizing either the object-ive methods or the societal uses of Western science; there have been critiques of science as far back as one wishes to go. It falls outside the scope of this chapter to say much about these earlier critiques, however. For our purposes, what is significant are the ways in which alternative scientific traditions have come to be rediscovered in recent years and applied to contemporary concerns. At least since the pub-lication of Thomas Kuhn's *The Structure of Scientific Revolutions* in 1962, the contemporary view of Western science has undergone what might be called a contextual revolution, as scientists and their dis-coveries have ever more come to be viewed in their historical and social contexts (for representative articles, see ref. 3, and for reviews, see ref. 12). Among anthropologists and other social scientists, as well as among philosophers and scientists themselves, the truth claims of Western science have been relativized (perhaps most dramatically and influentially in Feyerabend [21]), and for the past 15 years, it has become increasingly respectable to contrast Western science to other belief systems and ways of knowing [50]. Western science provides a kind of knowledge that works, but does it lead to wisdom or enlight-enment? The relativization of science involves an enquiry into its underlying premises and motivations [46] and into its psychological and more personal, subjective meanings [93].

On the one hand, there has been a rediscovery of the various spir-itual and holistic sciences and pseudo-sciences that have been based on different philosophical points of departure [18, 91]. Both alchemy and astrology, for example, have in recent decades come to be studied not merely by mystically minded initiates, but they have also been re-evaluated by historians and philosophers seeking to unravel the various crises of modern society [22]. There has also been a growing concern with the limited capacity of Western science to address moral and ethical issues and fulfil what might be considered the ideal of self-enlightenment that has often been traditionally associated with the pursuit of knowledge. In general, from the 1960s onward, there has been a marked "return to cosmology" and a rather widespread questioning of the previously hegemonic world-view assumptions of Western science [92].

Particularly influential have been the re-examinations of the role that magic, religion, and alchemy played in the formation of modern, Western science [101, 30]. The historical record has come to be re-written with increased emphasis given to figures like Paracelsus and Bruno, who had sought to give early modern science a far broader and more spiritual orientation than it ended up receiving. The hermetic and gnostic texts of the early modern period have come to be re-examined, and they have been seen to have played an important role in developing the more visionary, utopian sides of Western science [44]. Even Isaac Newton himself, the father of the mechanical philosophy, has been shown to have been a much more complicated personality than had earlier been imagined, as historians have investigated his alchemical research and his concern with Biblical cosmology.

Historians of later periods have also come to direct attention to the alternative undercurrents within Western science and philosophy. The history of Western science has, as it were, been broken up into distinct historical periods characterized by debates and even struggles between different approaches. Thus, Paracelsian medicine, Goethe's science of colours, and Whitehead's organicism have been re-evaluated and shown to offer explanations and approaches to natural phenomena that challenge the dominant approaches of Western science. Particularly with the advent of feminism, there are many who actively work to show that Western science has been limited and biased in significant ways, and the critiques that have emerged have come to exert a substantial influence in several scientific fields [94]. What has been at work, according to feminist critics, is a particularly masculine way of conceptualizing reality, which has superimposed socially constructed patterns and relationships onto natural processes [41].

It may be helpful to group the critiques in three main thematic categories, corresponding to the three dimensions of Western science that I discussed above. On the one hand, there is what might be termed a philosophical or romantic critique, which has rediscovered the critical writings of poets and artists about the "single vision" of

Western science, as well as reinterpreted the significance of mystical and occult traditions. Here attention is directed primarily at what I have called the philosophical or cosmological dimension of modern science, the world-view assumptions and methodological precepts that are seen as characteristic of modern science. A second category of critique can be labelled technological, taking its points of departure in the range of problems – from environmental destruction to structural unemployment and military escalation – that have been associated with science and technology. In relation to the discussion above, this category of criticism focuses more on the technological uses – and misuses – of modern science than on the scientific research activity itself. Thirdly, there is the growing feminist critique of science that has emerged during the past 20 years, focusing on the gender biases at work in both the institutions and concepts of scientific research. The feminist critique is the most vocal, and probably the most significant, kind of criticism directed against what I have termed above the sociological dimension of science, the ways in which research is organized and institutionalized in modern societies. In reviewing the feminist critique, I will briefly mention some of the other critical voices within the sociology of science.

Within each category, we can further distinguish between what might be termed "internal" and "external" types of criticism, the first coming from within the scientific community and thus proposing alternatives that fall within the overall framework of scientific thought and behaviour, the second coming from outside the halls of science and thus much more open to and supportive of non-scientific, even anti-scientific, paths to knowledge or wisdom.

The romantic critique
In this category, there are those who have sought inspiration in the alternative traditions of Western civilization, as well as in the spiritual approaches of non-Western traditions. Important sources have been the writings of Joseph Needham and his collaborators on the history of science in China and the works of S.H. Nasr on Islamic science. Both projects – and the further developments that they have encouraged – have shown, in impressive detail, how Western science of the modern era is based on the findings and the insights of non-Western scientific traditions. According to Needham, all the world's civilizations have contributed to modern science; it is a world science that needs to recognize the crucial importance of the contributions of the non-Western peoples for its development [60]. Needham has

never sought an alternative to Western science; his ambition has rather been to correct the sense of omnipotence and omniscience, in short the scientism, that has been part of a certain philosophical interpretation of Western science [61].

For Nasr, Western science has narrowed what was a far richer and more spiritual scientific quest in the Islamic world [59]. Western science is, for Nasr and other spiritual critics, a pale reflection of what was, in other cultures, a more integrated social activity based on an attitude of harmonious contemplation rather than exploitation of Nature. In the 1960s, the works of Nasr and Needham, and of Frances Yates and others, on the mystical and magical roots of Western science helped inspire the international "counter-culture" with its rather substantial interest in Eastern religions and other modes of consciousness. Also important were the explorations of magical and mystical traditions in the scholarly writings of Mircea Eliade [16] and the extremely popular books of Carlos Castaneda.

Theodore Roszak's *Where the Wasteland Ends* [78] is a good example of this genre of critique in combining a rejection of the mechanical universe with a resuscitation of romanticism. Jean-Jacques Rousseau's glorification of Nature and later William Blake's critique of the industrial spirit – as well as Goethe's holistic science – all contribute to Roszak's project. Romanticism, for Roszak, is not a lost historical tradition but a necessity for spiritual survival in a technological age; in Roszak's words, "romanticism is the struggle to save the reality of experience from evaporating into theoretical abstraction or disintegrating into the chaos of empirical fact. . . . Whatever we must leave behind of the Romantic style, we can scarcely afford to abandon its steady determination to integrate science into a greater vision of reality, to heal and make whole the dissociated mind of its culture" [78, pp. 256, 258]. The counter-culture of the 1960s, which had a profound influence on many literary intellectuals and artists, such as Roszak, can be seen as a kind of romantic renaissance, leading to the revival of occultism and mysticism that is such a noticeable presence in the world today. Much of what is left of this revival is degenerate in that it turns critique into sectarianism and a kind of escape from society; but, particularly in some of the so-called "new age" formulations of, e.g., Fritjof Capra [11], attempts are made to apply holistic and romantic approaches to physics and economics.

For the purposes of this chapter, the most significant contemporary versions of the romantic critique are those that have been directed against the (high) technological culture. Roszak himself has

143

criticized the "cult of information" that has, through the widespread diffusion of computers in education, sought to promulgate a new data processing model of knowledge upon the Western societies, and increasingly upon the non-Western world as well [79]. For Roszak, the information revolution has imposed a new level of machine dependence in both education and scientific, even humanistic, research, and, even more seriously, the information ideal tends to reduce human thinking to machine manipulation.

The romantic critique of Western science builds, of course, on a long legacy of thinkers; and, in their responses to the new advanced technologies, neoromantics such as Roszak and Langdon Winner have drawn on Lewis Mumford's ideas about the megamachine and "authoritarian technics," as well as Jacques Ellul's conception of an autonomous technology that has grown out of human and social control [17, 55, 99]. Other important sources of inspiration have been the critical social theorists and philosophers of the 1940s and 1950s – Heidegger in Germany, Sartre in France, Marcuse in the United States – who tried to apply new philosophical approaches to the postwar technological society.

The environmental critique
In the United States, Jeremy Rifkin has published a number of books (and held countless public meetings over the past 10 years) to oppose the technological applications of genetic manipulation. Rifkin has combined the romanticism of the counter-culture – with its poetic imagination and its distrust of modern technology – with a second category of criticism, which can be labelled environmentalism. While Roszak has questioned the information ideal of knowledge as a fundamental challenge to earlier conceptions of human thinking, Rifkin has seen the new biotechnological "products" as a challenge to earlier conceptions of Nature. "Two futures beckon us," according to Rifkin. "We can choose to engineer the life of the planet, creating a second nature in our image, or we can choose to participate with the rest of the living kingdom. Two futures, two choices. An engineering approach or an ecological approach" [73, p. 252].

What is at issue among environmental, or ecological, critics of Western science is not so much the power and control embodied in Western science and technology as the anthropocentrism and species reductionism of much of Western science. Ecology, as both science and philosophy, has been presented as an alternative way of approaching Nature and of managing the various crises of pollution, overpopulation, climatic change, etc. What ecology offers for its proponents is a systemic view of Nature, derived as much from field biology as from cybernetics [100, 65]. Nature is seen not in a reductionist way, in terms of its component parts, but in its interrelations and underlying patterns. Particularly in some of the newer formulations of the Green parties and groups, a so-called "deep ecology" of empathy for all living things has challenged many practices of mainstream Western science, such as animal experiments, genetic manipulation, and nuclear power. The alternative is a "kinder" science that draws on the organismic and even animistic philosophies of the past while making use of the feedback and systemic understandings of computer science [14]. An influential source of inspiration is Gregory Bateson, whose attempts to delineate the "ecology of mind" among both the Balinese and contemporary Western scientists, has provided insights for biologists, anthropologists, and psychologists.

In Norway, the philosopher Arne Naess has, under the influence of environmentalism, developed a new kind of ecological philosophy, based on the idea of species egalitarianism. Naess and the Australian Peter Singer, author of *Animal Liberation*, and the American anarchist Murray Bookchin, have been among those who have sought to take the environmental critique of Western science to what might be called a new metaphysical level [90, 9]. Also significant in this domain is the propagation of the so-called Gaia hypothesis [42], by which the Earth and its inhabitants are seen as part of one overall process of life. In our terms, they have criticized the philosophical dimension of Western science, while most environmental activists have criticized the particular technological uses or applications of Western science. Animal rights and the preservation of virgin natural regions are concerns that require a new attitude to Nature, a non-exploitative worldview that, in many ways, is similar to pre-modern and non-Western attitudes (for a critical review, cf. ref. 10). For many deep ecologists, American Indians and other "primitive" peoples offer alternative modes of interacting with the natural environment, both practically and cognitively. And, as we shall see, the rediscovery of more "ecological" traditions is also becoming significant within environmental movements in developing countries.

The environmental critique is not alone in opposing the uses to which modern sciences are put. After the Second World War, and the dropping of the atomic bomb over Hiroshima and Nagasaki, many scientists and ordinary citizens took to the streets to protest the new destructive weapons and try to "ban the bomb." The British philosopher Bertrand Russell was for many years a leader in the international efforts to oppose the increasing militarization of science and

145

technology and the consolidation of what, in the United States, was labelled a "military-industrial complex." The criticism of military technology remains significant in the 1990s; it reminds us of the fact that modern science is by no means a universally positive phenomenon. Compared with the other social and environmental problems that are, in part, caused by science and technology, military escalation has proven to be one of the most difficult to counteract. Indeed, many argue that science and technology are so thoroughly connected with military or aggressive intentions that only a moratorium on research or a slowing down of the rate of innovation would make a significant impact on world peace. On the other hand, the critique of military research has stimulated the development of science itself by spawning a number of peace research institutes around the world and thus generating a kind of "internal" process of reform or conversion of at least some portion of modern science from military and aggressive purposes to more idealistic or peaceful objectives.

The feminist critique

A third category of critique is associated with feminism and has come to exercise an ever growing influence on scientists, particularly women scientists, throughout the world, but perhaps especially in the United States. At issue here are both what is called the "gender bias" of Western science, as reflected in the concepts, theories, and even experimental methods of many sciences, and the overall philosophical or epistemological criteria that are used to validate scientific findings [27]. On the one hand, feminists claim that Western science portrays and investigates Nature in particularly aggressive and exploitative ways, following Francis Bacon in articulating a "masculine" conception of science and using a particularly sexist kind of rhetoric to portray both the natural world and technical artefacts [32]; on the other hand, Western science as such is seen as following a particular masculine form of logic, being competitive rather than dialogic, monopolistic rather than pluralistic, individualistic rather than collective [41]. The feminist critique thus becomes both epistemological and sociological and supports attempts to develop a social epistemology whereby the verification of truth claims is seen as dependent on the social contexts in which scientific results are produced or "manufactured." In this way, feminism has both fostered and been enriched by the more general social theorizing of science that has been growing among sociologists and philosophers in recent years.

In this social theorizing or sociological critique, attention has been

146

focused on the professional or institutional systems of modern science. Science and technology have been criticized for their hierarchical or authoritarian social relations, with a small number of leaders or managers dominating the majority of scientific workers [25]. Science has been seen in terms of its production organization, or labour process, and, particularly in the 1970s, when Marxism regained popularity within many academic fields, science and technology came to be criticized in class terms. It was the relations of science and technology to capitalist corporations that were questioned and challenged. In the 1990s, much of this sociological criticism has disappeared, while feminism has taken over and focused the critique on the particular sphere of gender relations.

The critiques of Western science are, of course, not limited to romanticism, environmentalism, and feminism, but the three categories do indicate both the range and the variety of contemporary critical voices. What might have been seen as conventional wisdom among philosophers and scientists themselves some 30 years ago – a more or less common "scientistic" belief that the methods, institutions, and technological applications of modern science were superior to other modes of knowledge production – has come increasingly to be challenged. These critiques have fostered a growing relativism or agnosticism among sociologists of science, who have increasingly come to see science as merely one form of social activity among others. For Latour and Woolgar [40], science is seen as a way of life rather than a path to truth, and for Mulkay [54], science is a kind of language game, constructing concepts and "discourses" like any other literary activity. The dominant sociological view of science today is that of social constructivism, whose practitioners are not so critical of Western science or anxious to provide alternative ways of producing knowledge as they are sceptical of its aims and social implications. The feminist and sociological critics seek to expand the scientific enterprise into something more pluralistic and variegated: sciences instead of Science [13].

The search for alternatives

It is as part of the efforts to achieve independence from foreign domination that non-Western intellectual traditions will be considered. Here it is possible to delineate two main approaches: a traditionalist approach, which has sought to revive the pre-colonial past in a more or less unadulterated form, and an integrative approach, which has sought to combine elements of indigenous traditions in one or another developmental framework. In all of the liberation struggles in the so-called third world, there has been a tension between the two approaches, and in most developing countries there continue to be conflicts over the most appropriate way to develop "non-Western" ways of doing science.

The communist model of development, first put into practice in the Soviet Union and then in China, Vietnam, Cuba, and, to varying degrees, in several African countries, tended to follow and pro-mulgate a weak integrative approach: traditional techniques in medi-cine, agriculture, and small-scale industry have been tolerated only when they could be combined with Western approaches in the aim of producing a new "socialist" or "people's" science of some kind. Although patterns of development varied from country to country, the standard procedure was to build up formal systems of science and technology based on Western approaches, while allowing some in-formal systems of training, diffusion, and service in non-Western approaches. The dichotomy has roughly corresponded to the division between the urban and rural economies. The general ideology of socialist development has been modernist, depicting Western science and technology as intrinsically progressive, and traditional belief sys-tems as belonging to a pre-modern past [5, 6].

In many of the non-communist developing countries, the scientistic value system associated with Western science has more explicitly been distinguished from the practice; certain elements of Western philosophy, religion, and belief have been characterized as "colonial mentality" or "Westernization," and attempts have been made to fos-ter and encourage indigenous religions and belief systems. At the same time, the natural and engineering sciences have been developed along Western lines, since most of the leading scientists in developing countries were, at least until independence, educated in Western countries. Usually, non-Western philosophy and art have been en-couraged alongside the Western sciences, which has meant that even though the formal systems are modelled on the West, the actual re-search and education are influenced in many ways by non-Western culture and beliefs. In a very real sense, all science in non-Western countries is non-Western science, since the institutional traditions and cultural patterns are different from those that produced Western sci-ence. At the same time, however, the official ambition in almost all non-Western countries has been to copy Western models and apply Western modes of knowledge production [34, 88, 71].

Post-colonialism

The assimilation of Western science can be seen, somewhat schematically, to have gone through a number of phases since the end of the Second World War. In a first phase that lasted in most countries at least until the second half of the 1960s, there was little concern with developing alternatives to Western science on either the sociological or technological level; it was usually only the Western philosophy that was challenged and countered by reinterpretations of traditional belief systems. In Africa, the attempts of Nkrumah, Senghor, and others to formulate an indigenous African philosophy involved both the reinvention of African tradition and also the conscious application of selected elements of that tradition to contemporary political and social projects: "Africanization" [47]. Such use of the past has been criticized for its irrationality and its confusion of philosophy with myth; for Paulin Hountondji, for example, African philosophy is based on "the myth of primitive unanimity, with its suggestion that in 'primitive societies' – that is to say, non-Western societies – everybody always agrees with everybody else. . . . African philosophy does exist, . . . but in a new sense, as a literature produced by Africans and dealing with philosophical problems" [28, p. 63].

For our purposes, the attempts to develop African philosophy and revive traditional non-Western religion are interesting in seeking to provide a different cultural framework for the development of science, not a different science. It is also important to note that they are the result, for the most part, of interaction with Western critical traditions; the Western-trained leaders and cultural spokesmen of the newly independent countries of the third world have applied or at least made use of certain tools of Western cultural criticism in seeking to foster the traditions of their own peoples. In Africa, the rediscovery of the past was inspired by Western anthropology [53]. Those who came to formulate African philosophy were influenced especially by the works of the anthropologist Lévy-Bruhl, and they were affected more generally by the cultural relativism that was a rather common feature of European philosophy and sociology between the First and Second world wars.

While some leaders of newly emerging countries thus sought to develop alternatives to what we have called the philosophical dimension of Western science, the articulators of socialist development strategies sought to impose a different agenda for putting science to use. The writings of Franz Fanon, which had a major influence in third world intellectual circles during the first period of independence, can be taken as representative of this socialist position. For Fanon in Algeria, much like Mao in China, Nehru in India, and Castro in Cuba, traditional approaches to knowledge were part of the pre-colonial, undeveloped, and backward society; the starting point was the observation that traditional society had been "thrown into confusion" by the experience of colonization. In his view, the liberation struggle in Algeria had helped solve the problem by taking sides for modern medicine. "Witchcraft, *maraboutism* (already considerably discredited as a result of the propaganda carried on by the intellectuals), belief in the *djinn*, all things that seemed to be part of the very being of the Algerian, were swept away by the action and practice initiated by the Revolution. . . . The notions about 'native psychology' or of the 'basic personality' are shown to be vain. The people who take their destiny into their own hands assimilate the most modern forms of technology at an extraordinary rate" [20, pp. 124, 126].

What liberation and independence provided was thus not a return to tradition but a different way to use Western knowledge, not only to benefit the previous élites and colonial rulers, but to "serve the people," as Mao put it in China. It is certainly no accident that it was Western-trained medical doctors, lawyers, engineers, and scientists who were among the leaders in most of the third world struggles for independence. They were modernists who had imbibed the teachings of Marxism and European positivism and who saw their revolutions, among other things, as a crucial step toward assimilating Western science and technology into their "underdeveloped" societies. Marx, in the nineteenth century, had of course been a critic of capitalism and its commodity fetishism, but his criticism had not been directed toward science and technology; indeed, central to his critique was the belief that capitalism could not make satisfactory use of the new productive forces that it had unleashed on the world. It was rather the task of the working class to put the revolutionary discoveries of modern science to more effective and widespread use. In the twentieth century, first in Russia and then in the colonies, Marxism was disseminated to other groups of oppressed peoples, but its attitude to science and technology was not particularly affected in the process. The revolutionary movements that came to power after the Second

World War, many of which explicitly identified themselves as Marxist, were thus propagators of Western science and technology, although traditional methods in medicine and agriculture were tolerated as long as they "worked."

Anti-imperialist movements

A second wave of opposition to Western science began to take shape as part of the widespread questioning of Western-style development that emerged in the anti-imperialist movements of the 1960s. What was at issue was not primarily the Western science and technology that was central to development but the orientation to the imperialist centre, the dominance that the imperialist countries continued to exercise over the newly independent countries of the third world. In order to continue the struggle beyond independence to a true national liberation, it was necessary among other things to take the pre-colonial past much more seriously and to question some of the Marxian and positivist assumptions that had hitherto guided the development of science and technology.

The Vietnam War brought these issues to a head. The United States, now seen as the dominant imperialist power, mobilized a massive destructive force in order to keep North and South Vietnam as separate nations. In response, the Vietnamese mobilized their indigenous skills and traditional knowledge and, in the process, came to stand for a new kind of popular approach to science and technology and military resistance. Mao in China had also come to launch his Great Proletarian Cultural Revolution, closing the universities and sending students to the countryside to learn from the people rather than from the "bourgeois" professors who were still supposedly in power in the cities. Where the Vietnamese people were forced to defend themselves by rediscovering methods of guerrilla warfare, the Chinese people were forced to take part in a massive and, it must be said, largely disastrous social experiment. For both countries, the experiments produced a great deal of suffering, wasted effort, and human and natural destruction; and yet they were none the less innovative attempts to impose a new social order of knowledge on large human populations. To speak in Karl Popper's terms, they were massive social experiments, which failed to falsify Western science. Indeed, in both countries, the enthusiasm for Western science and technology has, if anything, been greater after the revolutionary experiments than before [31]. In their time, however, both efforts provided models, at least at the rhetorical level, for other countries

151

to emulate, and contributed to a more general search for alternative or appropriate approaches to science and technology [15, 89].

In the 1970s, appropriate technology – by which was usually meant the creative combination in particular contexts of traditional and modern techniques to meet the problems at hand – developed into a multifaceted movement. In our terms, appropriate technology addressed or challenged the technological dimension of Western science and sought to break the link that had been formed already in the early modern period between the development of science and the development of practical techniques. Appropriate technologists argued for a return to an artisanal technology, a technical ideal that focuses on the craftsman rather than the scientist as the main source of innovation. Appropriate technology tended to be seen as a process of development from below, a non-scientific, locally based technical activity that made better use of the available human and natural resources than a technology development from above, directed by scientific experts with little awareness of local conditions and capabilities.

Appropriate technology had difficulty in meeting the challenges of the new advanced technologies of micro-electronics and biotechnology that began to appear in the international market-place in the late 1970s. These technologies were based on the latest scientific understanding and thus seemed to imply a re-Westernization; appropriate technology, in the course of the 1980s, tended to be marginalized, and now serves not so much as a real alternative to Western science and technology as a nostalgic memory. Part of the problem is that the alternatives quickly grew too specific. Rather than develop a comprehensive set of appropriate technologies and encourage each country to ransack its own traditions and find those ideas and approaches that seemed most fruitful to develop further, all too many appropriate technology enthusiasts wanted to develop immediate solutions, technical fixes to contemporary problems. The units that still survive are primarily those that have sought to stimulate appropriate processes for technological development and training rather than appropriate products. But what was also stimulated was a much more thorough historical reconnaissance than had ever been encouraged before [1, 24, 64].

Particularly important were the efforts made to reinterpret the pre-colonial scientific traditions. In Latin America, as part of the effort to save the tropical rain forests from extinction, the ethnobotanies of the Amerindians were rediscovered, and, by now, research institutes have been established to carry out agricultural programmes based on the revitalized traditional knowledges [69]. In China, acupuncture and herbal medicine have not only become fully legitimate parts of medical science and treatment but they have been transferred to the rest of the world as a visibly non-Western way to treat – and understand – the human animal. In Africa and central America, the pre-colonial astronomical and cosmological theories have been redis-covered, and some of the mysteries of modern astrophysics are beginning to receive different kinds of explanations when they are filtered through the non-Western paradigmatic and cosmological frameworks.

These ethnosciences have not merely been of interest to scientists; particularly in the Islamic world, they have given support to full-fledged traditionalist movements, which in countries like Iran and Pakistan have tried to develop more or less complete non-Western scientific institutions. Indeed, with the Iranian revolution in 1979, the search for alternatives to Western science can be said to have moved into a third and still unfolding phase. More polarized and explicitly conflictual, the new more fundamentalist tendencies in the anti-Western debate seek to revive a comprehensive alternative at once cosmological, technological, and sociological.

Fundamentalism and the return to tradition

In a recent book, Ziauddin Sardar, a spokesman for Islamic science, has identified four streams of thought among those who would de-velop an alternative to Western science in the Middle East [84]. One, which he identifies with the Persian scholar S.H. Nasr, is criticized for its reduction of Islamic science to what I have called the philo-sophical dimension; but even more seriously for Sardar is the tend-ency that he finds in Nasr's writings to equate Islamic science with a general, occultist interest in "gnosis." All too many spokesmen for Islamic science, according to Sardar, weaken their criticism by not satisfactorily specifying the alternative. Their position becomes merely another restatement of the old debate between religious ex-perience and scientific knowledge that merely seeks to replace one belief system with another.

A second group, composed primarily of people who are both Mus-lims and scientists, and often leaders within their own countries' sci-entific establishments, is one that continues to pursue business as usual. The critiques of Western science that have been promulgated over the past two or three decades are simply brushed aside, accord-

153

ing to Sardar, and the scientists in Islamic countries continue to live schizophrenic lives, Western scientists by day, practising Muslims by night.

Abdus Salam, one of the leading physicists of the Arab world, can be taken as a representative of this position [80]. His view is that science is universal, but, all too often, Muslims and people in developing countries are excluded from contributing to and participating in its development: "There truly is no disconsonance between Islam and modern science. . . . What gives one hope is that there *are* Muslim scientists working principally (though not exclusively) in developed countries who have registered the highest attainments in sciences. This implies that it is basically environmental factors in our societies which need to be corrected" [80, pp. 323, 348].

The third and fourth groups identified by Sardar are, in many respects, more interesting for the purposes of this chapter. They involve those who would establish a new metaphysical starting point for scientific enquiry that would have far-reaching consequences for the actual pursuit of scientific research. If I follow Sardar's argument, the difference is one of degree; the one group would alter the relations between scientific fields, the selection of problems, the depth of moral and religious reflection attached to scientific research; while the other group, to which Sardar himself belongs and which he calls the Ijmali position, would seek to create an entire new science, in which the very "facts" of nature would be different, derived solely from the ethical, value, and conceptual parameters of Islam [84, p. 155].

Islamic science, as perhaps the most ambitious ethnoscience tradition, has thus already spawned internal dissension and, judging from Sardar's treatment of his adversaries, a rather large amount of aggression in an enterprise that claims to be based entirely on a love of God, or Allah, the "one and only God." Indeed, in comparison with his first book, *Science, Technology and Development in the Muslim World* [83], the programme of Islamic science appears to have increased in rhetoric but lost something in practical achievement and focus. Indeed, in this respect the attempt to develop an Islamic science seems to be repeating much of the same process that the attempt to develop a "science for the people" went through in the early 1970s. In both cases, a critical identification of problems leads to an overly ambitious formulation of an alternative that has proved impossible to realize in practice. While the alternative becomes ever more extreme and absolute in terms of rhetoric, it thus fails to solve the particular problems that were initially attributed to Western science.

The four schools of thought that Sardar delineates can be taken as representative of the different alternative approaches to Western science that have developed, albeit in very different ways in different countries, during the past decade, as fundamentalist religious movements have exercised a growing political influence. On the one hand, there is what might be called a spiritualist position: the particular alternative teachings are not as important as the general ambition to counter materialism and "material" Western science with a revival of spirit, occultism, and religious faith. On the other hand, there are the realists, who continue to practise Western science while professing a set of moral values, as it were, on the side. Science and values continue to be separate spheres of existence for this second group, which still seems to include most of those who actually work as scientists and engineers in most developing countries.

It is among students that one might expect the strongest resonance for the other two, somewhat newer, schools of thought; and, as such, there seems to be a significant generational dimension to the ethnoscientific enterprise. The one, the critical school, sees the development of alternatives by taking the Western tradition seriously, pointing to its weaknesses, both methodologically and practically, and seeing a new ethnoscience as an explicit combination of Western and non-Western approaches. The other, more dogmatic, orientation sees the alternative, Islamic science as a self-enclosed activity that in some way can separate its own ethnoscience from others. In the next section, I look at how this tension between critical and dogmatic approaches has played itself out in India. The tension between a critical assimilation of Western science and a dogmatic reconstruction of non-Western tradition can be expected to increase in importance as today's students grow into the scientific cadres of many developing countries.

The example of India

Let us examine some of the pros and cons of Western science in the context of one particular developing country. India has been chosen not merely for the size and diversity of its population and the richness of its culture, but also because almost all of the themes that have been taken up in the general debates about Western science can be found there. Indeed, it could be argued that India's struggle for independence was, to a greater extent than elsewhere, also a struggle for the resurrection of Indian civilization. At the very least, it can be said that traditional techniques and non-Western beliefs and customs were mobilized in the political struggle more explicitly than elsewhere. Under the inspiration of Mahatma Gandhi the peoples of the Indian subcontinent were encouraged to revive traditional technical practices and even managed to put aside, for a time, some of their religious antagonisms in order to achieve national independence.

Gandhi, of course, was Western-trained and learned about Western philosophy and Western science while studying law in Britain. Perhaps most important for our purposes here is that Gandhi became acquainted with Western traditions of cultural criticism, associated with such names as Ruskin, Tolstoy, and Thoreau. The "experiments with truth" that made up Gandhi's life were, in large measure, a conscious effort to combine these critical Western ideas with a very personal interpretation of Hindu belief. Gandhi embodied an alternative science and technology in his own person, but he was not particularly successful in writing about it or in institutionalizing it. He has served, in post-independence India, as both a legend and personal model; and, as we shall see, his inspiration can be seen in a number of alternative activities in India today.

Gandhi was not alone in his attempts to develop alternative approaches to science and technology in colonial India, although it was his vision that has perhaps been most influential. Ashis Nandy has recently contrasted Gandhi's "critical traditionalism" to the more absolute glorification of tradition represented by the art historian and Buddhist scholar Ananda Coomaraswamy [57]. Where Gandhi made use of Indian traditions in an open-ended, reflective way, Coomaraswamy's "tradition remains homogeneous and undifferentiated from the point of view of man-made suffering. . . . Today, with the renewed interest in cultural visions, one has to be aware that commitment to traditions, too, can objectify by drawing a line between a culture and those who live by that culture, by setting up some as the true interpreters of a culture and the others as falsifiers, and by trying to defend the core of a culture from its periphery" [57, pp. 121, 122].

Gandhi's critique of Western science was fundamental and comprehensive. He rejected Western science in terms of all three of our dimensions, recombining the romantic or poetic critique of secularization with critiques of the institutionalized élitism and the "technicist" orientation of Western science. It was the lack of morality, the lack of idealism of Western civilization that Gandhi objected to; and Western science was, for him, a central part of that immoral value system.

156

The double nature of Gandhi's critique is important in understanding the subsequent Indian discourse(s) on Western and non-Western science. Unlike the Marxist or positivist leaders of most other independence movements in non-Western societies, Gandhi sought to develop an alternative way of life in which traditional techniques and non-Western beliefs had a central place. His critique of Western civilization was thus not merely a critique of its immorality, but also of its epistemology. "Traditional technology, too, was for him an ethically and cognitively better system of applied knowledge than modern technology. He rejected machine civilization, not because he was a saint making occasional forays into the secular world, but because he was a political activist and thinker with strong moral concerns" [57, p. 160].

India, of course, did not follow Gandhi's lead in the first two decades of independence. Instead, under the leadership of Jawaharlal Nehru, ambitious efforts were made to implant what Nehru called a scientific temper in Indian society. Nehru's scientism, and that of his leading scientific and political advisers, was deep and unambiguous. "It is science alone that can solve the problems of hunger and poverty, of insanitation and illiteracy, of superstition and deadening custom and tradition, of vast resources running to waste, of a rich country inhabited by starving people. I do not see any way out of our vicious circle of poverty except by utilizing the new sources of power which science has placed at our disposal" (Nehru, quoted in [35, pp. 7–8]).

For Nehru, Indian civilization, with its superstitions and religious strife, was in need of radical change; a "scientific temper" needed to be imposed on Indian society, and his governments did their utmost to develop both scientific institutions and also a popular understanding and appreciation for science. Like other post-independence leaders in the third world, Nehru's attitude to Western science was positive; if there was a "non-Western" component to his science policy, it was in seeking to apply scientific research in a planned, systematic way. From the late 1940s, scientific and technological research were organized roughly along the lines of the Soviet model, with central planning and strong state control over priorities and orientation. In a recent review, Krishna and Jain have written:

The Indian experience of science policy up to the late 1960s, which was based on the close alliance between elite scientists and the political leadership, had the major objective to expand the infrastructural base for science, technology and education. The leadership of Nehru provided the necessary political will and economic assistance to ensure continuous expansion of scientific organisations and funding of science and technology. [35, p. 15]

It would be an oversimplification to say that Nehru's death in 1964 led to a revival of Gandhian thought. But as the 1960s progressed, a number of challenges emerged to the developmental strategies and emphases that had guided India since independence. The wars with China and Pakistan fostered nationalistic tendencies, and a variety of popular peasant movements began to wage struggles against the central and regional authorities. The international wave of student and anti-imperialist protest also played its part, so that, by the early 1970s, India was a society torn by inner conflict. Most significant from our perspective was the revitalization of the Gandhian undercurrent, spearheaded by Jaraprakash Narayan, or JP as he came to be called, with his "total revolution" that aimed to revive village economic life and grass-roots initiatives. The revival of Gandhism was an important factor in the protests against the large dams and government-sponsored social forestry programmes as well as the emergence of environmental movements, especially the famous Chipko "tree-huggers" in northern India. In 1978, Prime Minister Indira Gandhi, after having ruled the country through an unpopular State of Emergency, was defeated by the opposition Janata party, which in many ways tried to apply Gandhian ideas during its few short years in power, before being torn apart by internal dissension.

It was in this general spirit of criticism and change that the political scientist Rajni Kothari gathered together a number of Western-trained humanists and social scientists at the Centre for the Study of Developing Societies (CSDS) in Delhi. Kothari had been the chairman of the Social Science Research Council and had been a key actor in the infrastructure building of the Nehru era. In the 1970s, however, Kothari and his colleagues at CSDS grew increasingly disillusioned with the path that Indian development had taken, and began to reconsider the Gandhian intellectual legacy. Indeed, throughout the country, perhaps particularly among science and engineering students, who were finding their knowledge increasingly irrelevant to the needs of their country, the received position about the crucial role of modern science in Indian development began to be questioned. It was particularly among engineering students that the appeal of appropriate technology seems to have been felt most strongly, and in the 1970s a number of different units were established [43].

At the end of the 1970s, three books appeared that served to articulate a new kind of intellectual critique of Western science in India. In 1978, J.P.S. Uberoi, professor of sociology at Delhi University, published *Science and Culture*, in which he developed an all-encompassing critique of Western science, or, more specifically, of the Western positivist tradition, which he traced back to the Reformation and the separation of subject and object. According to Uberoi:

I am persuaded that so long as the problem of the alternative is seen in India or elsewhere in purely practical extrinsic terms, whether political, social or economic, modern Western science itself will remain a stranger and liable to exploit us for its own ends. Its so-called diffusion, implantation or assimilation in the non-Western world will very properly remain a failure or turn into something worse. On the other hand, if the intrinsic intellectual problem of the positivist theory and praxis of science and its claims come to be appreciated by us, leading to a dialogue with native theory and praxis, whether classical or vernacular, then modern Western science will find itself reconstituted into something new in the process [95, p. 86].

The following year, 1979, the Bombay-based journalist and political activist Claude Alvares, who had gone to Holland to study philosophy, provided what would become a catalyst for much of the new critical thinking in his doctoral dissertation, *Homo Faber: Technology and Culture in India, China and the West 1500–1972*. Alvares's book opened up an arena for critical reappreciation, among intellectuals, of the non-Western scientific traditions in India. It presented what Alvares called a new anthropological model of technological development, and explicitly called for the integration of ethnosciences, or indigenous scientific traditions, in the development of appropriate technologies and developmental strategies. For Alvares, "the model of social and technological development idealized out of the industrial revolution in England, the United States and certain parts of Western Europe is no longer the sole means by which the Southern countries and nations of Asia, Africa and Latin America can hope to survive" [1, p. 45]. Alvares traced the historical development of technology in India, China, and England and sought to show how cultural traditions and, in particular, the experiences of imperialism and colonialism had affected all three countries in fundamental ways. Such historical relativization was necessary, according to Alvares, if the non-Western countries were to escape their historical dependency on the West. "The displacement of the West in its monopoly over the productive process will be accompanied by the displacement of its monopoly position as the arbiter of what is proper for the Southern nations in the realm of culture, ideas and ideals. The wider dispersal of the ability to produce goods will be accompanied by the wider dispersal of the ability to produce ideas" [1, p. 221].

A third book of the Janata period, Ashis Nandy's *Alternative Sciences*, brought the critique of Western science down to a micro, or individual, level. Nandy analysed the different ways in which Jagadis Chandra Bose, the plant physiologist, and Srinivasa Ramanujan, the mathematician, had become "alien insiders" in the world of Western science. His was not a straightforward critique of Western science, but rather a more subtle psychological critique that carried a number of different messages. On the one hand, Nandy showed how two Indian scientists had been constrained in their work by their Indianness, but he also indicated how Indian tradition had provided opportunities for creative "dissent" from Western science [56].

The theme of creative dissent has continued to concern Nandy in his more recent writings [57]. His discussion of Gandhi's "critical traditionalism" referred to earlier also stresses the psychological dimension of non-Western science. His criticism, like Gandhi's, has come to be directed ever more to the intrinsic violence of Western science – against Nature and against humanity. While Uberoi has tended to focus more of his attention on alternative traditions in the West – he has recently written on Goethe's "alternative" science [96] – Nandy has continued to explore the psychological tensions and conflicts at work in Indian science. His critique of a "statement on scientific temper" produced in 1981 by a group of distinguished Indian scientists led to a major debate between the proverbial two cultures in India – the humanists and the scientists; and the intellectual critique of Western science that Nandy and his colleagues at CSDS have produced [97] can be expected to grow ever more relevant to the future development of Indian science.

Even more significant has been the emergence of a critique of Western science in the various new social movements themselves. On the one hand, there are the so-called people's science movements that have been particularly active in southern India, beginning with the founding of the Kerala Sastra Sahitya Parishad (KSSP) in 1962. Here the emphasis has been on critical popularization, linking science in selective ways to popular myths and traditions and bringing scientific expertise to bear on protests against government-sponsored irrigation and forestry projects [19, 35]. The people's science movements are not critical of Western science; rather they are critical of the ways in which Western science has been misused in Indian society. Much like the Red Guard in China during the Cultural Revolution, but with less rhetoric and often, it seems, more popular support, the people's science movements are seeking to develop a socialist science, a "science for social revolution," according to the KSSP's main slogan.

What has emerged in other parts of India, as an outgrowth of the environmental movements in the forests and on tribal lands, has been a very different kind of alternative. Here the various critiques of Western science developed in the West have been "recombined" in the praxis of environmental activism. As articulated by the physicist turned Green activist Vandana Shiva, "maldevelopment is intellectually based on, and justified through, reductionist categories of scientific thought and action. Politically and economically, each project which has fragmented nature and displaced women from productive work has been legitimised as scientific by operationalising reductionist concepts to realise uniformity, centralisation and control" [87, p. 14]. In her book *Staying Alive*, Shiva combines an ecological and feminist critique of Western science and discovers alternative "feminine" principles and a feminine attitude to Nature in traditional Indian thought. "Contemporary Western views of nature are fraught with the dichotomy or duality between man and woman, and person and nature. . . . In Indian cosmology, by contrast, person and nature (Purusha-Pakriti) are a duality in unity" [87, p. 40].

Shiva's argument is that social forestry and the Green Revolution in agriculture have been masculine, reductionist projects that have separated women (and men) from their natural roots as well as destroying valuable natural resources. In the protests of rural women, especially the Chipko movement in northern India, Shiva sees the "countervailing power" of women's knowledge and politics:

Women producing survival are showing us that nature is the very basis and matrix of economic life. . . . They are challenging concepts of waste, rubbish and dispensability as the modern West has defined them. . . . They have the knowledge and experience to extricate us from the ecological cul-de-sac that the Western masculine mind has maneuvered us into. [87, p. 224]

Shiva and other scientists who have joined forces with the environmental movements in India have, by the end of the 1980s, developed a range of research institutions and alternative organizations for the dissemination of their ecological alternative. Particularly significant

161

has been the Delhi-based Centre for Science and Environment, which has produced widely read reports (in 1983 and 1985) on *The State of India's Environment* and produced a large number of magazine and newspaper articles through its press service. Together with the appropriate technology groups that still are dotted around the Indian countryside, the environmental movements represent a practical critique of Western science in India. Here, as elsewhere, the critique is Western-inspired and the critics Western-trained; but it has produced an ongoing dialogue with Indian traditions that is likely to grow in importance in the years ahead.

The significance of the alternatives

Until now, alternatives to Western science have tended to be partial and often self-defeating. One aspect of Western science has been criticized or challenged while other aspects have been accepted – even utilized – in mounting the critique. This is to be expected. Western science has developed its contemporary form and its impressive power through a long historical process and it is thus only to be expected that it cannot, in a short time, be replaced by a new form of knowledge production that is as effective and all-encompassing. On the other hand, the problems with Western science do not mean that the entire tradition is in need of overhaul. Very few of the critical viewpoints that have been discussed in this chapter reject the general ambition of modern science to provide a verifiable, even universal, kind of knowledge about Nature. Rationality itself is not the issue as much as the uses to which rationality is put and the institutional contexts in which it is organized.

In an article published in 1979, the German philosopher Gernot Böhme contrasted alternative approaches to science with alternative traditions in science [7]. For Böhme, the alternative to science is irrationalism or obscurantism; there had been, throughout modern history, sufficient alternative traditions within science to sustain visions of the good society. The difficulty was in realizing the good science while avoiding the "bad" applications and priorities. Over 10 years later, the situation is not much changed. There has been a much greater movement to address environmental issues in developing countries, and the rediscovery of non-Western idea traditions has, if anything, grown more intense. While the level of rhetoric has been raised, however, it is far too early to see a full-fledged alternative to Western science emerging in the efforts currently under way.

162

If the search for alternatives to Western science can lead to a more modest, even more humane, science, or if it can encourage a more open dialogue with other traditions of knowledge production, then much will be gained. At the very least, the critiques of Western science have raised some fundamental questions about the ways in which human societies make use of their creative resources, and out of that questioning, it is perhaps not too optimistic to think that the world's citizens might obtain a more variegated, even pluralistic, range of approaches to deal with the problems that confront them.

References

1   Alvares, C. *Homo Faber: Technology and Culture in India, China and the West 1500–1972*. Bombay: Allied Publishers, 1979.

2   Barber, B. *Science and the Social Order*. New York: The Free Press, 1952.

3   Barnes, B., and D. Edge, eds. *Science in Context: Readings in the Sociology of Science*. Milton Keynes: Open University Press, 1982.

4   Ben-David, J. *The Scientist's Role in Society*. Englewood Cliffs, N.J.: Prentice-Hall, 1971.

5   Bernal, J. *Science in History*. Harmondsworth: Penguin, 1969.

6   Blomström, M., and B. Hettne. *Development Theory in Transition*. London: Zed Books, 1984.

7   Böhme, G. "Alternatives in Science – Alternatives to Science." In: H. Nowotny and H. Rose, eds. *Counter-movements in the Sciences*. Dordrecht: Reidel, 1979.

8   Böhme, G. et al. "The 'Scientification' of Technology." In: Krohn et al., eds. *See* ref. 36.

9   Bookchin, M. *The Ecology of Freedom*. Palo Alto, Calif.: Cheshire Books, 1982.

10   Bramwell, A. *Ecology in the 20th Century: A History*. New Haven: Yale University Press, 1989.

11   Capra, F. *The Turning Point: Science, Society and the Rising Culture*. London: Wildwood House, 1982.

12   Chubin, D., and E. Chu, eds. *Science off the Pedestal: Social Perspectives on Science and Technology*. Belmont, Calif.: Wadsworth, 1989.

13   Cozzens, S., and T. Gieryn, eds. *Theories of Science in Society*. Bloomington: Indiana University Press, 1990.

14   Devall, B., and G. Sessions. *Deep Ecology*. Layton, Utah: George M. Smith, Inc., 1985.

15   Dickson, D. *Alternative Technology and the Politics of Technical Change*. Glasgow: Fontana, 1974.

16   Eliade, M. *The Forge and the Crucible*. New York: Harper, 1971.

17   Ellul, J. *The Technological Society*. New York: Knopf, 1964.

18   Elzinga, A., and A. Jamison. *Cultural Components in the Scientific Attitude to Nature: Eastern and Western Modes?* Lund: Research Policy Institute, 1981.

19 ———. "The Other Side of the Coin: The Cultural Critique of Technology in India and Japan." In: E. Baark and A. Jamison, eds. *Technological Development in China, India and Japan*. London: Macmillan, 1986.

20 Fanon, F. *A Dying Colonialism*. 1959. Reprint. Harmondsworth: Penguin, 1970.

21 Feyerabend, P. *Against Method*. London: New Left Books, 1975.

22 ———. *Science in a Free Society*. London: New Left Books, 1978.

23 Giere, R. *Explaining Science: A Cognitive Approach*. Chicago: University of Chicago Press, 1988.

24 Goonatilake, S. *Aborted Discovery*. London: Zed Books, 1984.

25 Gorz, A. "On the Class Character of Science and Scientists." In: H. Rose and S. Rose, eds. *The Political Economy of Science*. London: Macmillan, 1976.

26 Hacking, I. *Representing and Interpreting*. Cambridge: Cambridge University Press, 1983.

27 Harding, S. *The Science Question in Feminism*. Ithaca: Cornell, 1986.

28 Hountondji, P. *African Philosophy: Myth and Reality*. Bloomington: Indiana University Press, 1983.

29 Hull, D. *Science as a Process*. Chicago: University of Chicago Press, 1988.

30 Jacob, M. *The Cultural Meaning of the Scientific Revolution*. New York: Knopf, 1988.

31 Jamison, A., and E. Baark. *Technological Innovation and Environmental Concern: Contending Policy Models in China and Vietnam*. Lund: Research Policy Institute, 1990.

32 Keller, E. *Reflections on Gender and Science*. New Haven: Yale University Press, 1985.

33 Kolakowski, L. *Positivist Philosophy*. Harmondsworth: Penguin, 1972.

34 Kragh, H. *On Science and Underdevelopment*. Roskilde: RUC Forlag, 1980.

35 Krishna, V., and A. Jain. "Country Report: Scientific Research, Science Policy and Social Studies of Science and Technology in India." Paper presented at the First Workshop on the Emergence of Scientific Communities in the Developing Countries, 22–27 April 1990, Paris: ORSTOM.

36 Krohn, W. et al., eds. *The Dynamics of Science and Technology*. Dordrecht: Reidel, 1978.

37 Kuhn, T. *The Essential Tension*. Chicago: University of Chicago Press, 1977.

38 Lakatos, I., and A. Musgrave, eds. *Criticism and the Growth of Knowledge*. Cambridge: Cambridge University Press, 1970.

39 Landes, D. *The Unbound Prometheus*. Cambridge: Cambridge University Press, 1969.

40 Latour, B., and S. Woolgar. *Laboratory Life*. Beverley Hills: Sage, 1979.

41 Longino, H. *Science as Social Knowledge*. Princeton: Princeton University Press, 1990.

42 Lovelock, J. *Gaia*. Milton Keynes: Open University Press, 1987.

43 MacRobie, G. *Small is Possible*. New York: Harper & Row, 1981.

44 Manuel, F., and F. Manuel. *Utopian Thought in the Western World*. Cambridge, Mass.: Harvard University Press, 1979.

45 Mason, S. *A History of the Sciences*. New York: Collier Books, 1962.

46 Maxwell, N. *From Knowledge to Wisdom*. Oxford: Basil Blackwell, 1984.

47   Mazrui, A. *Political Values and the Educated Class in Africa*. London: Heinemann, 1978.
48   Medawar, P. *The Limits of Science*. Oxford: Oxford University Press, 1984.
49   Mendelsohn, E. et al., eds. *The Social Production of Scientific Knowledge*. Dordrecht: Reidel, 1977.
50   Mendelsohn, E., and Y. Elkana, eds. *Science and Cultures*. Dordrecht: Reidel, 1981.
51   Mendelssohn, K. *Science and Western Domination*. London: Thames and Hudson, 1976.
52   Merton, R. *Social Theory and Social Structure*. New York: The Free Press, 1957.
53   Mudimbe, V. *The Invention of Africa: Gnosis, Philosophy and the Order of Knowledge*. Bloomington: Indiana University Press, 1988.
54   Mulkay, M. *Opening Pandora's Box*. Cambridge: Cambridge University Press, 1984.
55   Mumford, L. *The Pentagon of Power*. New York: Harcourt Brace Jovanovich, 1970.
56   Nandy, A. *Alternative Sciences*. New Delhi: Allied Publishers, 1980.
57   ———. *Traditions, Tyranny and Utopias*. Delhi: Oxford University Press, 1987.
58   Nasr, S. *Man and Nature: The Spiritual Crisis of Modern Man*. London: George Allen and Unwin, 1968.
59   ———. *Islamic Science: An Illustrated Study*. London: World of Islam Festival, 1976.
60   Needham, J. *The Grand Titration*. London: Allen and Unwin, 1969.
61   ———. "History and Human Values: A Chinese Perspective for World Science and Technology." In: H. Rose and S. Rose, eds. *The Radicalisation of Science*. London: Macmillan, 1976.
62   Northrup, F. *The Meeting of East and West*. New York: Macmillan, 1946.
63   Ornstein, M. *The Role of Scientific Societies in the 17th Century*. Chicago: University of Chicago Press, 1928.
64   Pacey, A. *Technology in World Civilization*. Oxford: Basil Blackwell, 1990.
65   Pepper, D. *The Roots of Modern Environmentalism*. London: Croom Helm, 1984.
66   Popper, K. *Conjectures and Refutations: The Growth of Scientific Knowledge*. London: Routledge and Kegan Paul, 1963.
67   ———. *Objective Knowledge: An Evolutionary Approach*. Oxford: Clarendon Press, 1972.
68   ———. *Unended Quest: An Intellectual Biography*. LaSalle, Ill.: Open Court, 1976.
69   Posey, D. "Alternatives to Forest Destruction: Lessons from the Mebengokre Indians." *The Ecologist* 19 (1989), no. 6.
70   Putnam, H. *Realism with a Human Face*. Cambridge, Mass.: Harvard University Press, 1988.
71   Raj, K. "Images of Knowledge, Social Organisation and Attitudes to Research in an Indian Physics Department." *Science in Context* 2 (1988).
72   Ravetz, J. *Scientific Knowledge and Its Social Problems*. Harmondsworth: Penguin, 1971.

73 Rifkin, J. *Algeny*. Harmondsworth: Penguin, 1983.
74 Rodney, W. *How Europe Underdeveloped Africa*. London: Bogle-L'Ouverture Publications, 1972.
75 Rosenberg, N., and L. Birdzell. *How the West Grew Rich*. New York: Basic Books, 1986.
76 Rossi, P. *Francis Bacon: From Magic to Science*. London: Routledge and Kegan Paul, 1968.
77 ———. *Philosophy, Technology and the Arts in the Early Modern Era*. New York: Harper Torchbooks, 1970.
78 Roszak, T. *Where the Wasteland Ends*. New York: Doubleday, 1972.
79 ———. *The Cult of Information*. New York: Pantheon, 1986.
80 Salam, A. *Ideals and Realities: Selected Essays*. Singapore: World Scientific, 1989.
81 Salomon, J., ed. *Science, War and Peace*. Paris: Economica, 1990.
82 Salomon, J.-J., and A. Lebeau. *Mirages of Development*. Boulder, Colo.: Lynne Rienner, 1993. Originally published in French as *L'écrivain public et l'ordinateur*. Paris: Hachette, 1988.
83 Sardar, Z. *Science, Technology and Development in the Muslim World*. London: Croom Helm, 1977.
84 ———. *Explorations in Islamic Science*. London: Mansell, 1989.
85 Scheler, M. *Problems of a Sociology of Knowledge*. 1923. Reprint. London: Routledge and Kegan Paul, 1980.
86 Shapin, S., and S. Schaffer. *Leviathan and the Air Pump*. Princeton: Princeton University Press, 1986.
87 Shiva, V. *Staying Alive: Women, Ecology and Development*. London: Zed Books, 1988.
88 Shiva, V., and J. Bandyopadhyay. "The Large and Fragile Community of Scientists in India." *Minerva* 18 (1980), no. 4: 575–594.
89 Sigurdson, J. *Technology and Science in the People's Republic of China*. London: Pergamon, 1980.
90 Singer, P. *Animal Liberation*. London: Croom Helm, 1976.
91 Tambiah, S. *Magic, Science, Religion, and the Scope of Rationality*. Cambridge: Cambridge University Press, 1990.
92 Toulmin, S. *The Return to Cosmology*. Chicago: University of Chicago Press, 1982.
93 Traweek, S. *Beamtimes and Lifetimes: The Worlds of High Energy Physics*. Cambridge, Mass.: Harvard University Press, 1988.
94 Tuana, N., ed. *Feminism and Science*. Bloomington: Indiana University Press, 1989.
95 Uberoi, J. *Science and Culture*. Delhi: Oxford University Press, 1978.
96 ———. *The Other Mind of Europe: Goethe as a Scientist*. Delhi: Oxford University Press, 1984.
97 Visvanathan, S. *Organizing for Science*. Delhi: Oxford University Press, 1984.
98 Weingart, P. "The Relation Between Science and Technology – A Sociological Explanation." In: Krohn et al., eds. *See* ref. 36.
99 Winner, L. *Autonomous Technology*. Cambridge, Mass.: MIT Press, 1979.

100   Worster, D. *Nature's Economy: The Roots of Ecology*. San Francisco: Sierra Club Books, 1977.

101   Yates, F. *Giordano Bruno and the Hermetic Tradition*. London: Routledge and Kegan Paul, 1964.

102   Zilsel, E. *Der Sozialen Ursprunge der Neuzeitlichen Wissenschaft*. Frankfurt: Suhrkampf, 1976.

103   Ziman, J. *Reliable Knowledge: An Exploration of the Grounds for Belief in Science*. Cambridge: Cambridge University Press, 1978.

5

The institutionalization process

Hebe Vessuri

Overview

In the process of transplanting Western science into developing countries, the scientific institutions of the most advanced nations became "models" to be reproduced. The presence of Western-type scientific institutions in the developing world has been widely accepted as an indication of modernity. But this notion, embodied in endless projects of institutions created throughout the modern history of developing countries, has been accompanied by very unequal success and, in general, by difficulties of consolidation. It has often been argued that the social weight of scientific institutions in developing countries is very small, derived from the low prestige and marginality of science in those countries; that scientific institutions tend to suffer from premature obsolescence; that it is very difficult for them to survive their creators; that they have difficulties in adjusting to the transformations of society; that their excessive bureaucratization detracts from their original aims. In short, scientific institutionalization in developing countries as depicted in the literature frequently appears as characterized by fragility, fragmentation, and incoherence.

How true are these generalizations? What was the historical process of scientific institutionalization in developing societies? Did different (national) Western models lead to different "styles" of scientific institutionalization? How did receiving cultures perceive and respond to Western science? What was the local scientific structure, if any, that received it? How was it used in the knowledge transfer, or was it disregarded as simply backward? What is specifically "European" about Western science [45]? India, Japan, China, and Islam had well-developed scientific traditions, elaborate and firmly established theories of life, and rich traditions of education that drew the admiration of many in the West. The high cultures of Latin America, like the Mayan, Aztec, and Incan civilizations, also surprised Westerners because of their achievements. Australia, North America, and most of South America and Africa, with smaller populations, had their cultures pushed aside and destroyed by Europeans. The enormous differences in the frequency and nature of the contacts between the West and what eventually came to be categorized as the developing countries and the recent renaissance of historical scholarship about the non-Western world invite a reassessment of prevailing approaches to the institutionalization of science in developing countries. However, received opinion about the spread of Western science has been so one-sided and prejudiced since the heyday of the European-centred world of the nineteenth century [10] that comparatively little progress has been made towards the resolution of the Western science–backward cultures dichotomy.

Within the larger disciplines of history or sociology, the subject appeals to only a handful of devotees – most of them practitioners of a still unfashionable social or institutional study of science. Most existing literature merely sketches the terrain, using scientific institutions as markers and identifying significant social forms upon which more interpretative studies may be based. What follows is a reconnaissance that highlights some of the themes and concepts that have received attention from scholars.

Scientific institutionalization in the present analysis is the process by which modern national scientific traditions have emerged in the varied social contexts in the post-colonial nation-states, and where scientific institutions have represented at different times the multifarious manifestations of specific patterns of cultural and economic response to the complex combination of ideas and developments identified as Western science. Stress is laid on the diversity of forms in the social organization of science, on the contextual definition of norms and of the establishment of social control, and on the provisions made to ensure the continuity of scientific activity in peripheral settings lacking scientific traditions or in cultures that accommodated Western science with rich, non-Western traditional sciences.

The Pandora's box of "colonial science"

"Colonial science" is a blanket term, supposed to cover a variety of situations. It has been described as "low science" (limited to data gathering, while the theoretical synthesis was supposed to take place in the metropolis); "derivative" (working on problems set by savants in Europe); "dependent" on metropolitan recognition [44, p. 221]; "a lodge in the wilderness," the product of expatriate Europeans for European consumption [62, pp. 1–16]. The term has even been used to refer to an indigenous population that was itself European in culture and outlook, like French Canadians and the Irish, but who for different reasons left the cultivation of science in the nineteenth century to the colonizers of British stock [39, p. 339].

The question of colonial science is relatively new, dating back only to 1967, when Basalla wrote his by now classic paper on the global spread of Western science [11]. He proposed a simple three-phase evolutionary model, very much in tune with the conceptual framework of developmentalism and international cooperation of the 1960s. Not only has this model been very much discussed and disputed since its publication, but its "colonial phase" in particular has attracted a good deal of attention in recent years, stimulating a continuous flow of empirical research that reveals that the phenomena involved are much more complex than originally thought. The question may be more profitably looked upon as a complex power relationship involving a metropolis, a colonial or semi-colonial territory and social structure, scientists of European descent living overseas, and non-Western people involved in scientific research. Western science developed a most powerful assemblage of social devices for validating knowledge and indeed for *moving* knowledge among localities. The key dilemma of modernity for those who do not belong merely by birth, primary socialization, and intellectual training to some version or other of the hegemonic culture of the modern world, is, in Dunn's words [23, p. 5], how to distinguish those aspects of the culture that genuinely exemplify the capacity to know better from those that exemplify instead only its brazen and deceptive claim to do so. For it is the ability to draw this distinction, continues Dunn, that alone makes it possible to discriminate an extension of cognitive capacity that no human agent or human society could have good reason to reject in itself from a cognitively arbitrary erosion of personal or social identity by the action of alien force. The problem comes, of course, when – as is usually the case – culture contains un-

170

mistakable elements of both. There has been continuous negotiation and redefinition over who gains local control over useful knowledge institutions. If controlled from *outside* the national boundaries, then local knowledge and local interests are condemned to marginality. If controlled from *within*, there are potentialities but also dangers of other kinds.

It is worthwhile keeping a double approach to this subject. On the one hand are the strategies of the major powers for the export of Western science to their colonial outposts and zones of influence. There has been substantial variation among the strategies and at different periods. This chapter examines only the last 150 years, although of course colonialism goes back much longer, but such de-limitation makes it possible to consider processes that have a direct bearing on contemporary arrangements. A characteristic figure of "colonial science" linked to colonial administration was the individual or institution that was basically a "gatekeeper" of colonial science, actually blocking the advancement of scientific research by keeping an image of "low science," for activities useful to the colonial ad-ministration, although the picture would be incomplete without men-tioning the "scientific soldier," for whom the work ethic was of para-mount importance and who did his best in the given circumstances [40, p. 58]. On the other hand are the views and interests of indi-viduals in societies other than Western ones towards scientific de-velopments occurring beyond their frontiers and/or towards the emer-gence of national scientific traditions in the new nations resulting from the often traumatic experience of colonialism. Typical figures in this other perspective were the groups of scientists – mostly non-European but also some Western settlers – who were basically part of the emerging nationalism and who were also partners in the freedom movement in colonial outposts and in semi-colonies or zones of in-fluence. Let us look first at the metropolitan powers.

Strategies and styles of the major powers

Academic, administrative, and commercial interests were involved in different combinations in colonial science as practised within British, Dutch, French, Spanish, Portuguese, German, Belgian, and Amer-ican frameworks. More is known about the role of science in British colonialism than in the other colonial powers. MacLeod has made a sweeping overview of the evolution of scientific institutions and re-search in their relationships to imperial rule and the distribution of

171

power between Great Britain and other parts of the British Empire. His theoretical construct highlights the *function* of empire (a moving metropolis), selecting, cultivating intellectual and economic frontiers. "In retrospect," he says, "it was the peculiar genius of the British Empire to assimilate ideas from the periphery, to stimulate loyalty within the imperial community without sacrificing either its leadership or its following" [44, p. 245]. Pyenson [62–64] embarked on an ambitious comparative project between France, Germany, and the Netherlands concerning strategies of scientific expansion in the exact sciences associated with the history of cultural imperialism. In France there is a tradition of colonial history, in which attention was occasionally drawn to the role of science in colonization, but its impact on sociology or the history of science has been minimal. Since 1984, the REHSEIS (Recherches épistémologiques et historiques sur les sciences exactes et les institutions scientifiques) team has been working on the subject of science and empires, and has organized an international colloquium on the topic that resulted in the most recent addition to the literature [57].

On reaching the culminating stage of imperialism in the early twentieth century, the major powers had evolved their international strategies in line with changing policies of colonial development. Conveying scientific practice from metropolis to periphery grew more intense and was marked by rivalry. It had two main aims: cultural influence and competition with other nations, although formally it was possible to identify the need to support science as an inherently international activity [82]. In the period preceding the Second World War, similar agencies and policy instruments were established in the major countries.

Probably one of the better known examples of colonial science is the one resulting from the actions of the British in India. Britain organized scientific activity in India from the time the Crown took over the country from the East India Company (1857), in an effort aimed primarily at meeting the strategic needs of the empire – army, trade, and the welfare of European inhabitants. The social structure of "colonial science" seriously discriminated against natives. For a long time Indians were denied access to scientific departments. At the end of the nineteenth century a Board of Scientific Advice was constituted in India to coordinate the activities of various scientific services, but its purview was limited to the governmental sphere and thus kept separate from Indian society. The requirements of colonial government made science dependent on the British metropolis and limited the scope of the board in India [44]. The characteristic British policy was not to encourage technological development but to increase the productive resources of the country through the agency of imported technology. Whatever information Indians gathered regarding the technology was therefore a result of their quest for it [80, p. 222].

In the nineteenth century, France revamped its overseas empire. The "functionary" style scientist, in terms of nineteenth-century science, evokes the behaviour of the French scientist, for a French scientist was a federal civil servant, usually employed as a teacher in a secondary school or institution of higher learning, and his assignments could turn out to be anywhere in greater France or the colonial outposts. He operated within a state bureaucracy. Pyenson argues that this kind of scientist intended to interweave his research institution into the social fabric of his setting. The *institution*, such as it might be defined by either local or metropolitan directives, was paramount, and the criteria that determined progress up the ladder paid little attention to scientific research. To secure new territories in such places as China, Madagascar, and the Near East, France relied on the enormous resources and organizational talents of French Catholic missionaries. It can be argued, as indeed Pyenson does [62, p. 391], that the most successful overseas French research institutions in exact sciences were those conceived and staffed by Jesuit functionaries.

The beginning of the twentieth century saw a new awareness by the French state of the importance of science and culture for international relations [52]. The "Groupement des Universités et Grandes Ecoles de France pour les Relations avec l'Amérique Latine," created in 1907 at the initiative of French scientists, for example, was aimed at developing university cooperation and at competing with Germany in a strategy of "intellectual expansion" [54, pp. 428–442]. It led to the establishment of French institutes in several Latin American capital cities, and of the journal *Revue d'Amérique Latine*, through which the notion of "latinité," as opposed to North American panamericanism, was pushed forward. Probably the most important Latin American institution in whose creation and early years the French were involved was the University of São Paulo, founded in 1934. A considerable number of first-rate European scientists and intellectuals gave shape to its central unit, the Faculdade de Filosofia, Ciências e Letras [55, pp. 339–362].

The Dutch colonial expansion began during the first half of the seventeenth century. Only with the last quarter of the eighteenth century, however, did scientific institutions come to Batavia, the commercial centre of the principal overseas colony. Particularly from the middle of the nineteenth century the colonial ministry pushed ahead to transform Java into a vast tea and sugar plantation. Those who supervised the extraction of natural resources for the needs of European markets subsequently set up institutions shadowing metropolitan models. The dominant interests in nineteenth-century Indonesia related to geographical botany and ethnology. New institutions of higher learning emerged over the next decades.

The development of the Batavia observatory illustrates that the practical demands placed on colonial scientists related directly to commerce and agriculture. Rainfall had to be measured, weather predicted, and the time of day established. Planters and merchant princes went beyond supporting research into practical problems. They believed not so much that pure science would solve the colony's practical problems, but rather that pure science could bring lustre to them and their bourgeois confederates in the metropolis [63, p. 183].

In their struggle for the partition of the world, the European powers tried to penetrate China from an early date. Specialists usually recognize two main waves of introduction of Western science there: first through the Jesuit missions in East Asia from the sixteenth century on spurred by the Portuguese and later French Asian expansion; then in the nineteenth century, when Britain was the leading colonial power in the opening of China to international trade, through the Protestant missionaries, whose tools of conversion were mainly institutions of higher education and Western medicine [43]. It is interesting to note that by the late seventeenth century, scientific activity in China evinced many characteristics of a continuous and systematic social activity, although it remained weak with respect to the social perception of its value. Unlike Europe, science in China did not achieve the momentum of a radical social and intellectual movement within the larger social system. One answer to why a sustained scientific movement began in China only when the West again intruded in the nineteenth century is provided by Porter [59, pp. 529–544], who refers to the broader political and ideological climate prevailing in China after the mid-seventeenth century.

Although American imperialism, in its early phases, concentrated on ensuring US control of the North American continent, it was never exclusively continental in outlook. From the beginning it looked out across the Pacific to Asia. This expansive imperialism that recognized no geographical limits brought the US into contact, and often into conflict, with the other major powers of the nineteenth century.

174

Prominent American private foundations were instrumental in creating and maintaining an economic and political order of international scope, increasingly interconnected, with the US in its centre. Among them, the Rockefeller, Carnegie, and Ford foundations invested in the growth of institutions of higher education, think-tanks, and research centres all over the world. They were the main architects of international networks of research and agencies involved in the production and diffusion of knowledge, networks connecting talented individuals and their institutional bases among themselves as well as with their benefactors. They stretched from colonial China (the Peking Union Medical College) and Argentina's Institute of Physiology led by Nobel prizewinner Bernardo A. Houssay to W. Cannon's lab at Harvard University, which received scores of Latin American physiologists.

Patterns of metropolitan involvement in scientific cooperation in the early twentieth century did not necessarily reflect historical – usually colonial – relationships or were circumscribed by the language the recipient country adopted for "higher culture." The example of German cultural imperialism reflects the convergence of a rich flowering of science and governmental policies to acquire territories and influence abroad. The dominant style for intrusion in foreign lands in this case seems to have been the actual production of scientific knowledge, following the tradition that had characterized its wandering scholars, far in excess of the numbers that could be absorbed at home and thus seeking their fortune abroad, through the tangible proof of their scientific talent: scientific publishing [62]. Thus in a society like Argentina, with extensive cultural and economic influences from Spain, France, Italy, and England, strikingly German interests became prominent in several sectors, such as education, the army, and the electrical utilities firms around Buenos Aires. Between 1904 and 1913, the Prussian Kultusministerium (that is, the Prussian Ministry of Spiritual Affairs, Instruction, and Public Health) planned and staffed the Instituto Nacional del Profesorado Secundario in Buenos Aires, which had a lasting impact on the training of secondary school teachers [61]. With the active support of the imperial foreign office, German learning was implanted in the new University of La Plata in open competition with North American interests, and the German tradition in exact sciences came to dominate twentieth-century Argentine research until mid-century.

Other small European countries, like Spain or Italy, which at the beginning of the century were experiencing a renaissance in science and the arts, also endeavoured to reinforce their linkages with the colonies or the post-colonial nation-states. Thus the Institución Cultural Española, created in 1914, had as its goal to diffuse Spanish learning in Hispanic America, through the establishment of chairs to be filled by Spanish intellectuals and the development of other activities directly related to the intellectual exchange between Spain and the region [70, pp. 217–260].

Cultural responses to Western learning

The implantation of Western learning as an integral part of imperial strategies had its counterpart in a multiplicity of cultural responses by which such learning was assimilated or rejected. From the nineteenth century onwards, strong feelings of cultural nationalism, sometimes expressed in social movements like the 1899–1900 I-ho t'uan Movement in China (known in the West as the Boxer Rebellion), were apparent throughout the colonial and post-colonial world, including in some cases the revaluation of traditional modes of understanding. In this respect, distinguishing colonial from national periods of a scientific tradition has once again proved to be of doubtful validity. Not only is there usually a considerable overlap between the supposedly colonial institution and the supposedly national institution, but also colonial science is seldom in any significant sense transformed into national science following political independence [16]. The metropolitan-colonial dialectic is a complex process, which by no means moves through logical stages to a preordained dénouement. In some instances the formation of a sense of identity may well precede independence, but in others, as in Canada, the search for identity may follow rather than precede effective emancipation [26, pp. 4–5].

The very concept of "identity" in a colonial society is fraught with ambiguity [27, 18]. Whose identity is at issue and what determines it? No culture, no society has ever possessed a single and comprehensive identity. In this essay we are basically concerned with the development of a self-image among the colonial élites. But it is possible to recognize the development of a sense of identity and even other scientific traditions among the less privileged groups in a colonial society. In Mexico, for instance, there existed at least two distinct cultures: that of the criollos and their descendants, and that of the Indians and mestizos, whose aspirations first emerged in the Hidalgo-Morelos revolt of 1810 and then again, with more lasting results, in the revolution 100 years later [51, pp. 51–93]. Western science, often adopted by the cultivated élites in search of "modernity," helped to reinforce an identification with European culture, which in the nineteenth and twentieth centuries often assumed the guise of "cosmopolitanism," variously regarded positively or negatively in the political struggle for self-assertion of the new nations [79]. Probably the feeling of distinctiveness, a lack of identification with Europe, was present very early, most among the Blacks and the children of miscegenation, those for whom the colony was the only "mother country."

Colonial societies, like all societies, were in a constant process of defining and redefining themselves. But, owing their existence to a distant mother country, they found themselves trapped in the dilemma of discovering themselves to be at once the same and yet not the same as their country of origin. The dilemma was made all the more acute in that metropolitan contempt for provincial cousins seems to have known no bounds. The continuous bombardment of calumny to which settler communities were subjected gave them an early and powerful incentive to develop a more favourable image of themselves, if only in self-defence. Where the settlers lived in the midst of an allegedly "barbaric" native population, as in Ireland or Mexico, this meant in the first instance differentiating themselves from these alien peoples, to whose characteristics they were assumed by misguided Europeans to have fallen victim. But the actual relationships between the various social groups making up the new societies told a complicated history of the triangular relationship of mother country, colonists, and subject populations. European ethnocentrism was present not only in the knowledge of the man in the street, but also in the "scientific" knowledge that Europe presumed to have of non-Western cultures. And as such it is to be found in the action of government officials, "experts," and businessmen in their encounters with cultural diversity [53].

The weaknesses of colonial science can be illustrated by the example of the egg-laying mammals taken from the history of Australian zoology during the nineteenth century [22]. Data from the periphery, transferred to Europe by colonial collectors and observers, was interpreted within a theoretical framework provided by European professionals. It took more than 85 years after the first platypus specimen arrived in Europe for European scientists finally to accept that the monotremes laid eggs. The delay reveals the weaknesses of the system of collecting information in colonial science and the resistance of the European scientific community to evidence that violated their theoretical preconceptions. European biologists made a serious mistake and for a considerable time persisted in it. Australian scientists were more easily convinced by European authority than by the empirical evidence available in Australia. Yielding to the scientific judgement of renowned British biologist Sir Richard Owen, most Australian scientists agreed that monotremes gave birth to live young. Meanwhile, aborigines and other Australians not educated within the European scientific tradition had the necessary knowledge, but they were not heard by the scientific community because of the blatant racism of colonial science.

By the mid-nineteenth century, Indian zeal for the learning of European sciences was explicitly demonstrated through actions such as the opening of the Anglo-Indian College in Calcutta by local inhabitants for the promotion of the teaching of European science in India, "in a manner, forcing upon the British" [80, p. 217]. Many of the country's leading scientific and technical institutions were established from the late nineteenth century onwards, at the height of colonialism, by Indian philanthropy. In 1876, as a reaction to British colonial science by the Indian political and scientific intelligentsia, the Indian Association of Cultivation of Science (IACS) was inaugurated, thus giving birth to "national" colonial science. Among its aims, it was explicitly stated that Indians "should endeavour to carry on the work with [their] own efforts, unaided by government. [It ought to be] entirely under [their] management and control. [They wanted] it to be solely native and purely national" (M.L. Sircar, quoted in [41, p. 6]).

By the 1920s, as part of an emerging nationalism, the efforts of eminent individual Indian scientists such as J.C. Bose, C.V. Raman, and C.P. Ray led to the creation of basic research institutions in physics, chemistry, mathematics, and plant physiology, which were the genesis of Indian science. A common platform for the small teams and scientific societies spread all over India was provided by the launching of the Indian Science Congress Association (ISCA), in 1914. Mathematical and engineering societies were established in the 1920s. During the next quarter-century, about 10 professional societies were established, along with scientific periodicals and professional journals. *Current Science* from Bangalore and *Science and Culture* from Calcutta, and two scientific weeklies patterned on *Nature* were launched in the mid-1930s. By the 1940s there were at least six universities established by Indians and more than one hundred colleges where science and technical teaching was introduced. The demand to

Indianize the colonial scientific organizations was an important plank of the political agenda to mobilize mass support. Thus, when India achieved its independence in 1947, Nehru could launch an ambitious programme in science and technology.

If in the contacts that China had kept with the West since the sixteenth century, Chinese interest in Western science was linked to a renewal of "concrete studies" within the Chinese tradition, by the end of the nineteenth century, the learning of Western science and technology had become a necessity because they were regarded as a key to military power by a country in havoc after facing the successive attacks of Britain, the United States, Japan, Russia, Germany, France, Austria-Hungary, and Italy [38, p. 86].

The Japanese adoption of Western science and technology with the Meiji Restoration offers a case-study of historical discontinuity [36]. It would appear to be a strong argument in favour of the peripheral nature of science in Japanese society prior to industrialization, at the same time that there was a social ability to *respond* to new influences coming from outside. The presence of forces exogenous to the nation was significant, and the receptiveness of Meiji society and the economy to those external forces helps to explain the radical nature of the social transformation. Among the external forces that stand out are foreign teachers of new technologies. The most prestigious scientific centres were serviced by foreigners. There were also foreigners employed as technicians and applied scientists or general advisers. One Japanese publication cites 1,392 as the number of foreigners employed by Japanese industry and government between 1860 and 1912, at least 900 of whom were invited. Many young Japanese officials and businessmen were sent to Europe during the early Meiji years. Scientific and technical works in Western languages were published in Japan, and foreigners dominated the major science-cum-technical associations formed for the diffusion of knowledge, such as the Tokyo Academy or the Electrical Society [36]. Thus the resulting Japanese science involved a discontinuity with regard to its own past and to past European experiences of industrialization. The scientific community in Japan, in contrast to Europe at the time, had a "planned character," planned for the set purpose of catching up with the Western standard of science as quickly as possible [50].

However, the institutionalization of science and technology by government initiative, which was very efficient for transplanting and introducing foreign science and technology, was not good for the purpose of fostering original creative activity. The Meiji government paid little attention to scientific research. From 1886 a reorganization of institutions took place, accompanying the maturation of the industrialization process. Up to the 1880s, the scientific institutions created by government were mostly of the geophysical kind for survey work, typical of a non-industrial modern state. Starting in the 1890s, however, many national research institutes were established for fostering industrial development. The war mobilization of research in Europe and the USA during the First World War led government and scientists in Japan to think about financing scientific research. The creation of the Riken (Institute for Physical and Chemical Research) in 1917 was a landmark of this change, since the major source of funds was the industrial sector (85 per cent) rather than the government. During and after the First World War, several private firms, notably in the chemical industry, established their own industrial laboratories. Another unique arrangement was the creation of university-affiliated research institutes and of government research funds [49].

The pinnacle of the colonial hierarchy was reserved for Europeans, and even in the nations that achieved independence in the nineteenth century, the arrogance and rigidity of European teachers often created conflicts with the nationals who wanted to make a career in science [74, p. 427]. This the Argentine evolutionary palaeontologist Florentino Ameghino learnt bitterly in his confrontation with German creationist zoologist Carl Burmeister in Buenos Aires during the last decades of the nineteenth century [46, 9, 71]. Burmeister never recognized Ameghino's value as a scientist and tried to block his career. Even at Burmeister's death, Ameghino was prevented from being appointed to the directorship of the Buenos Aires Museum because the German professor had left it in the charge of another European, Carl Berg. However, Argentine palaeontology as led by Ameghino had managed to constitute a critical mass, creating an original disciplinary approach in evolutionary studies. Among the signs of maturity was the presence of an interconnected disciplinary group, the control by Darwinians of two first-rate local museums, the support of the Ministry of Education, and broad contacts with the European research front. The earliest works by Ameghino were published in France and the United States, and he kept intense contact (even an active collaboration with Henri Gervais) with the great figures of French transformism. Ameghino attracted to his cause nationalist forces that helped to rally support and at the same time to reduce the efficacy of traditionalist opposition [30].

The disciplines and institutions of colonial science

By the end of the eighteenth century, agricultural and mineral sciences were employed more systematically to exploit the resources of the colonies. New soil conditions, surveying, pests, weather conditions, transportation, and communication required scientific inputs. Economic and geobotany acquired enormous importance. Every new plant was scrutinized for its use as food, fibre, timber, dye, or medicine. Among their central tasks, professional botanists sought the best techniques for transplanting commercially viable species from one part of the world to another. Their technical work was closely linked to the establishment of plantation economies on conquered land, as reflected in the history of sugar, cocoa, coffee, tea, rubber, quinine, and sisal. These activities were best carried out on institutional locations *in situ*. There was a proliferation of institutions from the last quarter of the eighteenth century onwards in many different latitudes. Botanic gardens consciously served the state as well as science, and shared the mercantilist and nationalist spirit of the times. Initially intended for the introduction and acclimatization of plants, like the Real Hôrto of Rio de Janeiro, many grew into institutes for serious experimentation and study. Kew Gardens, the Calcutta Garden, Peradeniya Garden on Ceylon, and Buitenzorg Garden on Java became important research centres [12]. Major agronomical research institutions also emerged in the colonies and other tropical nations in the second half of the nineteenth century, as in Campinas in Brazil, Buitenzorg on Dutch Java, and at Amani in German East Africa.

Other institutions that witnessed significant growth throughout the world during the nineteenth century were the museums of natural history. Successful metropolitan museums served as inspiration and example, not only for materials but also for architectural designs, organizational models, and qualified personnel. In Africa, museums of natural history were concentrated in the extreme southern portion of the continent. South African and Rhodesian museums survived only in centres with large White populations, such as Cape Town, Durban, Pietermaritzburg, and Grahamstown. A common feature was the exclusion of Blacks every day but Thursday, when admittance depended on wearing boots or shoes. In India, museums are said to have counted for little, were meagrely supported, and few and far between. As in Africa, widespread illiteracy, extreme poverty, and patterns of rural settlement made museums irrelevant to the vast majority of the populace [86]. Elsewhere, however, the museum movement was more successful. A handful of active, enthusiastic men directed museums located in the principal urban centres of Canada, Argentina, Australia, and New Zealand. South American museums tried to function both as research institutions and as instruments of popular enlightenment. Supported by national and provincial governments, important museums could be found in every capital city of the new republics. Rio de Janeiro, Buenos Aires, Santiago de Chile, and Montevideo built autonomous museums of natural history. While not reaching funding equivalents to that of the top museums in the world, their budgets rank with those of the better European institutions.

In Mexico, as mining was the principal source of income for the Spanish Crown, this activity received special attention in the context of social and economic renewal of the late eighteenth century. An impressive Royal School of Mining was founded in 1792 as part of a larger project, sponsored by Charles III of Spain, with the purpose of preparing individuals to direct the work of the mines and the exploitation of metals in those metal-poor minerals normally thrown away [1, pp. 137–146]. In Brazil, the School of Mines, built in the colonial town of Ouro Preto, close to the country's richest mineral deposits, was created only in 1875 [14, 85]. Alongside mining and engineering schools, we find astronomical observatories and meteorological stations. The scientific instruments available in some of those institutions suggest an interest in the basic sciences that occasionally went beyond practical concerns.

The intense competition between European colonial powers in seeking cures for the major tropical diseases that hindered the further colonization and exploitation of the tropics led to the emergence of tropical medicine as a distinct scientific specialty around the turn of the century, first in Britain, then in France, Italy, Belgium, Germany, the Netherlands, and, somewhat later, in the USA. European doctors were posted to the four corners of the world to service the imperial outposts that secured markets, trade, and raw materials for the imperial economies. A School of Tropical Medicine was needed to increase the quantity and quality of Colonial Medical Officers as an integral part of late-nineteenth-century British imperialism, the strengthening of political control, and attempts at more systematic exploitation [97, p. 93]. However, tropical medicine became a legitimate metropolitan scientific specialty and not merely a satellite activity instrumental to general public health in the colonies. Also in the French case, although relatively little attention has been paid yet to the export of pasteurianism to the tropics and around the world (an exception being Arnold [7]), it has been argued that the scientific imperialism of the Pastorians cross-reacted with colonial imperialism without being absorbed into it [47, pp. 307–320].

In the educational division of labour between metropolis and colonies, little importance was usually placed on developing local training and research capacities beyond those in applied fields, but differences were remarkable. In cases like that of the backward Portuguese Empire, timid, unstable, and bureaucratized attempts were made simply to train cadres for the state administration and the discovery of new wealth in the huge possessions of Brazil. But there was an absence of a social sector with greater interest in the development of education and science locally [85]. One may ask to what extent, for example, did the absence of universities in Brazil or the West Indies, and their continuing dependence on the home country for higher education, reduce those societies' chances for establishing a firm sense of their own identities in comparison with Mexico or New England, which had their own universities [26, p. 12]. In Africa, too, the establishment of governmental scientific institutions usually preceded the founding of universities in many countries by several decades [24].

By contrast, universities arrived in Spanish America with the Spanish conquerors as a conscious administrative expression of the will of the Crown and the Church. Thus they remained linked from their very inception to the powers of the audience and the viceroy or of the Church and the monastic orders. Until independence, the 33 existing universities led a precarious existence, mainly devoted to the training of priests, lawyers, and administrators. Current systems of higher education have little to do with colonial institutions. Instead, there is a more direct genealogical linkage with the public universities created during the nineteenth century, when the classic Latin American "lawyers' university" emerged, exemplified by the University of Chile, created by Andrés Bello [91]. With the advent of the new republics, being a lawyer became the main socialization and access channel for the national political élites, ensuring at the same time the necessary training for positions within the state apparatus, to which normally one arrived through family or political patronage. The University of Buenos Aires was founded in 1821, the Republican statutes of the Central University of Venezuela were approved in 1826; the University of Chile was established in 1842, that of Uruguay in 1860, Asunción in 1889. In Mexico, after independence the old colonial university was suppressed by government on the grounds of its being "useless, irreformable and pernicious," and it was reopened several times (1833, 1857, 1861, and 1865), reaching consolidation only after 1920 [13, pp. 13–106]. However, higher education continued to be relatively simple until 1950. By then, in half the 20 countries total enrolment did not reach 5,000 students, and in seven of them was less than 2,000. Argentina, Brazil, and Mexico accounted for 64 per cent of the regional enrolment. The total number of students in Latin American institutions of higher education in 1950 was less than the number of students currently enrolled in just one of the region's universities, the Universidad Nacional Autónoma de Mexico (UNAM).

Throughout the nineteenth century, Americans expanded all levels of their educational system. Typically, four-year residential colleges provided education to a growing portion of the population. A few of the colleges were associated with professional schools of law and medicine. Conventional wisdom equated the first two college years with the work in the gymnasium, the grammar school, and the lycée, and the last two years, hopefully, with the European university level. Elevating the entire college course to university level often appeared to be beyond attainment. The assumption and reality was that hardly anybody would attend a real university. For that reason, the colleges had to provide what society required in the way of both cultivation and preparation for selected occupations. The motivation of some of the US professoriate for higher courses and for research opportunities is usually stressed in the historical literature [67]. Apparently around 1900 the United States had surpassed Germany, at least in numbers; quality was another question [66, p. 17].

In Australia, the government supported most scientific and technological research. The second half of the nineteenth century saw the establishment of a number of higher learning institutions that provided technical or practical information. The first two Australian state universities were established in the early 1850s, one in Sydney and one in Melbourne. Four other universities were created before 1914. The incumbents of chairs and directorships often came from England. They did little research or practical work. With the waning of European interest in this natural wonderland, there was little in the local scene to sustain intellectual endeavour. Indeed, almost everything was against the development of science: isolation from Europe, the small size of the population, colonial economies based on the exploitation of readily available natural resources, as well as the export of a narrow range of staple commodities, and an emerging egalitarian and anti-intellectual tradition more intense than in North

America. As late as 1937, the various Australian governments and even the Council for Scientific and Industrial Research (CSIR), created in 1926, hesitated to finance pure research into anything beyond local knowledge, for it was believed that it would "be made available by close liaison between Australian scientists and the great overseas laboratories," meaning those of Britain [81, p. 188].

Institutional growth in the moulds of "national science"

Development patterns of the former colonial world have been enormously varied, making generalization difficult. Whatever the origin of the ambitions of different colonized groups, by the mid-nineteenth century the nation-state had become the only acceptable frame of reference. After that, self-perception was a question of nationhood and of very little else. Nevertheless, the idea of the nation in the twentieth-century movements aimed at liberation from colonial domination had specificities that force us to look at them separately from the independence movements of the Atlantic world, both from British and Spanish domination, in the late eighteenth and early nineteenth centuries. But relevant differences are not only those of a temporal kind. A recent classification produced on the basis of the most readily available indicators of S&T capabilities distinguishes between three broad groups of developing countries:

1. Those with no S&T base, with an extremely fragile economy, and where much of the population lives in extreme poverty. The low level of provision for education and training is both a cause and a consequence of this situation. (Most African countries are included in this category.)
2. Those with the fundamental elements of an S&T base, thanks to past (mostly foreign) investment. Little attention is paid to domestic economic problems, often resulting in serious imbalances. They have established a certain industrial basis, with moderate GDP per capita. Some have a relatively high percentage of potential S&T manpower, but absolute numbers are low. (Includes countries of various sizes and ranging from East Asia to the Middle East and North Africa.)
3. Countries with an established S&T base. This highly heterogeneous group has an industrial basis with a higher percentage of potential S&T manpower and relatively high GDP per capita. Because of past achievements, and because they are more integrated into international trade than others, most are highly vulnerable to

185

international trends. (Includes some Asian countries, including Pakistan, India, and the newly industrialized countries, and most Latin American countries, particularly Argentina, Brazil, and Mexico.) [35]

Typologies such as this – although they permit comparisons between different areas – have a number of limitations for understanding the dynamics of science in specific national contexts. A scheme is needed that successfully contains both the local and the metropolitan factors operating in developing country science, rather than a static descriptive set of indicators. A comparative treatment of the scientific enterprise must account for differences *between* and *within* typical groupings [37].

For example, if we take the so-called areas of recent settlement, a restricted category that includes Canada, Australia, New Zealand, and Argentina, by the end of the nineteenth century all were characterized by an abundance of land relative to labour and capital, and they all managed to develop capitalist economies that were highly integrated into the world market through the export of staples. It might also be said that the development of the scientific enterprise in these areas falls into some pattern that allows us to speak of a common "type" of scientific community, on account of the presence of three factors common to all of these countries: immigrants of primarily European origin, foreign capital, and a reduction in the cost of ocean transport. However, the history of Canadian science is not precisely that of Argentine science, not even that of Australian science. Features of the socio-economic bases of such nations emerge as of prime importance, operating on different levels of the scientific enterprise, from the societal support base to the scientific community itself.

The most striking contrast within this group is between Argentina and the others [58, 89]. During the twentieth century, most of these countries became modern industrial societies. In Argentina, however, economic and political performance since the 1920s has clearly differed from the achievements of the other countries, making of it a remarkable case of failed development. That is, although it possessed some requisites for becoming a modern country – not the least being its high rate of economic growth between 1880 and 1930 – those conditions did not suffice. Government policies during the 1940s were short-sighted, and the political instability that followed made matters worse. Moreover, past economic growth helped raise expectations that could not be met. Among the complex subjects that still await further analysis are the characteristics of the Argentine dominant landowning class and of the scientific community that grew, sometimes supported by government patronage, sometimes persecuted by it. But also the different economic policies and factor endowment relative to other countries of recent settlement require deeper study.

One might thus begin to explain national scientific institutionalization as a product of both local conditions and metropolitan relations. Changing conditions are differentially absorbed within the socio-economic base for science and are transmitted to the scientific enterprise through the medium of the institutional infrastructure, in which the activity developed during the colonial period may have left a strong imprint. The debate in Nigeria about the reorientation of higher education after independence in order to respond to changing demands related to manpower and economic development is illustrative of the difficult balance of forces. The 1960 report of the Commission on Post-School Certificate and Higher Education in Nigeria – the well-known Ashby Report – was the point of departure for analysis and conceptualization [3, pp. 1–20]. Based on the theory of human capital [32], it argued that the economic development of every country is ultimately the result of the trained effort of its citizens. Therefore, the building of a broad reservoir of highly educated persons was the key to Nigeria's development [8]. These ideas, however, ran against the deeply rooted colonial tradition according to which the schools and colleges that emerged in Nigeria were developed as much as possible into replicas of similar institutions in England, i.e. "emphasizing in standard and curriculum, the thin stream of excellence and narrow specialism. In social function . . . [they were] restricted to an elite" [17]. The initial optimism and enthusiasm of the early days of independence obscured the signs of conflict that emerged from the effort to change the much criticized but highly revered system of higher education. The traditional "classical" orientation of Nigerian higher education has, in effect, been maintained, to the dismay of the initiators of reform. The pre-existing curriculum of higher education, with its primacy of academic subjects over science and technology, has been preserved.

The persistence of colonial relations and/or institutional networks, or the excessive bureaucratization of the scientific enterprise as the result of its close bondage with government in search of economic development, may hamper the ability of developing country science to adapt to changes at the local level. Scientific institution building in India from independence up to the late 1960s was based on a close and easy alliance between élite scientists and the top political leadership,

187

represented by Nehru. In contrast to Gandhi's anti-modern techno-logy stance, Nehru's modern, secular image and most of all his ideo-logy of "scientism," made him a "messiah" for Indian science [41, p. 12]. The importance of the personal linkage between Homi Bhabha in Atomic Energy, S.S. Bhatnagar and H. Zaheer in CSIR, P.C. Mahalanobis in the Planning Commission, and J.C. Ghosh with Nehru was crucial. The locus of scientific research shifted from the private research institutions characteristic of the previous period to government science agencies [5].

The shift in the locus of science also meant a shift in the power base of scientists and their career structure and status in a socially and culturally stratified society. Science agencies became subor-dinated to overarching political structures. Major decisions on science came to be determined by or within the political structures. A very narrow élite made up of the heads of government agencies and de-partments has played multiple roles, keeping control over a large and fragile scientific community [88, pp. 575–594]. Although in theory representatives of the scientific community, they turned out to be ser-vants of government and part of its hierarchical system. The power base and career path that draws more and more power and status has been located in the mission-oriented science agencies sector rather than in the academic, university sector, which is marginal to decision-making processes.

With the emergence of local scientific communities in developing countries, the assertion of a national identity in science grew at differ-ent times, assuming specific features in each case. In Australia, for in-stance, during the 1920s the imperial ideal predominated, the vision of a self-sufficient British Empire of which Australia would be a lead-ing part as an exporter of commodities and an importer of surplus labour and capital. Australian scientific resources were to be mobil-ized and linked with British science in pursuit of the economic in-tegration of the empire [81, p. 104]. By contrast, the arguments of the anti-*dependencia* scientists in Latin America emphasized the global nature of development and underdevelopment, placing them in the context of the "centre/periphery" model. Among the most commonly cited economic features of "dependency" were several connected with technology. These included the presence of heavy foreign investment and foreign capital-intensive technologies, which forced countries to specialize in exporting raw materials or labour-intensive manufactured goods; consumption patterns among the élites that were determined by the "centre," unfavourable exchange terms, and an increasing concentration of wealth and growing unemployment, particularly in the cities [94, p. 527]. The challenge was to find alternative courses of action for Latin American science in order to respond effectively to the problems of poverty, malnutrition, disease, and the increasingly unequal terms of international development [93, 33, 72, 75].

With all the differences of national contexts, the fact remains that a tension was always present in the developing world between the assertion of national identity and autonomy and the socio-psychological feelings of peripherality, marginality, or invisibility. That local science and metropolitan science share the same institutional model, while being widely separated (mentally and spatially), has led to ambivalence and tensions. Real or perceived intellectual isolation, a felt lack of recognition, debates about standards and quality of results, the claims to design alternative indicators for the scientific production in developing countries on account of the contextual variables that condition and determine their potential productivity – not the least important being the fact that developing country science is usually understood as "science for development" – these are all elements that have characterized the debates in the process of scientific institutionalization in the national "mould" in recent decades.

The groups of developing country scientists, engineers, and government officials who, at one time or another, managed to put their projects for autonomy into practice also achieved something else in the process. For a while they managed to change the conditions of the competitive game by their unexpected achievements [19]. The development of local capabilities in science, technology, industry, management, and labour skills introduced significant changes in the local social structures, created new sets of actors with technical and managerial skills, and gave them a better understanding of the art of negotiation. But the changes they produced have been usually insufficient to alter the background of social and economic conditions characteristic of underdevelopment.

The role of government science policy

The history of developing country science is full of examples of attempts at institutionalization followed by collapse, unbounded optimism followed by pessimistic indifference, and a lack of public trust in long-term intellectual endeavour. In general, government support for science on a significant scale by formation of concrete institutions and programmes was provided by the different countries in the context of changing conditions in the international markets.

Important cases of technological development in Latin America and South-East Asia illustrate the crucial role that tactical alliances have between the scientific élites and the state. The state has time and again revealed itself the most important factor in developing countries' successful or failed use of S&T for industrial development. Recent research contributions show the systemic and comprehensive state intervention in the economies of the newly industrialized countries, as well as the state's strategic guidance of the performance of national and multinational companies located in their territory [15]. The "developmental state" has been a fundamental factor in creating the conditions for economic growth in South-East Asia, as well as ensuring the transition of their industrializing economies to each of the different stages in their evolving articulation to the world economy [28].

At different times since the 1930s, but more systematically since the 1950s, most countries established national councils of science and technology or specialized units in planning agencies; the numbers of research centres grew, accentuating the fragmentation of the scientific and technological effort; new public institutions were created to promote and carry out scientific and technological activities, as were government units to regulate the importation of technology and to provide service to industry, mining, and agriculture. The rapid industrialization of the largest Latin American countries produced a demand for science and engineering graduates to handle operational and service problems of the new assembly industries; research funding mechanisms, which so far had operated according to the "little science" model, began to be transformed. New demands forced the emergence of intermediaries in the form of research managers, project administrators and negotiators in the funding agencies, with increasing formalization of research activities. The mechanisms and criteria adopted were not always compatible with the experience and tradition accumulated until then through isolated and small group efforts. Indeed, in the 1970s the bureaucratization of the state apparatus was visible in a country like Brazil. Active lobbying by groups of technocrats and intellectuals succeeded in convincing policy makers and then in creating the bureaucratic apparatus and the financial devices to enable the idea of autonomous scientific and technological development to prosper. Brazil's economy has been a mixture of

190

market mechanisms, state intervention, and planning. The state has played an important economic role through guidelines and planning, incentives and controls – establishing the objectives and the means to achieve progress [29]. Nevertheless, the recent difficulties of the Brazilian economy and therefore of its S&T base show the fragility and vulnerability of the whole enterprise.

Recognition of the role of government science planning in developing countries should not give the impression that results were always positive. In fact, national experiences have been subject to strong criticisms on various accounts. As a result of vigorous promotion by Unesco of "science planning" in the 1960s and 1970s, many African countries also created national science units, but most of their objectives have not been achieved [21]. Several such units have been abolished in recent years. Others have been absorbed into ministries of education. Most surviving science policy units engage in such activities as passing on requests for the clearance of foreign research workers who wish to work in the country. A few national units have acquired important but more modest functions: the Kenyan National Council on Science and Technology administers a fund that supports university research.

A rearrangement of the international system eased the transition to a new world order in which international cooperation with the third world contributed to shaping the modern process of scientific institutionalization in the developing countries. Until the 1960s, a high degree of congruence between the policies of industrialized country donors, as they came to be called, and the needs articulated by developing country recipients characterized international educational and scientific assistance. The most influential (Western) studies of the day demonstrated the productivity-raising effects of investments in education [e.g. 84], and showed that the magnitude of effects increased with educational level. Major Western universities were "twinned" with developing country "sister institutions" in a pattern closely resembling the affiliation of colonial institutions with metropolitan ones, as in the case of the London University network for British colonial dependencies and the Université de Bordeaux for the French ones. A large number of industrialized country universities and other institutions became involved in institution building overseas. But results often differed significantly from the international technocratic rhetoric. For example, the purpose for which universities were established in Africa was the indigenization of educational, technical, scientific, and administrative services, to contribute to economic development. Progress was measured in terms of the supply of trained manpower chiefly to the public sector. However, what did expand, generally speaking – because it lay within the power of politicians to create them – were not industrial jobs as expected of "high-level manpower planning in practical subjects," but appointments in the public services. Surely, technological imperatives arise from localized socio-economic forces and are not especially subject to pressure emanating from scientists. An increased number of graduates was what politicians wished to see so that they could fill the civil service. What the graduates studied was of less interest to them than the fact that they had academic degrees. An effective demand was thus created (although not the one served by public rhetoric) and African universities responded accordingly.

Regional institutional collaboration has been recommended as one of the most effective ways to build up rapidly a better S&T base, but the actual concentration of advanced scientific training in regional centres has often been resisted. Regional centres of scientific training and research have been moderately successful when institutions are supported by international organizations and by foreign governments in bilateral arrangements with particular governments. Instances of such successes have been the Latin American Centres of Mathematics in Buenos Aires, of Physics (CLAF) in Rio de Janeiro, and of Biology (CLAB) in Santiago, or the international agricultural research institutes established in Africa, like that of Tropical Agriculture in Nigeria, the International Livestock Centre for Africa in Ethiopia, and the International Laboratory for Research on Animal Disease in Kenya. However, their connections to national scientific institutions, especially with African universities, have always been fragile and have been weakened in recent years by the declining research efforts of many countries. When a developing country government and its universities are asked to share the costs of regional institutions, or to contribute to the development of institutions outside their country, these efforts have usually failed.

The economic crisis that affected most of the developing world in the 1980s increased the difficulties of the best research institutions. A growing flow of diagnostic studies confirm the deterioration of working conditions and the increasing alienation of researchers as a result of greater financial restrictions and physical and intellectual isolation. In a world in which academic networking is rapidly expanding, developing countries remain as poorly connected areas. Marginality in science and technology is increasing, both in quantitative and in qualitative terms in many developing countries. Given the difficulties in which academic science is immersed in Latin America and the serious threat of an intensified "brain drain," several countries have tried to implement programmes to minimize this process. Argentina, Mexico, Venezuela, and Brazil have implemented programmes to supplement the salaries of the élite S&T cadres, aimed at preserving the core of the national stock of researchers and fostering their improvement and productivity and the participation and self-evaluation of the research community.

The interface between higher education and research capabilities

As with other elements linked with the idea of modernization, like technology, economics, and the values of progress, the force of science lay for some developing countries more in its abstract symbolic power than in its actual practice. It has been argued that often attention was placed on collateral features of this science rather than on the means of producing it, that is to say, it was a science devoid of its crucial *research* component. The founding of schools for the training of engineers, medical and pharmaceutical doctors, and other specialists without the provision of specific institutional space for research has been common.

Even after independence, no provision was made in most new African nations for training in research as part of the work of the universities [87]. A research capability came to be demanded when governments decided that they could not endogenize their teaching and research staff without providing local graduate training. Master's programmes began to be hurriedly introduced in the most advanced developing countries in the late 1960s and 1970s. Brazil and Mexico are the Latin American countries that have developed the most this educational level, with over 1,500 M.A. programmes each, although doctoral programmes continue to be few and unproductive. In fact, graduate professional training was overwhelmingly a prolongation of undergraduate courses, with accreditation being linked to the access to and/or promotion in bureaucratized labour markets or with conspicuous cultural consumption, having little to do with scientific education for the preparation of researchers.

The introduction of research in the professional universities of developing countries was often the result of technical assistance received from the more advanced countries or of professional and scientific training obtained by individuals abroad. A forceful description of social conditions, which could be easily extended to many a national tradition in the developing world, has been given by Araújo e Oliveira with regard to Brazil: "in such a turbulent and hostile environment, where constancy, quality, excellence, seriousness, obstinacy and the disinterested search are of little or no value at all, those scientific groups that are implanted, prosper, and bear fruit, can best be considered as *islands of competence*" [6]. Such islands of competence are found particularly in countries like India, Pakistan, Argentina, Brazil, or Mexico, which at one time or another seemed on the verge of making the last leap forward to become truly independent centres of scientific creativity but have not quite made it. Although the historical record reveals examples of research groups or isolated scientists who managed to do competent science in the most difficult conditions, our subject of institutionalization of science refers to social conditions of receptivity and stimulus for the modern scientific enterprise. The emergence of scientific communities (of relatively autonomous scientific fields in developing countries), as we have seen, has had unequal results, and it is an area that still poses more questions than it answers.

For all the diversity of institutions and national and cultural contexts, develping country universities share a number of problems:
1. High cost.
2. Explosion of numbers. Latin American higher education experienced an unprecedented expansion from 1950 to 1980, increasing 20-fold, from about 250,000 students in 1950 to 5,380,000 in 1980. In most African countries, too, enrolment has increased enormously since independence, but the total number there is still comparatively small, less than 20,000. Part of the reason is that universities are unable to absorb more students because, being residential institutions, they have reached the upper limits of their intake capacity.
3. Scarcity of trained human resources. In some African countries such shortages had serious consequences for the operation of government scientific services. Sometimes one hears the criticism that a developing country has "over-invested" in or has "over-qualified" its human resources. But in societies that combine mass cultures in an accelerated process of formation with a weak, heterogeneous, and dependent base of externally conditioned prosperity and crisis cycles, the relationship between education and work posts becomes loose and tenuous.

4. Poor quality of instruction at all levels.
5. Privatization of higher education. The developing world shows almost the maximum possible range in the private sector's proportion of total enrolments by nation. In many countries, most private growth has occurred in recent decades. In terms of its impact on the whole system, the performance of privatization in Asian and Latin American higher education to date leaves considerable room for debate [42].

Concluding remarks

The institutionalization of Western science in the developing world proceeded as both an instrument of the interests of the most advanced countries and a result of active attempts by underdeveloped nations to master the knowledge that was the promise of modernity. At different times the major colonial powers and the new independent nations established S&T institutions, but it has been difficult for science to take root, particularly since it was expected to produce economic growth.

Emphasis on organizations and institutions has forced us to focus attention on scientific and technical élites. Scientific institutions were seen as the formulated and communicated outcomes of thought, as manifested in institutional ideologies, roles, and functions, "carriers" for particular collective understandings [2]. Specific scientific institutions represent ideals in operation, serving as channels for the realization and transmission of personal and intellectual will. At different times and places, institutional leaders have provided the beliefs, expectations, and goals that show the way, a particular way of identifying problems and their solutions. Of course the summation of individual scientific institutions does not necessarily result in the institutionalization of science in a particular country. They are necessary, albeit not sufficient, conditions for success or failure.

References

1   Aceves, P. "The First Chair of Chemistry in Mexico (1796–1810)." In: Petit-jean et al., eds. *Science and Empires*. Dordrecht: Kluwer Academic Publishers, 1992.
2   Adler, E. *The Power of Ideology: The Quest for Technological Autonomy in Argentina and Brazil*. Berkeley: University of California Press, 1987.
3   Ahmed, A. "The Asquith Tradition, the Ashby Reform, and the Development of Higher Education in Nigeria." *Minerva* 27 (1989), no. 1: 1–20.

195

4   Alvares, C. *Homo Faber: Technology and Culture in India, China and the West from 1500 to the Present Day*. The Hague: Martinus Nijhoff, 1980.

5   Anderson, R.S. "The Government of Scientific Institutions (Case Studies of Research Laboratories in India in the late 1960s)." *Contributions to Indian Sociology* (NS) 11 (1977), no. 1: 137–168.

6   Araújo e Oliveira, J.B. *Ilhas de Competência: Carreiras Científicas do Brasil*. São Paulo: CNPq/Brasiliense, 1985.

7   Arnold, D., ed. *Imperial Medicine and Indigenous Societies*. Manchester: Manchester University Press, 1989.

8   Ashby, E., and M. Anderson. *Universities, British, Indian, African: A Study in the Ecology of Higher Education*. London: Weidenfeld and Nicolson, 1966.

9   Babini, J. *La evolución del pensamiento científico en la Argentina*. Buenos Aires: La Fragua, 1954.

10   Barraclough, G. *An Introduction to Contemporary History*. Harmondsworth: Penguin, 1967.

11   Basalla, G. "The Spread of Western Science." *Science* 156 (1967): 611–622.

12   Brockway, L. *Science and Colonial Expansion: The Role of the British Royal Botanic Gardens*. New York: Academic Press, 1979.

13   Brunner, J.J. "Educación superior, investigación científica y transformaciones culturales en América Latina." In: BID-SECAB-CINDA, *Vinculación Universidád Sector Productivo*. Santiago: CINDA 1990, pp. 11–106.

14   Carvalho, J.M. de. *A Escola de Minas de Ouro Preto. O peso da glória*. Rio de Janeiro: FINEP/Editora Nacional, 1978.

15   Castells, M. *Four Asian Tigers with a Dragon Head: A Comparative Analysis of the State, Economy, and Society in the Asian Pacific Rim*. Working Document no. 14. Madrid: Universidad Autónoma de Madrid, Instituto Universitario de Sociología de Nuevas Tecnologías, 1991.

16   Chambers, D.W. "Period and Process in Colonial and National Science." In: Reingold and Rothenberg, eds. *See* ref. 68.

17   Colonial Office. *Report on Higher Education in the Colonies*. Cmnd. 6647. London: HMSO, 1945.

18   Constantino, R. *Neocolonial Identity and Counter-Consciousness: Essays in Cultural Decolonization*. London: Merlin Press, 1978.

19   Cueto, M. *Excelencia científica en la periferia: Actividades científicas e investigación biomédica en el Perú, 1890–1950*. Lima: GRADE-CONCYTEC, 1989.

20   Cueto, M. "Visions of Science and Development: The Rockefeller Foundation and the Latin American Medical Surveys of the 1920s." Paper presented at the conference, Science, Philanthropy, and Latin America: Cross Cultural Encounters in the Twentieth Century. North Tarrytown, New York: Rockefeller Archive Center, 1991.

21   Davis, C.H. "L'UNESCO et la promotion des politiques scientifiques nationales en Afrique subsaharienne, 1960–1979." *Etudes Internationales* 14 (1983).

22   Dugan, K.G. "The Zoological Exploration of the Australian Region and Its Impact on Biological Theory." In: Reingold and Rothenberg, eds., pp. 79–100. *See* ref. 68.

23   Dunn, J. *Identity, Modernity, and the Claim to Know Better*. HSDRSCA-103/
      UNUP-441. Tokyo: The United Nations University, 1982, pp. 1–23.
24   Eisemon, T.O., and C.H. Davis. "Can the Quality of Scientific Training and
      Research in Africa Be Improved by Training?" *Minerva* 29 (1991), no. 1.
25   Eisenstadt, S.N. "Transformation of Social, Political, and Cultural Orders in
      Modernization." *American Sociological Review* 30 (1965): 659–673.
26   Elliott, J.H. "Introduction: Colonial Identity in the Atlantic World." In: N.
      Canny and A. Pagden, eds. *Colonial Identity in the Atlantic World, 1500–
      1800*. Princeton: Princeton University Press, 1987, pp. 3–13.
27   Fanon, F. *Les damnés de la terre*. Paris: Maspero, 1961.
28   Gereffi, G. "Rethinking Development Theory: Insights from East Asia and
      Latin America." *Sociological Forum* 4 (1989), no. 4.
29   Gereffi, G., and P. Evans. "Transnational Corporations, Dependent Develop-
      ment, and State Policy in the Semiperiphery: A Comparison of Brazil and
      Mexico." *Latin American Research Review* 16 (1981), no. 3.
30   Glick, T. "Perspectivas sobre la recepción del darwinismo en el mundo hispa-
      no." In: M. Hormigón, ed. *Actas del II Congreso de la Sociedad Española de
      Historia de las Ciencias*. Madrid: SEHC, 1982.
31   Goonatilake, S. *Aborted Discovery: Science and Creativity in the Third World*.
      London: Zed Books, 1984.
32   Harbison, F.H., and C.A. Myers. *Manpower and Education: Country Studies
      in Economic Development*. New York: McGraw-Hill. 1965.
33   Herrera, A. *Ciencia y política en América Latina*. Mexico: Siglo XXI, 1971.
34   Home, R., ed. *The Shaping of Australian Science*. Sydney: Cambridge Uni-
      versity Press, 1988.
35   ICSPS (International Council for Science Policy Studies). *Science and Tech-
      nology in Developing Countries: Strategies for the 90s, A Report to UNESCO*.
      Paris: Unesco, 1992.
36   Inkster, I. "Meiji Economic Development in Perspective: Revisionist Com-
      ments Upon the Industrial Revolution in Japan." *The Developing Economies*
      17 (1979), no. 1: 45–68.
37   ———. "Scientific Enterprise and the Colonial 'Model': Observations on
      Australian Experience in a Historical Context." *Social Studies of Science* 15
      (1985), no. 4: 677–704.
38   Jami, C. "Western Mathematics in China, Seventeenth Century and
      Nineteenth Century." In: Petitjean et al., eds., pp. 79–88. *See* ref. 57.
39   Jarrell, R.A. "Differential National Development and Science in the
      Nineteenth Century: The Problems of Quebec and Ireland." In: Reingold and
      Rothenberg, eds., pp. 323–350. *See* ref. 68.
40   Krishna, V.V. "The Colonial 'Model' and the Emergence of National Science
      in India: 1876–1920." In: Petitjean et al., eds., pp. 57–72. *See* ref. 68.
41   Krishna, V.V., and A. Jain. "Country Report: Scientific Research, Science
      Policy and Social Studies of Science and Technology in India." Paper pre-
      sented at the First Workshop on the Emergence of Scientific Communities in
      the Developing Countries, 22–27 April 1990. Paris: ORSTOM.
42   Levy, D. "Problems of Privatization." Paper prepared for the World Bank's
      Worldwide Seminar on Innovation and Improvement of Higher Education in

Developing Countries, Kuala Lumpur, 30 June–4 July, 1991.

43   Lutz, G.W. *China and the Christian Colleges 1850–1950.* Ithaca: Cornell, 1971.

44   MacLeod, R. "On Visiting the 'Moving Metropolis': Reflections on the Architecture of Imperial Science." In: Reingold and Rothenberg, eds. *See* ref. 68.

45   ———. "The Contradictions of Progress." Sydney: University of Sydney, 1991. Unpublished manuscript.

46   Márquez Miranda, F. *Ameghino: Una vida heroica.* Buenos Aires: Nova, 1951.

47   Moulin, A.M. "Patriarchal Science: The Network of the Overseas Pasteur Institutes." In: Petitjean et al., eds., pp. 307–320. *See* ref. 57.

48   Moyal, A. *Scientists in Nineteenth Century Australia.* Sydney: Cassell Australia, 1987.

49   Nakayama, S. *Characteristics of Scientific Development in Japan.* New Delhi: CSIR, 1977.

50   Nakayama, S., D.L. Swain, and Y. Eri, eds. *Science and Society in Modern Japan: Selected Historical Sources.* Cambridge, Mass.: MIT Press, 1974.

51   Pagden, A. "Identity Formation in Spanish America." In: N. Canny and A. Pagden, eds. *Colonial Identity in the Atlantic World, 1500–1800.* Princeton: Princeton University Press, 1987, pp. 51–93.

52   Paul, H.W. *From Knowledge to Power: The Rise of the Science Empire in France 1860–1939.* Cambridge: Cambridge University Press, 1985.

53   Perrot, D., and R. Preiswerk. *Ethnocentrisme et histoire: L'Afrique, l'Amérique indienne et l'Asie dans les manuels occidentaux.* Paris: Anthropos, 1975.

54   Petitjean, P. "Le Groupement des Universités et Grandes Ecoles de France pour les Relations avec l'Amérique Latine, et la création d'Instituts à Rio, Sao Paulo et Buenos Aires (1907/1940)." *Anais do Segundo Congresso Latino-Americano de História da Ciência e da Tecnologia.* São Paulo: Nova Stella, 1989, pp. 428–442.

55   ———. "Autour de la mission française pour la création de l'Université de Sao Paulo (1934)." In: Petitjean et al., eds., pp. 339–362. *See* ref. 57.

56   ———. "La coopération France–Amérique Latine." Paper presented at the Third Latin American Congress for the History of Science, Mexico City, 12–16 January 1992.

57   Petitjean, P., C. Jami, and A.M. Moulin, eds. *Science and Empires: Historical Studies about Scientific Development and European Expansion.* Dordrecht: Kluwer Academic Press, 1992.

58   Platt, D.C.M., and G. Di Tella, eds. *Argentina, Australia, and Canada: Studies in Comparative Development, 1870–1965.* New York: St. Martin's Press, 1985.

59   Porter, J. "The Scientific Community in Early Modern China." *Isis* 73 (1982), no. 269: 529–544.

60   Pyenson, L. "The Incomplete Transmission of a European Image: Physics at Greater Buenos Aires and Montreal, 1890–1920." *Proceedings of the American Philosophical Society*, vol. 122 (1978), pp. 92–114.

61   ———. "*In Partibus Infidelium*: Imperialist Rivalries and Exact Sciences in Early Twentieth-Century Argentina." *Quipu* 1 (1984).

62   ———. *Cultural Imperialism and Exact Sciences: German Expansion Overseas 1900–1930*. New York: Peter Lang, 1985.

63   ———. *Empire of Reason: Exact Sciences in Indonesia, 1840–1940*. Leiden: E.J. Brill, 1989.

64   ———. "Pure Learning and Political Economy: Science and European Expansion in the Age of Imperialism." In: R.P.W. Visser et al., eds. *New Trends in the History of Science*. Proceedings of a conference held at the University of Utrecht. Amsterdam/Atlanta, Ga.: Rodopi, 1989.

65   Raj, K. "Hermeneutics and Cross-Cultural Communication in Science: The Reception of Western Scientific Ideas in 19th Century India." *Revue de Synthèse*. 4ème série, nos. 1, 2 (January–June 1986): 107–120.

66   Reingold, N. "Reflections on 200 Years of Science in the United States." In: N. Reingold, ed. *The Sciences in the American Context: New Perspectives*. Washington, D.C.: Smithsonian Institution Press, 1979.

67   ———. "Graduate School and Doctoral Degree: European Models and American Realities." In: Reingold and Rothenberg, eds. *See* ref. 68.

68   Reingold, N., and M. Rothenberg. eds. *Scientific Colonialism: A Cross-Cultural Comparison*. Washington, D.C.: Smithsonian Institution Press, 1987.

69   Ríos, S., L. Santaló, and M. Balanzat. *Julio Rey Pastor matemático*. Madrid: Instituto de España, 1979.

70   Roca Rosell, A., and J.M. Sánchez Ron. *Esteban Terradas: Ciencia y técnica en la España contemporánea*. Madrid and Barcelona: Instituto Nacional de Técnica Aeroespacial/Ediciones del Serbal, 1990.

71   Romero, J.L. *El desarrollo de las ideas en la sociedad argentina del siglo XX*. Mexico/Buenos Aires: Fondo de Cultura Económica, 1965.

72   Sábato, J., ed. *El pensamiento latinoamericano en la problemática ciencia-tecnología-desarrollo-dependencia*. Buenos Aires: Paidós, 1975.

73   Safford, F. *The Ideal of the Practical: Colombia's Struggle to Form a Technical Elite*. Austin: The University of Texas Press, 1976.

74   ———. "Acerca de la incorporación de las ciencias naturales en la periferia: el caso de Colombia en el siglo XIX." *Quipu* 2 (1985), no. 3.

75   Sagasti, F., and C. Cook. *Tiempos difíciles: ciencia y tecnología en América Latina durante el decenio de 1980*. Lima: GRADE, 1985.

76   Sagasti, F. et al. *Conocimiento y desarrollo: ensayos sobre ciencia y tecnología*. Lima: GRADE-Mosca Azul, 1988.

77   Saldaña, J.J. "La ciencia y el Leviatán mexicano." *Actas de la Sociedad Mexicana de Historia de la Ciencia y de la Tecnología*, vol. 1 (1989).

78   Salomon, J.-J., and A. Lebeau. *Mirages of Development*. Boulder, Colo.: Lynne Rienner, 1993. Originally published in French as *L'écrivain public et l'ordinateur*. Paris: Hachette, 1988.

79   Salomon, N. "Cosmopolitismo e internacionalismo (desde 1880 hasta 1940)." In: L. Zea, ed. *América Latina en sus ideas*. México/Paris: Siglo XXI/Unesco, 1986.

80   Sangwam, S. "Indian Response to European Science and Technology." *British Journal of History of Science* 21 (1988).

81   Schevdin, C.B. "Environment, Economy, and Australian Biology, 1890–1939." In: Reingold and Rothenberg, eds., pp. 101–128. *See* ref. 68.

82   Schroeder-Gudehus, B. "Science, Technology, and Foreign Policy." In: I. Spiegel-Rösing and D. de Solla Price, eds. *Science, Technology and Society*. London: Sage, 1977.

83   ———. "Sciences exactes et politique extérieure." In: Petitjean et al., eds. *See* ref. 57.

84   Schultz, T.W. *The Economic Value of Education*. New York: Columbia University Press, 1963.

85   Schwartzman, S. *A Space for Science: The Development of the Scientific Community in Brazil*. University Park: Pennsylvania State University Press, 1991. Originally published in Portuguese in 1979.

86   Sheets-Pyenson, S. *Cathedrals of Science: The Development of Colonial Natural History Museums during the Late Nineteenth Century*. Kingston and Montreal: McGill-Queen's University Press, 1986.

87   Sherman, M.A.B. "The University in Modern Africa." *Journal of Higher Education* 61 (1990), no. 4.

88   Shiva, V., and J. Bandyopadhyay. "The Large and Fragile Community of Scientists in India. *Minerva* 18 (1980), no. 4.

89   Solberg, C.E. *The Prairies and the Pampas: Agrarian Policy in Canada and Argentina, 1880–1930*. Stanford, Calif.: Stanford University Press, 1987.

90   Stavrianos, L.S. *The Global Rift: The Third World Comes of Age*. New York: William Morrow & Co., 1981.

91   Steger, H. *Las Universidades en el desarrollo social de América Latina*. Mexico: Fondo de Cultura Económica, 1974.

92   Stepan, N. *Beginnings of Brazilian Science: Oswaldo Cruz, Medical Research and Policy, 1890–1920*. New York: Science History Publications, 1981.

93   Varsavsky, O. *Ciencia, política y cientificismo*. Buenos Aires: Centro Editor de América Latina, 1969.

94   Vessuri, H. "El proceso de profesionalización de la ciencia venezolana: la Facultad de Ciencias de la Universidad Central de Venezuela." *Quipu* 4 (May–August 1987), no. 2: 253–281.

95   Vessuri, H.M.C. "O Inventamos o Erramos: The Power of Science in Latin America." *World Development* 18 (1990), no. 11: 1543–1553.

96   Vessuri, H., ed. *Ciencia académica en la Venezuela moderna*. Caracas: Fondo Editorial Acta Científica Venezolana, 1984.

97   Worboys, M. "The Emergence of Tropical Medicine: A Study in the Establishment of a Scientific Specialty." In: G. Lemaine, R. MacLeod, M. Mulkay, and P. Weingart, eds. *Perspectives on the Emergence of Scientific Disciplines* Paris/The Hague: Mouton/Aldine, 1976.

98   Worsley, P. *The Three Worlds: Culture and World Development*. London: Weidenfeld, 1984.

99   Yudelman, M. "Imperialism and the Transfer of Agricultural Techniques." In: P. Duignan and L.H. Gann, eds. *Colonialism in Africa*, vol. 4: *The Economics of Colonialism*. Cambridge: Cambridge University Press, 1975.

6

The behaviour of scientists and scientific communities

Jacques Gaillard

There is a relatively large storehouse of documents and reports on science and technology policies in the developing countries, often prepared for international conferences such as the United Nations Conference on Science and Technology for Development in 1979 in Vienna. That said, it must be recognized that these official documents mainly contain statements of intent and that our knowledge of science, scientists, and scientific communities in developing countries is very incomplete. There is also a fairly abundant literature on third world science scattered through numerous journals, seminar reports, and proceedings, but there are far too few empirical studies. The late Professor Moravcsik [61] provided one of the most complete bibliographies on this subject, which covers literature up to the early 1970s, but research on science in developing countries is still an unexplored and fruitful area [62]. A decision was taken in October 1990 at the International Conference on Science Indicators for Developing Countries in Paris to create an international association interested in science in developing countries; one of its tasks would be to update Moravcsik's bibliography [5].

Generally on the basis of national statistics compiled by organizations such as Unesco and the OECD, certain authors have emphasized the shortcomings of the research systems in developing countries and the shortage of available resources [71]. Other authors have matched socio-economic conditions against the level of scientific development in these countries. In some of his writings, Price [66] gives quantitative indicators for the developing countries. Research by Garfield [39] and his Institute for Scientific Information in Phil-

201

adelphia (ISI) points to the low productivity of third world scientists, the difference in productivity levels, and degrees of dependency (articles by scientists from developing countries have greater impact when co-authored by scientists from industrialized countries). Using the ISI database, quantitative analyses of mainstream scientific literature, i.e. articles in internationally read publications, were made at the continental and national levels. One such bibliometric study on mainstream science in Singapore recently showed that articles written by national scientists and published in international journals were very rarely cited [2]. Other authors, such as Frame et al. [30], provide interesting general information on the respective ranking of various developing countries and on the distribution and orientation of scientific disciplines there.

There has also been relatively little research on the scientists who make up the third world scientific communities, or on how these communities are emerging and reproducing. The comparative study made by T.O. Eisemon [26] was until recently among the few exceptions. He interviewed teachers/scientists in the mathematics and zoology departments at the universities of Ibadan and Nairobi in 1978 and concluded that:

the achievements of Nigerian and Kenyan science are primarily quantitative and in the sphere of construction of an institutional framework for scientific research. Science teaching programmes have been developed, scientific societies established, publishing institutions formed. These are not trivial accomplishments in my view. Nevertheless, it is also true that scientific work – in a more substantial sense – has not been much advanced. . . . Nor have hopes for rapid scientific development been realised. A much longer time will be required before a conclusive judgement can be passed on the effective implementation of the scientific "ethos" in Black Africa.

Thus, most developing countries are still in the institutionalization and professionalization stage. The initial institutional structure has been created, but its material expression – scientific research – has not yet been institutionalized, i.e. has not been recognized as a fully fledged component of society. Another study on science in Mexico suggests that, as in the other developing countries, there are fewer scientists than alleged. The study concludes by saying that the "Mexican scientific community is like an army which has too many generals and too much equipment but which lacks soldiers, particularly well trained soldiers" [73, p. 404]. During the 1980s a number of additional comprehensive studies were carried out and published [50], particularly on Latin American scientific communities, e.g. on Peru [42], a very detailed survey of 218 scientists, unfortunately still unpublished, Brazil [74, 10–12, 16, 17], and Venezuela [1, 81, 4, 69].

In reality most of these studies tend to show that none of the developing countries has a genuine scientific community, not even India, which numerically is among the five largest scientific communities in the world [77], or Brazil [74]. What an Indian scientist had to say about research practices in his country is most revealing: "There is no scientific community in this country. . . . I meet my colleagues only abroad. I meet my colleagues even from Delhi abroad. . . . In a well-knit community, where you are exchanging preprints, things are happening and there is excitement. There is no excitement here. Our excitement comes by mail from outside. It depends on the postal system. This is the worst part; the spirit is dead" [77, p. 587]. This dependence on the outside environment, in other words the West, is a constant refrain among many third world scientists. Consequently, since the knowledge formation process in their countries is largely influenced and determined by the West, a considerable part of their scientific output is foreign to the area where it is produced. Another common feature of these scientific communities is the difficulty for many of them to reproduce themselves [35].

These recent studies confirm what Stevan Dedijer wrote in the early 1960s, when he said that,

in the underdeveloped countries scientists suffer from isolation from each other, and thus they do not have the benefits of the stimulation of the presence of persons working in closely related fields. They are in danger, a danger to which they too often succumb, of losing contacts with their colleagues in the international scientific community. They feel peripheral and out of touch with the important developments in science unless they can visit and be visited by important scientists from the more developed countries; they feel inferior and neglected because their own journals and organs of publication, where they exist at all, are seldom read by foreign scientists, seldom quoted in the literature and are indeed often neglected by their own colleagues at home. They have little contact with their colleagues in neighbouring underdeveloped countries. They are in brief not fully-fledged members of the international scientific community and their work suffers accordingly. [25, pp. 80–81]

Close to 30 years have gone by since these words were written, and we have to admit that they still ring true.

The scientific communities in developing countries

Scientific community: A concept open to challenge

The concept of scientific community is today widely used by philosophers, historians, and sociologists of science, as well as, though to a lesser extent, administrators. "It is probably no exaggeration to say that the notion which has been most frequently associated with the social organization of science is that of the scientific community" [49, p. 164]. It is, however, of rather recent origin, appearing in the context of industrialized countries in the early 1940s. Its meaning varies for different authors and in different contexts, and its use, as a methodological tool, has recently been challenged.

The concept of scientific community, as a community in the sociological sense, clearly has a variety of meanings. In its broad sense, it refers to a group of scientists sharing the same attitudes, norms, and values. It is also being used, in a narrower way, to characterize a group of scientists active in a specific field of science. All the scientists active in a country are said to form a "national" scientific community, whereas most scientists claim to belong to the "international" scientific community. Thus, the same concept is used for various levels and with different meanings to describe the world, or international, scientific community and down to small groups of specialists. Furthermore, as correctly stated in an article by Struan Jacobs [47], the existence of scientific communities is assumed without question or argument in the works of leading contemporary philosophers of science (Kuhn, Popper, Toulmin, Lakatos, Hacking et al.), as in many historical studies.

According to several authors, the concept of scientific community was explicitly defined for the first time in the lecture by Michael Polanyi in 1942 to the Manchester Literary and Philosophical Society [43]. For Polanyi, the members of the scientific community, or "Republic of Science," should be given the maximum of liberty: "The Republic of Science is a Society of Explorers. Such a society strives towards an unknown future, which it believes to be accessible and worth achieving. In the case of scientists, the explorers strive towards a hidden reality, for the sake of intellectual satisfaction. And as they satisfy themselves, they enlighten all men and are thus helping society to fulfill its obligation towards intellectual self-improvement" [65, p. 19]. The concept was then used in the 1950s by other authors

204

(e.g. Barber [6]; Shils [76]; Kuhn [53]) and it became a key concept in the sociology of science in the 1960s [8].

It was also in the early 1940s that the sociology of science started to emerge as a discipline, soon dominated by the functionalism of Robert Merton and his school. The Mertonian "normative structure of science" defines a number of ideal norms and values (universalism, communism, organized scepticism, and disinterestedness) that scientists are believed to share [60]. For Merton, science is organized according to an idealized model. Most of its work tends to view the scientific community as if it were a separate social system, without taking into account its relations with other elements of the society to which it belongs. Most of the works until the 1970s, while criticizing the normative Mertonian approach, have also tended to consider the scientific community as an autonomous entity.

Warren Hagstrom [44] analysed the mechanisms of social control, and particularly the reward systems, acting to ensure the autonomy of the scientific community and its reproduction and growth. It is one of the most important and comprehensive works on the subject. Ben-David [8, p. 4], while recognizing that "science is conceived as the activity of a human group (the scientific community or, rather, communities specialized by fields)," suggests that "this group is so effectively insulated from the outside world that the characteristics of the different societies in which scientists live and work can, for many intents and purposes, be disregarded." I shall come back to this question when presenting some case-studies of developing countries to show that the emergence and the functioning of a given scientific community is strongly influenced by the society in which it takes shape.

Since then the concept of scientific community as defined above has been challenged (e.g. [22, 48]). Bourdieu [13] introduced the notion of the scientific field as a space where struggles for the monopoly of the scientific authority or credit take place. Bourdieu's model has been modified and further developed by Latour and Woolgar [55] among others. Others have gone beyond the credibility model to introduce the notion of networking [84] and translation [18]. While the latter studies have been useful to pinpoint the limits and the irrelevance of an internalist conception of the notion of scientific communities, no appropriate alternative concept has yet been proposed. Thus, it has been used in more recently published works [19, 31, 33, 64] and is still considered as a useful methodological tool in ongoing research programmes (e.g. [83]).

The widening gap and the need for a revised typology

Most of the studies reviewed at the beginning of this chapter tend to conclude that none of the developing countries has a genuine scientific community. Care must be taken, however, to avoid overgeneralizing. The last decade has made it increasingly clear that it was impossible to treat the developing countries as if they were a homogeneous entity. The gap is clearly widening between the "least-developed countries" and the "newly industrialized countries." The latter have reached a fair level of technological and scientific research, industrial capacity, and domestic sales that justifies their hope to better capitalize on new scientific development and technology [85], while most of the least-developed countries have unproductive, inadequate scientific research systems and lack an industrial base, qualified personnel, and capital.

Seven developing countries (Taiwan, Korea, Hong Kong, Singapore, Brazil, Mexico, and Argentina) account for almost 90 per cent of the total manufactured exports of the developing world, and the four in Asia account for 77 per cent. Although the development of endogenous scientific communities has not been the impetus for development in most of the Asian newly industrialized countries (in particular in Singapore and South Korea, where steady growth has been supported by acquiring techniques and transforming imported resources, together with staff training), these countries are now trying harder than ever to develop their national science and technology activities. In the late 1970s and early 1980s, when most developing countries were generally devoting between 0.1 per cent and 0.4 per cent of their GNP to research, Korea, for example, was already spending over 1 per cent and is today spending close to 2 per cent. Singapore, which is lagging behind slightly – starting from an average level of spending characteristic of most developing countries in the late 1970s and early 1980s (between 0.2 per cent and 0.3 per cent) – is now spending more than 1 per cent and is planning to catch up with Korea and Taiwan before the end of the century [41]. A similar development might take place in the South-East Asian countries such as the Philippines, Thailand, Malaysia, and Indonesia during the coming decade, although their economies will no doubt remain more dependent on agriculture. The situation is clearly different for the remaining Latin American industrializing countries (Brazil, Mexico, and Argentina), which, unlike their four Asian counterparts, belong in the category of large countries.

The question of the large countries is more difficult because of the size of their scientific communities and because most of them can hardly be considered as single entities but rather as several countries in one. One should, however, distinguish here between the two giants (China and India) and the other countries (Indonesia, Brazil, Mexico, etc.). India, which has been described as "excellence in the midst of poverty," has today among the five largest scientific communities in the world and accounts for 50 per cent of the scientific production of the developing countries. China, like India, also has a very high scientific and technological manpower potential in absolute terms due to its huge population, but low as a percentage of the total population. Both countries have vast regional disparities. The development of the scientific community in Brazil, the largest scientific community in Latin America, also illustrates the profound regional imbalance between the southern states (and more specifically the state of São Paulo), and the rest of the country. But large often goes together with fragile [77], and the economic difficulties recently experienced by most of these countries, plus the political events that arose in China, remind us that the future of their scientific communities is far from secure. They still have to struggle to create a space for science [75].

But let us remember that the majority of developing countries are small and very small countries. In 1985, about 67 per cent of all developing countries had a population of less than 10 million, and 52 per cent had less than 5 million. Botswana, Lesotho, Vanuatu, Swaziland, and Chad are typical examples. Although size measured in absolute terms is not an adequate indicator of the prospects of developing a science and technology base, it is more difficult to establish one in the smaller countries. Due to resource constraints, small developing – and developed – countries cannot solve all their problems alone. Major decisions have to be made as to what should be attempted using their limited research capabilities and what can be borrowed from elsewhere. This also requires adequate access to information and participation in research networks.

Thus, it is no longer possible to consider the developing countries as a single entity, and there is an obvious need to establish a typology reflecting the level of development in science and technology and the problems described above. An analysis of the different typologies available shows that the most common are linked to economic indicators, especially per capita GNP, and suggests a classification based on thresholds, e.g. the World Bank typology, which recognizes low income countries (US$0–US$400 per inhabitant per year), medium income countries (US$400–US$1,700), and oil-exporting, high income countries. The United Nations system, especially UNCTAD, makes a distinction between newly industrialized, oil-exporting, and least-developed countries.

But, as correctly stressed by Salomon and Lebeau [72], "purely economic definitions of developing countries tend to be distorting mirrors." Based on science and technology resources, they proposed a classification with five categories of developing countries. A recent report presented to Unesco by the International Council for Science Policy Studies [46] proposes an aggregate typology of "science and technology capabilities." Excluding the industrialized countries, three groups are identified: those with almost no science and technology base; those with fundamental elements of such a base; and those with an established science and technology base. Most African countries belong to the first group.

The latter classifications are the most interesting ones for our purpose, but a number of misgivings suggest that further research and efforts are needed to produce a more dynamic typology that takes account of recent set-backs and fluctuations. The main reason for the misgivings is the lack of reliable, comparable, and recent data on some of the basic indicators, including science and technology activities, in many developing countries. The adequacy of some of the science and technology indicators (in particular output indicators, which are controversial – even for industrialized countries) for measuring or evaluating third world science is also very much open to question (e.g. [5]).

Furthermore, many of the crucial factors that affect a society's ability to take advantage of modern science cannot be measured and translated into indicators. The search for a more "explicative" typology must extend beyond quantifiable indicators to include social structures, political systems, and national history. In order to go beyond the question of indicators, country case-studies are presented in the next section to pinpoint similarities and differences and to show that the conditions surrounding the emergence of given national scientific communities are producing different styles of science [83].

National scientific communities and styles of science

Before illustrating the different styles of science by referring to studies on specific countries, it may be useful to look at the way national scientific communities have emerged or are still struggling to emerge in smaller, lesser-known countries that are representative of many developing countries.

(a) Costa Rica, Senegal, and Thailand

These three countries are representative of a number of small (Costa Rica and Senegal) or medium-size (Thailand) developing countries. All three are agricultural, although the economies of Costa Rica and Thailand have been changing structurally as industry accounts for an increasing part of GNP. They are characteristic examples of young scientific communities. The first traces of research institutes and schools of higher learning started appearing at the end of the nineteenth century and became really visible during the first part of the twentieth century. And it was not until the 1960s and even more the 1970s that national scientific communities started taking root and becoming bona fide institutions.

As universities grew, research activities went through a process of institutionalization, and science policy-making bodies were created. This process started in Thailand at the end of the 1950s, in Senegal in the 1960s, and in Costa Rica at the beginning of the 1970s. Their scientific communities are small and difficult to evaluate. The Senegalese and the Costa Rican scientific communities probably numbered 800–1,000 scientists and the Thai community just over 5,000 in the mid-1980s [33]. The scientific potential is concentrated in universities and focused on agriculture, social sciences, and health. University scientists are the best trained and have the highest qualifications. The number of scientists in engineering is low in all three countries. Unlike the other two, the large number of scientists in Senegal working on what could be called "general advancement of knowledge" is rather unusual and can only be explained by historical reasons going back to the colonial period: the Senegalese scientific community is still heavily dependent on expatriate scientists 30 years after independence [33, 35].

Of the three, Costa Rica offers its scientists, especially those working at the University of Costa Rica (UCR), the best salaries and the most incentives to be involved in scientific activities and to publish their findings. Proposals have recently been made through the UCR internal promotion system to encourage academic staff members to do more research. Obviously there is still room for improvement, and different institutes offer different opportunities. The major devaluation of the cólon early in the 1980s seriously affected salaries and lowered their purchasing power. But this situation was not unique to people working in science.

In Thailand, despite a political decision to support scientific work and an atmosphere propitious to its research development, there are inherent difficulties in the profession that are far from being solved. In the public sector, Thai research scientists and university teachers are badly underpaid. Promotions and salary increases depend almost exclusively on seniority, with very little credit for educational level, work performance, and services rendered, be it in research, education, or administration. This situation is of course deleterious to normal research schedules, since far too many research scientists and teachers doing research have to supplement their salaries by accepting unrelated jobs on the side.

The status of the research scientist in Senegal is precarious. Since there is no career stability, a future in this profession is fraught with uncertainty. Institute regulations make no allowances for the unique characteristics of the profession. Career paths and positions are very different from each other. The scientists who are classed as civil servants are probably the worst off, since the professional scale in the government civil service does not provide for a Ph.D. level. This means that scientists with a doctorate, a master's, or a bachelor's degree are all in the same category and move up the seniority-driven scale at the same pace. Attempts to provide the research scientists with common professional statutes were nearing completion when the Ministry of Research was dissolved in 1986. In view of the present institutional structures, the economic crisis, and the budgetary restrictions in Senegal, there is little chance that research will be given statutes of its own in the near future.

(b) Different styles of science

How can the profession of scientist distinguish itself in society and how do scientific communities gain legitimacy? The history of science in a number of developing countries shows that the professionalization of third world scientists, the emergence of national scientific communities, and the legitimation of scientific activities are often associated with the creation of an active professional association. This is the case in India, where the launching of the Indian Science Congress Association (ISCA) "afforded a widely scattered scientific community a much needed common meeting ground" [51]. In Brazil, the Brazilian Society for the Progress of Science (SBPC) played a crucial role in the professionalization of Brazilian scientists between

1950 and 1960. In the case of the SBPC, "scientific interests and political legitimacy were closely imbricated" [11]. Modelled on the American Association for the Advancement of Science, the Venezuelan Association for the Advancement of Science (ASOVAC), created in 1950, has brought a stronger cohesion to the young, emerging Venezuelan scientific community while legitimating its role [82]. It is also responsible for a distinct, rather academic, style of science.

The charismatic role of leading political and/or scientific figures must also be taken into consideration. Thus, in India, the important role of Nehru – "a passionate believer in modern science" and "the main architect of India's science policy" – is widely recognized (e.g. [52]). Other, lesser known examples are Clodomiro Picado in Costa Rica, Cheikh Anta Diop in Senegal, and King Mongkut, considered as the father of science in Thailand. All of them are venerated by their national scientific communities; universities, research centres, and scientific awards have been named after them [33]. Conversely, dictatorships and anti-science political systems always have devastating effects on the emergence and growth of scientific communities everywhere. One of the most extreme examples is the Chinese Cultural Revolution (1966–1976), during which civilian scientific research came to a halt. The situation hopefully changed radically in 1978, when the Party declared science and technology to be one of the four modernizations, but much time had been lost while Chinese researchers were sent to the countryside and isolated from the rest of the world.

The (social) origin of the scientists may also be linked to the emergence of different styles of science. In Africa, for instance, the social origin of the scientist is less distinctive than in the other continents, although many come from rural areas. In Latin America, the scientific communities mainly recruit among the middle classes and in particular among immigrants.

By contrast, in many Asian countries, scientists are often drawn from groups traditionally associated with learning and/or power. In Thailand, as already mentioned, the royal family played, and is still playing, a very important role in the birth, growth, and dissemination of science. Princess Maha Chakri Sirindhorn, for instance, who recently received her Ph.D. from Srinakharinwirot University, is considered a symbol and often participates in public scientific events [35]. In India, ever since the British brought Western science to Bengal in the nineteenth century, the scientific community seems to have been dominated by the upper Hindu castes, especially the Brahmans.

Kapil Raj describes how the Brahmans in their own way "appropriated occidental ideas and science to give credence to their new – dominant – status in the Indian society" [67]. This, according to Raj, explains why Indian science tends to pay more attention to basic disciplines and to be what he calls a "clean" science. Krishna [51, 52] also describes how an Indian national science developed in the 1920s in opposition to the British colonial science around a few scientific leaders, mainly in "pure" sciences, such as physics, chemistry, and mathematics. This may also partly explain why the Indian scientific community is clearly influenced and attracted by the international scientific community and why Indian scientists tend to publish in mainstream journals.

Conversely, Brazil, the second largest scientific community of the third world after India, is characterized more by an "inward-looking" or "inbred" research approach, and the Brazilian scientists tend to publish, particularly within agricultural sciences, in Portuguese and in local journals [78, 19]. The development of the Brazilian scientific community in the field of informatics illustrates how young, jobless Brazilian university graduates succeeded in convincing national banks and the government to launch a national plan to develop a Brazilian computer industry [10]. The future of the Brazilian scientific community is, however, far from secure. "There is certainly more science and technology in Brazil today than only twenty years ago; and yet, it is clear that a space for science, in terms of socially defined, accepted and institutionalized scientific roles, is barely there. What we have at most are islands of competence, niches where science was able to develop for some time, but always precariously, and threatened by an unfriendly environment" [74].

The example of Singapore, where R&D activities are now developing under the influence of a strongly technocratic political approach, is attractive to many developing countries because of its impressive results, but it leaves us with a number of unanswered questions. In a recent article, Goudineau [41] shows that Singapore reached a rather advanced stage of socio-economic development before thinking of developing a national science and technology policy. This occurred in a context where a technological potential existed but where local scientific élites had not been trained nor had a true national scientific community emerged. This is compensated for by a massive importation of knowledge, experts, and know-how in a limited and carefully selected number of areas (informatics, biotechnology, and new materials). A more and more vigorous attempt to encourage Singaporean scientists to return from abroad is also taking place.

The origins, behaviours, and conditions of scientists

The results reported below are mainly derived from a survey carried out during 1985 and 1986. A questionnaire was sent to the 766 scientists in 78 countries who had received research grants from the International Foundation for Science (IFS) between 1974 and 1984. (The IFS, which was founded in 1972 in Stockholm, is a non-governmental organization, multilaterally funded by a number of countries and development agencies. It provides support and guidance to young scientists in and from developing countries. To date, over 2,000 scientists in more than 90 countries have benefited from IFS support.) The results obtained from the questionnaire survey were supplemented by interviews.

All the respondents (489 in 67 countries) are third world scientists, and all are working in their home country. The majority of them (71.4 per cent) work for universities or other academic institutions; 83.4 per cent are men, and 80 per cent are between 30 and 45 years of age. More than 60 per cent of the respondents have a Ph.D. or the equivalent, for which most of them (76 per cent) have studied in an industrialized country. They are mainly working in the agricultural and biological sciences, which are high priority and dominant areas of research in developing countries. The most distinctive feature of the survey population may be that it is composed of internationally *selected* scientists, chosen according to criteria that have become ever stricter over the years.

The main conclusion of the study [33, 36] is that scientists from developing countries find themselves faced with a dilemma: whether to participate in solving local problems or to follow the models and reference systems more or less imposed by the international scientific community. They are highly dependent on countries in the centre, as well as on the international scientific community. They often rely upon outside sources for education and training, institution building, research financing, etc. To a large extent, third world scientists use international scientific literature as their reference, choose research topics on the basis of essentially the same criteria as their colleagues in the centre, and tend to select the same equipment that they grew accustomed to during their Ph.D. studies in the laboratories of the industrialized countries.

But importing equipment manufactured in the North into the developing countries of the South, even with clear instruction manuals, is not enough to ensure equal quality service [37]. Similarly, scientists who studied in the North often discover that the subject of their thesis, their course curricula, knowledge, and experience are not directly applicable upon return to their home country. It is becoming increasingly obvious that applying major international criteria on scientific communities of the periphery, especially in the developing countries, will not guarantee the latter's integration into the international scientific community. Furthermore, it may detract from the relevance of research to local needs and problems.

Origins

Close to one-third of the researchers who responded to our questionnaire come from farming families (many of them from small subsistence-level farms), and one-fifth spent their childhood in a small village. This rural background is even more widespread among the African scientists. Eisemon provided further evidence for these results through interviews he conducted in Kenya and in Nigeria in 1978: "African scientists, like most other Africans with higher education, are usually the first in their families to receive secondary and higher education. Many, particularly in Kenya, come from rural backgrounds" [26, p. 512]. Thus a relatively large number of scientists from developing countries have experienced a rapid social rise, going from a small village to a big (capital) city. At the end of this socio-intellectual adventure, they go on to become members of the intelligentsia, leaving their home village behind them.

For the other categories of social origin, results unquestionably prove that the grade system and then the university system have selection criteria that are hardest on the least favoured classes, albeit without totally excluding them. The intermediate categories (especially crafts and commerce), with approximately one-fifth, are rather well placed. Close to one-tenth have a father in the "office staff" category, whereas the percentage of sons and daughters of "labourers" was lower (3.7 per cent). This last low percentage reflects the inequality of opportunities for the lower social classes; it can also be partly explained by the lower rate of industrialization in most of these countries. The high percentage of researchers (close to one-fourth) whose parents are in liberal professions or senior management positions – a social category that accounts for a small percen-

214

tage of the population in most developing countries – confirms the inequality of opportunity.

With 16.6 per cent of the overall population, women appear to be underrepresented. However, a quick comparison with the situation in the industrialized countries of the world softens this initial reaction. For example, in 1982 only 13 per cent of the scientists and engineers in the United States were women, and this was a 200 per cent increase over the 1972 figure. In a country like Sweden, which is well known for its efforts in favour of equality of the sexes, women accounted for only 12 per cent of the research scientists in 1982. The use of averages obscures regional disparities and important differences between countries. Women researchers in our population figure as follows: 9 per cent for Africa, 15 per cent for Latin America, and 23 per cent for Asia. The Philippines (36 per cent) and Thailand (33 per cent) had the highest percentages. Some African countries such as Tunisia (27 per cent) and Tanzania (23 per cent) have a laudably high percentage compared to the continent as a whole, while countries like Burkina Faso, Morocco, and Senegal rank far below the average. Our results also brought out a strong degree of disciplinary specialization: women tend to choose disciplines that involve laboratory work and that offer jobs in the capital. Women are often reluctant to live outside urban areas, not only because of their discipline; other factors such as marital status, the number of children to support, and the spouse's profession can also affect the researcher's region of residence.

Compared with the national average in their countries, third world scientists marry late. In our study population, 70 per cent in the 25–29 age group were unmarried, as are close to one-third in the 30–34 age group, and one-fifth in the 35–39 age group. One reason may be that many of them had long years of schooling and extended journeys abroad. Another reason may be the contact with Western models during their studies outside their home countries. The Western standard also seems to have been adopted for the number of children, since two-thirds of the scientists in our population had at most two children. Close to half the scientists in the 30–34 age group and over one-fourth in the 35–39 age group had no children at all. Who do the scientists marry? There is a strong endogamous trend, since half of the spouses are scientists and teachers. The marriage strategy (late marriage, strong endogamy, Malthusian behaviour) seems to characterize a very rational approach to reproduction. Under the influence of the Western model, which holds that small families are more mobile and do better socially than large families, the scientists produce as many children as they think they can establish at a level they would be satisfied to occupy themselves. The investment required for research quite clearly implies postponing marriage and the first child. Since in research the social status that accompanies the profession seems to take more time to acquire than in other professions, scientists have to – and seem prepared to – make the relevant sacrifices.

Higher education and research training

Student populations stayed small and relatively few diplomas were awarded by higher learning institutions in most developing countries until the end of the 1960s. During the 1970s student enrolment grew substantially in all countries. By the early 1980s a number of countries, mainly in Asia and in Latin America, boasted a student population comparable with the OECD countries in relative terms (2 per cent or more of the total population). Part of the explanation lies in the creation of many new public universities (most of them outside capital cities) during the 1970s and in the overpopulation of most universities. The proliferation of private universities, mainly specializing in business and administration, also contributed to this spectacular development. The student boom and the large number of graduates produced, combined with the economic crisis and the budgetary cuts, gave rise in the late 1970s to a new phenomenon: unemployment among the intellectuals. This does not mean that all the employment needs have been fulfilled. The situation is quite the opposite. But the key employer, i.e. the state, is no longer able to keep up with the need to create new posts. This is particularly true in Africa, where nearly all branches suffer and where associations of unemployed university graduates have been created.

Until relatively recently – except for certain countries such as India – many third world students had to leave their home countries to attend a university and obtain the education needed to become scientists. Studying abroad is nothing new and is not limited to young people from developing countries, but it is noteworthy that the percentages of such students in the total foreign student population has increased considerably in most Western countries since the 1960s [58, 59, 63, 32]. During the colonial period, most of the (very few) students who were sent abroad for their education studied in the colonizers' country. During the pre-independence years, increasing numbers of students applied to study abroad, and the number of scholarships made available by industrialized countries rose considerably. This showed increased awareness of the importance of higher education in development-oriented science; it also reflected the donor countries' desire to maintain – or acquire – political and economic influence in newly independent states.

At the time of independence, there were some universities in the developing countries, but they did not go as far as the doctoral level and did not offer a full range of science and technology courses. In some countries, after the first university was created, change took hold very quickly, especially in the 1960s. By way of illustration, a country like Brazil now offers hundreds of graduate programmes in some 30 independent institutions and universities. Two-thirds of the programmes lead to a master's degree, one-third to a doctorate. At the end of the 1980s, the University of São Paulo alone offered 100 master's programmes and 66 doctoral programmes in a great variety of disciplines.

Although the proportion of doctorates conferred in the developing countries has been constantly increasing since the beginning of the 1970s, research scientists, especially the most active ones, still rely heavily on foreign education. Among the countries that train third world students to the doctoral level, three stand out on the international scene: the United States, Great Britain, and France [32]. A student who has the choice between studying at home or abroad will generally choose the latter. Besides the economic benefits that accompany a stay in an industrialized country, a diploma obtained there is usually rated higher than a diploma from a developing country. The quality of the doctoral programmes in third world universities is also often questioned by the officials of the same countries. It is also claimed that their graduate programmes provide a very slow rate of training; chemistry training in Brazil, for example, requires on average 4.5 years to complete a master's programme and a further 6.5 years to take a Ph.D. [17]. In Thailand, too, it takes an abnormally long time to finish a Ph.D., because many Ph.D. candidates work at the same time, and it is often difficult for them to meet their supervisors, who have many other commitments outside the university [35].

Studying abroad is expensive, though the cost obviously varies depending on the country. In the late 1980s it ranged from US$3,000 per annum in the USSR to US$7,400 in the United States and US$10,800 in Japan. These figures do not include registration and tuition fees nor travel expenses. With US$4,500 as the average tuition fee for a semester in the United States and annual living expenses of $7,440 (adjusted for inflation between 1985 and 1988), two and a half years of study for a master's degree in the United States would cost about $44,000; a doctorate requiring four years of study would "cost" $70,000. By way of comparison, a "maestria" at UCR in Costa Rica would cost only 227,600 colones (including living expenses), which, at the 1987 exchange rate, comes to slightly less than US$4,000, in other words, one-tenth of the cost in the United States [68]. For African countries with small budgets, and small developing countries in general, the cost of scholarships for training abroad represents a high proportion of the budget for higher education. Thus, in Mali up to 85 per cent of the higher education budget is spent on scholarships for training abroad [24].

Nevertheless, research training is too heavily reliant on foreign facilities and countries, and training abroad usually does not satisfy the needs of the third world scientists. It would be more realistic, efficient, and, in time, productive to allocate the considerable sums of money now being used to train these scientists abroad to reinforce and establish doctoral programmes leading to a Ph.D. in priority fields within the national universities. Doctoral programmes could also be organized on a regional basis. Strengthening national academia would contribute to improving the structure of scientific communities in developing countries, thanks to added input from both the national scientific potential and the student body. This implies that the countries of the North would have to remodel their educational aid policy, but obviously does not mean cancelling all opportunities for doctoral or post-doctoral education abroad in certain very highly specialized fields. Another important aspect that has to be taken into account is that studying abroad for a long period increases the risk of not returning to the home country.

Brain drain and brain gain

The survey showed that, logically enough, offers for positions abroad were made more frequently to scientists who had spent longer periods of time for training abroad. There is also a clear correlation between acceptance of the job offer and the number of years spent studying abroad. In other words, the longer one has studied abroad, the greater the chance of receiving an offer to work abroad and the greater the tendency to accept the offer. The motives for returning home or remaining abroad after studying there for several years are diverse. The full potential for emigration is not fulfilled, however, because most of the scientists are attached to their country and home environment. As Bernardo Houssay, the Argentinian Nobel Prize recipient, said, "Science does not have a country, but the scientist does . . . the country where he was born, or raised and educated, the country that gave him a place in his professional career, the country of his friends and family" (quoted in [20, p. 450]). This confirms some of the most important findings of a UNITAR study on emigration and return, namely that "the most common pulls back home are family, friends, and patriotic feelings" [40].

Paradoxically, economic and material factors – even if they may influence the outflow of scientists – are not the strongest determinant of a decision to emigrate. The possibility of obtaining a much higher income may, however, cause a scientist to emigrate for a short period of time. Family ties and children's future are believed to play an even stronger influence than salary or working conditions upon an individual's decision to return to the home country or to emigrate. Another important finding of the UNITAR study, which might have direct policy implications, is that students with scholarships or special grants from their home countries are more likely to return home than those who study abroad with a foreign grant or privately [40]. Similarly, having made the decision to go back home, few scientists plan to emigrate again. Out of close to 500 scientists supported by the International Foundation for Science to carry out a specific research programme in their home country, more than 95 per cent were still active within their national scientific communities in 1985, i.e. 15 years after the first grant was given [33]. Racial, ethnic, and political discrimination may also strongly influence the decision to emigrate or to return home.

In addition, the UNITAR study suggests that "in forecasting whether nationals of a particular country might become part of the brain drain, a more important factor than the stage of development is the extent to which a country trains an excess of professionals in a particular field" [33, p. xxvi]. If the country of the scientist is "the country that gave him a place in his professional career," it is at the same time clear that if the key employer, i.e. the state, does not offer him a position, he will not have much choice but to leave his country if he wants to remain a scientist. A heavy load of teaching and administration, not enough time for research, poor equipment and facilities, and isolation from the international scientific community are among the most important factors in a decision to emigrate, particularly among experimental (biological) scientists and engineers. When planning to emigrate, scientists always prefer to go to the industrialized countries they know best, i.e. the one where they studied. Thus, the United States is clearly the favourite, followed by Great Britain, France, Canada, and Australia. But there are also developing countries to which third world scientists migrate: Nigeria in Africa, some oil-producing countries, Singapore, etc.

The situation in the United States has reached such an extreme, particularly in engineering, that a few people are starting to wonder if the presence of foreign graduate students is a boon or bane [7]. Beginning in 1981, and for every year since then, more than one-half of the engineering doctorates awarded in the United States have been to foreigners, nearly 70 per cent of them Asians. Furthermore, foreigners, and particularly Asians, comprised about two-fifths of total post-doctoral employment in 1985, up from one-third in 1979 [63]. Thus, in given scientific fields (chemistry, physics, mathematics, and computer science) there is a clear shortage in the supply of high quality US applicants and a surplus of high quality foreign (mainly Asian) applicants [21]. A potential reinforcement of the rapatriation schemes in some Asian countries and the possible subsequent return of scientists to their home country may pose a threat to the long-term competitiveness of US universities and firms. Thus, a number of countries in Asia and to a lesser extent in Latin America have started to rethink the problem of brain drain and tend to consider that working abroad for a while can represent a gain to the home country, rather than a loss, if the scientist returns with increased skills directly related to the needs of national research groups. Measures should be taken to identify these needs and the scientists concerned. Mechanisms should also be implemented to bring them back home, with attractive research careers and proper professional status.

Research scientists in search of statutes and status

Thus far, research scientists in developing countries long to have a proper professional status; draft texts have often been prepared and then stored away in anticipation of better times to come. Research is often carried out as part of some profession or system designed to uphold professional standards or value systems that are not specific to research. Furthermore, in most developing countries, research scientists do not have high social standing or prestige. Doctors and lawyers and other professionals of that level, with at most the same amount of education as the research scientists, are not only better paid but also enjoy a much higher social status.

Speaking about Venezuela, Roche said, "I know many examples of young people whose rich parents forbade them to major in sciences or to devote themselves to research often because of the low salaries or uncertain career opportunities. The bourgeois attitude to careers in science is much the same as the attitude to professions in the arts; success is reserved to very outstanding people alone, all the others being condemned to a Bohemian life of uncertainty. The profession has probably changed since the Sputnik was invented, but research is still not seen as a fully acceptable profession" [70]. The low wages explain why many of them supplement their incomes by working overtime on side jobs that include anything from working as a consultant, a teacher, or a taxi driver. Anyone who has spent time with third world scientists quickly realizes that a second (or even third) job and income are vital. These additional jobs are of four main types: consultancy, teaching, agriculture, and commerce. Consultancy is nearly always related to the expertise developed in the research activity. Agriculture can mean anything from working on a coffee plantation to raising layer hens. Scientists working in commerce usually help in a family business.

How attractive research can be as a profession depends very much on the country. In Kenya, according to Eisemon, scientists have enjoyed a place of special pride in society since the European colonization period, when close relationships were established between the scientific and the politico-economic circles. Thus, a scientific career, which brings an individual into proximity with the élite, is pursued for social advancement [27]. In India, although the scientific community seems to be dominated by the upper Hindu castes [67], there is paradoxically not much prestige attached to the profession of research scientist; and, like most intellectual positions in the public sector, it is poorly paid [27]. In an effort to better understand the professional choice made by the research scientists in our population, we found out that social status ranked very low, whereas intellectual stimulation and social utility were rated first and second respectively. Actually, an a priori, carefully considered choice seemed to explain a career in scientific research after higher education less than the fact that students were selected or had access to a scholarship at the right stage of their education, even when it meant studying subjects that initially did not interest them [36].

The strategies adopted by the scientists are the result of negotiations carried out in a socio-economic, cultural, and political environment that is not always conducive to scientific perspectives and societal recognition of research science as a profession. Up to the present, science in the developing countries, especially in Africa, has been essentially controlled by government. The first step for the newly independent countries was to build up the state and its institutions. Education was given top priority in order to train civil servants for the state. Careers have, however, often been constructed without considering diploma qualifications. Success in the power struggle has been given more importance than professional specialization. Because of this situation, it has often been difficult to develop research science as a profession, or even as a vocation. As a career, it is not very appealing, and urgently needs statutes.

Choosing research topics and practising research

The conditions described above affect the way research subjects are chosen and research activities in general are practised. In an attempt to determine the different factors that may play a part in the choice of research subject, I found that third world scientists have more or less adopted the same reference systems as American researchers working in comparable fields. (For the comparison, I adapted a list of criteria tested in the United States: [15, p. 45].) The leading criterion, "importance to society," takes us back to the criterion that was in second position in the list on choosing research as a profession, namely "social utility." When I asked the scientists what this concept meant to them, they indicated that social utility was more or less the capacity of research to solve the economic and social problems facing their country. The fact that the criterion "demand raised by clientele" is at the bottom of the list no doubt reflects the marginal position of science in developing countries and supports the theory that research scientists and scientific institutions are kept out of the economic and production system.

The findings also confirm the fact that a choice of subject depends more on a series of factors – some of which are external to the science involved – than on any single factor. While saying that third world scientists have more or less adopted the same reference systems as their American colleagues, I do not mean that they orient their scientific activity toward research problems defined in the industrialized countries. On the contrary, they do tend to choose research topics

222

that they perceive to be relevant to local problems. This is demonstrated by the fact that a large number of third world scientists who studied in an industrialized country had to change research subjects when returning to their home countries so as to match research work with perceived national needs.

Thus, a researcher who had to work on the problem of nutrition linked to obesity in the United States quite obviously had to change subjects upon her return to Thailand; she decided to work on controlling the thiamine (vitamin B1) deficiency caused by consuming too much tea and tannin. The degree of relevance of the selected research topic to local problems may, however, vary among disciplines. While Lea Velho presents evidence in a recent paper that agricultural scientists in Brazil select research topics directly relevant to local agricultural problems [80], this may not be the case for other scientific disciplines such as physics. More studies would be needed on criteria for choice of research topic in developing countries to come to more definite conclusions.

Time devoted to research depends on various factors. One of them is the nature of the researcher's home institution. Obviously the researchers with the heaviest teaching load work in universities. This is the case of the majority of third world scientists. In an attempt to compare the sample scientist population with their American colleagues, I found that American university researchers on average spent less time teaching than their colleagues in the third world (27 per cent as against 37 per cent), and, above all, more time doing research (57 per cent as against 34 per cent). The differences are much less significant for researchers working in research institutes, although American researchers again spend more time (77 per cent) doing research in these institutions than do their third world colleagues. As for the size of their research budget, the differences are of another magnitude. While American researchers in government research institutes have an average annual budget of US$209,000 and their university colleagues have US$68,000, I found that researchers in developing countries on the average have only between US$5,000 and US$15,000 depending on the level of foreign funding. Even if we are dealing with estimates given by the researchers themselves, who very often do not know the precise total of their budgets, the differences observed are such that they require no further comment.

Other disparities also bring out the fact that third world scientists are at a significant disadvantage compared with their colleagues in scientifically more advanced countries. Lack of equipment, vehicles, technicians, and scientific documentation are among the most frequently observed and described. Another disadvantage that is perhaps even more critical and at the very centre of the scientific enterprise is communication. Many scientists suffer from a feeling of isolation, especially when they have just returned from studying abroad and are trying to fit into the scientific community at home. Moravcsik [61] describes how difficult, and in some cases impossible, it is for scientists in developing countries to communicate with their peers and colleagues by drawing a comparison with birds whose wings have been clipped. The feeling of isolation is probably heightened by the fact that these scientists have been trained in a large variety of universities located throughout the industrialized countries. Furthermore, during this early period, when the young national scientific communities are just "taking off," the scientists often have to cope with being the only specialists in their field within their institution, or even within their country. All the authors agree, however, that science cannot exist without communication, and that a colleague's criticism is vital to progress in any scientific endeavour: "an isolated person builds only dreams, claims and feelings, not facts" [54, p. 41]. Here again these scientists are enduring a handicap little known to their colleagues in the industrialized countries. Other handicaps relate to the visibility and the recognition of their scientific production.

Scientific production: Not very visible

Developing countries are credited with approximately 5 per cent of the world's scientific production. But science produced there is inadequately reflected in the international databases. International databases, and particularly that of the Institute for Scientific Information (ISI), are very selective and screen only the world's most prestigious scientific journals, the ones that publish the most frequently cited articles. Thus, the Science Citation Index (SCI), developed by ISI, focuses on what has become known as "mainstream science," i.e. the most internationally visible science carried in about 4,000 scientific journals. Since we know that there are about 70,000 scientific journals in the world, we can measure the ISI's selectivity in building up a database; less than 2 per cent of the scientific journals selected come from the developing countries. In general, journals that are not in English are at a disadvantage.

224

The place of third world science in mainstream science

The question of adequately representing third world science in international databases was the main issue at a 1985 conference organized at the ISI in Philadelphia. The title of the final conference report, "Strengthening the Coverage of Third World Science," pointed to a glaring gap [62]. The conference participants estimated that only about half of the science produced in the third world that meets international standards of excellence is included in the ISI database.

In fact, as D.J. Frame [29] so correctly wrote, it all depends on what you are trying to assess. "If the purpose of the bibliometric indicators is to help in the building of a national scientific inventory, telling us what kind of research is being performed at different institutions, then coverage of local as well as mainstream publications would seem important. On the other hand, if one is primarily interested in investigating Third World contributions to world science, then publication counts taken from a restrictive journal set would seem most appropriate." Thus, when Garfield prepared his "Mapping Science in the Third World" [39], he was actually measuring the impact of third world scientific output on the international scientific community, using, as his only criterion, the part of the third world scientific output that was cited and used by the international scientific community. For this reason, it is not surprising that the impact was found to be slight.

Mainstream science production is even more narrowly concentrated than is national wealth expressed as GNP. Ten countries produce more than 80 per cent of the international scientific literature. Except for India, which has maintained a steady ranking of eighth place since the beginning of the 1970s, all the countries are members of the industrialized world [30, 14]. Between 1981 and 1985, the developing countries produced 5.8 per cent of the world's mainstream scientific output, of which 3.7 per cent came from Asia, 1.1 per cent from Latin America, 0.4 per cent from sub-Saharan Africa, and 0.6 per cent from the Middle East [14]. Even if we challenge the representative value of these estimates, especially considering the database used, we still have to accept that mainstream science from the third world is marginal compared with the rest of the world.

Among the developing countries, India, the uncontested leader, produces five times more mainstream scientific publications than the People's Republic of China. The table lists the top 15 producers of mainstream scientific literature in the third world for 1973 and for the

225

Fifteen leading developing countries, ranked according to number of mainstream publications produced

	1973[a]		1981–1985[b]	
Rank	Country	Number of publications	Country	Number of publications (annual averages
1	India	6,880	India	10,978
2	Argentina	764	People's Rep. China	2,146
3	Egypt	683	Brazil	1,498
4	Brazil	573	Argentina	1,124
5	Mexico	368	Egypt	1,029
6	Chile	356	Nigeria	790
7	Nigeria	280	Mexico	709
8	Venezuela	200	Chile	590
9	Taiwan	186	Taiwan	509
10	Iran	174	Hong Kong	365
11	Malaysia	138	Saudi Arabia	319
12	Kenya	125	South Korea	312
13	Singapore	120	Venezuela	311
14	Thailand	117	Kenya	248
15	Lebanon	114	Singapore	214

Sources: a. ref. 30, table 4, pp. 507–508; b. ref. 14.

period 1981–1985. This list changed considerably during the reference period. Production in certain leading countries in 1973, like Brazil and Nigeria, rose sharply. Some countries with small – even very small – scientific output in 1973 started climbing, e.g. Hong Kong, Saudi Arabia, and South Korea. Other countries, like Iran and Lebanon, in the throes of political and military unrest, lost their standing. Most of the countries on the list produced substantially more in the years following 1973, but the per country mainstream scientific production remained small, even in countries at the top of the list, like Egypt, Mexico, and Nigeria.

A comparison with the production of scientific institutions in the OECD countries shows that a country such as Egypt produces less than the Harvard University Medical School [29]. The total production of sub-Saharan Africa, excluding South Africa, at present represents about one-tenth of the scientific production of a European country such as France [38].

Referring to the ISI and other international databases, recent studies have provided interesting information on the position of the various countries on the mainstream science supplier list and their impact on world science, but the description of how science is constructed in these countries, the researchers' scientific strategy, and their participation in national and international science is incomplete and often inaccurate. These studies, moreover, tend, either implicitly or explicitly, to assign research scientists of the peripheral scientific communities to two distinct categories: scientists who "really count," in other words, who are known to the international scientific community since they publish overseas in influential international journals; and the others, whose "local" science lacks originality and, at best, is published in low circulation local journals.

Mainstream science and local science: A needed revision

Several other recent studies justify a revision of this exaggerated – but widely held – caricature of science production in the periphery [19, 23, 34]. They substantiate the thesis that the bibliometric indicators based on an international database do not accurately assess the scientific output from the periphery, especially from the developing countries. International databases do not provide enough information to measure accurately the science produced in these countries and assess the scientific thrust of the countries of the periphery in general. Combining and comparing several international databases can improve the relevance of bibliometric indicators but will not tell the whole story. The international databases need to improve their coverage of science produced in the developing countries, and local databases need to be created and consulted. Databases at the local level, accompanied by periodic production and dissemination of documented analytical bulletins, would not only serve to better measure scientific output in the third world, but would also in time enhance South-South and North-South documentation exchange, as well as both the visibility and accessibility of developing countries' scientific output.

Given these handicaps, it is not surprising that third world scientific production and its impact are slight. The following analysis compares overall figures on numbers of publications per researcher with the findings of the survey of the lists of publications of 213 third world scientists who received grants from the International Foundation for Science [33, 36]. The latter produced on average 0.5 publications per year as sole author and 0.7 as co-author – that is to say slightly more than half that of American researchers working in related scientific disciplines [15]. Furthermore, half (55 per cent) of their total scientific production was published in local journals. Asian scientists tend to publish more than African scientists. In addition, Asian and Latin American scientists publish more locally (approximately 60 per cent) than African scientists (approximately 40 per cent). These percentages are exceptionally high in comparison with industrialized countries: in western Europe, scientists publish 12 per cent of their work in foreign journals, while the figure for Japan is 25 per cent [39].

When reflecting on these percentages we should remember that there are many more local journals in Asia and Latin America than in Africa. Logically, the more the scientists publish abroad, the more they work in collaboration with foreign scientists. Garfield [39] has shown that articles by researchers in developing countries have a greater impact (on the international scientific community, measured in terms of number of citations per article) when they are co-authored by researchers from industrialized countries. Here we come up against the dilemma of the strategic scientific choices that researchers in developing countries, in common with most researchers in peripheral scientific communities, have to make between participation in mainstream science (the most used, most visible, and most frequently cited) and the resolution of local problems through "inward-looking" research. Co-authoring with foreign scientists is the most prevalent among scientists who studied or worked in post-doctoral positions abroad. In most cases, however, these publications are produced in the years immediately following the stay abroad; sustained active collaboration with foreign scientists is rare if not reactivated by frequent stay abroad. The fields in which they publish most, such as chemistry, are also the fields in which they publish most abroad. We have also observed a relatively significant difference in productivity by gender, men publishing more than women. Women also tend to publish more in local journals than men.

With very few exceptions, English-speaking scientists publish in English, whereas more than one-third of the publications by Latin American scientists and almost one-fifth of those of French-speaking scientists were found to be in English. A case-study conducted in a French-speaking African country (Senegal) showed that English was increasingly used as a language of publication. I also found a relatively significant use of local languages in certain Asian countries, e.g. Indonesia, where more than half (52 per cent) of the published works of scientists appear in Indonesian languages, Thailand (28 per cent in Thai), and South Korea (18 per cent in Korean). Publication strategies differ greatly, depending on both the country and the discipline. Unlike South Korea, in Singapore all the scientific journals are in English. A glance at the lists of references consulted and cited confirms the hypothesis that the different linguistic worlds are almost "language proof," especially between the English and French languages. Spanish- and Portuguese-speaking scientists often cite literature in English; this is rarely the case for French-speaking scientists. And references by English-language scientists are drawn exclusively from literature written in English.

Most of the scientists publish in both national and international journals. Publication strategies differ according to country and to scientific discipline. Third world scientists cite references essentially (78 per cent) from mainstream scientific literature, which they seem to receive later than their colleagues in the centre, since nearly half the references are over 10 years old, as against 29 per cent of the references cited by scientists from the centre countries. An analysis of the citations indicates that third world scientists use articles from national journals in smaller proportion but much sooner than articles from international journals.

Citation modes usually work against third world scientists in particular and scientists at the periphery in general because, as we have seen above, much of the work is published in local journals that are only circulated within the country. The third world scientists are caught in an especially vicious circle, because even when their findings are published in highly influential, prestigious scientific journals in the centre, they are far less often cited than writings by their colleagues in the centre [2]. Recent work on referencing within the Brazilian scientific community showed that "citation patterns are significantly influenced by factors 'external' to the scientific realm and thus reflect neither simply the quality, influence, nor even the impact of the research work referred to" [79]. The place of publication strongly influences the number of times a publication is cited [56]. Arunachalam and Manorama [3, p. 395] explain that many leading Indian scientists have had the irritation of seeing work published by Western scientists after theirs had been cited; the Western scientists got the credit and their own original work remained unacknowledged. I also found that third world scientists often cite colleagues in industrialized countries, but rarely cite other third world scientists, even when their works are published in well-read international journals. This behaviour seems to be the result of a rather widespread, although difficult to prove, conviction among them that quoting works published by colleagues in industrialized countries brings more credit to their own work.

In sum, third world scientists often cite their colleagues from the developed countries, but their own work – being relatively "invisible" – is seldom cited. They often feel caught in a dilemma: either adopt the habit of scientists from industrialized countries and publish in international journals to become more "visible" and gain international standing, or else seek national recognition by publishing in local journals, and sometimes in local languages, thus being condemned to non-existence, or at best, marginal existence in mainstream science. The general trend is to adopt the two strategies together.

Concluding remarks

Considerable efforts have been made, particularly during the 1960s and 1970s, to develop a science and technology potential in many developing countries. Most countries have experienced a boom in student enrolments, particularly during the 1970s and 1980s, while many new universities were created outside the capital cities. The number of scientists has also increased significantly during the latter period, with annual increase rates often higher than in industrialized countries. Substantial efforts have also been made to build up research institutions and to support the emergence of national scientific communities. Yet the results are not always satisfactory. It was long believed that the accumulation of adequate resources (scientists, institutions, and funding) would automatically generate productivity. We now know that the availability of such resources, although necessary, is not sufficient to guarantee achieving the scientific results needed for development. It is not enough just to build institutions, train good scientists, and provide them with proper supplies.

Going beyond the availability of resources, research activities need a certain permanency through greater recognition by society. The scientists need to be able to find their place in a scientific community that has its own legitimate place in society. Wherever scientific communities are emerging, the debate henceforth centres on the professionalization of their scientists, the conditions under which scientific activities are performed, and the capacity of the scientific communities to reproduce themselves and sustain their activities. Therefore, a number of conditions should be fulfilled for supporting the emergence and reproduction of endogenous scientific communities in developing countries.

The strategies adopted by the scientists are the results of negotiations carried out in a socio-economic, cultural, and political environment that is not always conducive to a scientific outlook and societal recognition of research science as a profession. In addition to proper status and better salaries and working conditions, the emergence of tight-knit and lively scientific communities should be promoted, for example by establishing active Academies, professional associations, and scientific journals. Encouragement should also be given to activities such as national science days, science awards, science weeks for young people, annual conferences of national science associations, and also exhibits, science museums, and clubs that attract young people to science and scientific careers. Education is also important in shaping attitudes and scientific minds.

The dependency of most developing countries on (above all) Europe and the United States to train their scientists is not compatible with the creation of an independent scientific tradition and the emergence of a truly autonomous scientific community. It is becoming increasingly urgent to shift the "centre of gravity" of doctoral level education from the North to the South. This would require a revised cooperation between the Northern host countries (which often offer scholarships) and the developing countries themselves. The process will entail redefining aid policies (and the risk for the North of losing some of its influence) and also, in many cases, developing countries' education policies. The substantial sums that are still being spent by the countries that offer scholarships could be used by the universities in the South to establish or strengthen doctoral courses in disciplines of national priority. Strengthening national academia would contribute to improving the structuring of the emerging scientific communities as the result of added input from both the national scientific potential and the student body. This is essential if all the actors, from confirmed senior scientists to Ph.D. candidates to regular students, are to keep up with science in the making and remain up to date on progress in their disciplines.

To gain in legitimacy, the strengthened national universities should also be better linked not only to the other research and higher learning institutions but also to the society as a whole. New answers should be found to sustain the university as a socially relevant institution and to transform its attributes beyond the neoclassical university [45]. In many countries the situation of the national universities is so

231

critical that it may very well lead to curtailment of university research. Many countries have found no other solution than to circumvent the problem by creating specialized research institutes outside the university, usually with no responsibility for graduate and postgraduate education. (This is not the case in India, where the technological institutes, renowned as poles of excellence, provide close interaction between education and research.)

More historical and sociological research is also needed to achieve better understanding of the conditions that need to be fulfilled for a scientific community of the periphery to emerge, develop, and reproduce. The respective role of the different actors involved also requires further investigation. There is in particular a lack of studies on the roles and professions of engineers and technical workers in both the public and private sectors in developing countries (an exception is Longuenesse et al. [57]). More studies are also needed on the transfer of successful models of institutions and/or on institutional innovations such as the institutes of technology and the fashionable technopoles to improve understanding of the extent to which they could contribute to better linkages between the academic world and the productive sector, as well as to the reproduction of the national scientific communities.

References

1   Alvarez, R.D. *Universidad: Investigación y Productividad*. Caracas: Universidad Central de Venezuela, 1984.

2   Arunachalam, S., and K.C. Garg. "A Small Country in a World of Big Science: A Preliminary Bibliometric Study of Science in Singapore." *Scientometrics* 8 (1985), nos. 5, 6: 301–313.

3   Arunachalam, S., and K. Manorama. "Are Citation-based Quantitative Techniques Adequate for Measuring Science on the Periphery?" *Scientometrics* 15 (1989), nos. 5, 6: 393–408.

4   Arvanitis, R. "De la recherche au développement: les politiques et pratiques professionnelles de la recherche appliquée au Vénézuéla." Ph.D. diss., University of Paris VII, 1990.

5   Arvanitis, R., and J. Gaillard, eds. *Science Indicators for Developing Countries*. Proceedings of the International Conference on Science Indicators for Developing Countries, Paris, Unesco, 15–19 October 1990. Paris: ORSTOM, Colloques et Séminaires, 1992.

6   Barber, B. *Science and the Social Order*. New York: The Free Press, 1952.

7   Barber, E.G., and R.P. Morgan. *Boon or Bane: Foreign Graduate Students in U.S. Engineering Programs*. New York: Institute of International Education, 1988.

8   Ben-David, J. *The Scientist's Role in Society*. Chicago: University of Chicago Press, 1971.

9   Blickenstaff, J., and M.J. Moravcsik. "Scientific Output in the Third World." *Scientometrics* 4 (1982), no. 2: 135–169.

10   Botelho, A.J.J. "L'émergence de la communauté informatique au Brésil." Paper presented at the First Workshop on the Emergence of Scientific Communities in the Developing Countries, 22–27 April 1990, Paris: ORSTOM, 1990.

11   ———. "The Professionalization of Brazilian Scientists, The Brazilian Society for the Progress of Science (SBPC), and the State, 1948–60." *Social Studies of Science* 20 (1990): 473–502.

12   ———. "La construction du style scientifique Brésilien: institutionnalisation et croissance de la science et de la technologie au Brésil (1950–1990)." In: Arvanitis and Gaillard, eds. *See* ref. 5.

13   Bourdieu, P. "The Specificity of the Scientific Field and the Social Conditions of the Progress of Reasons." *Social Science Information* 14 (1975), no. 6: 19–47.

14   Braun, T., W. Glänzel, and A. Schubert. "The Newest Version of the Facts and Figures on Publication Output and Relative Citation Impact of 100 countries, 1981–1985." *Scientometrics* 13 (1988), nos. 5, 6: 181–188.

15   Busch, L., and W.B. Lacy. *Science, Agriculture and the Politics of Research*. Boulder, Colo.: Westview, 1983.

16   Cagnin, M.A. "Patterns of Research in Chemistry in Brazil." *Interciencia* 10 (1985): 64–77.

17   ———. "The Conditions of Scientific Research in Chemistry: Perceptions from the Brazilian Community." In: Arvanitis and Gaillard, eds. *See* ref. 5.

18   Callon, M., and J. Law. "On Interests and Their Transformation." *Social Studies of Science* 12 (1982): 615–625.

19   Chatelin, Y., and R. Arvanitis. *Stratégies scientifiques et développement: Sols et agriculture des régions chaudes*. Paris: ORSTOM, 1988.

20   CIMT (Committee on International Migration of Talent). *International Migration of High-Level Manpower*. New York: Praeger, 1970.

21   Coward, H.R., C.P. Hayles, and S.J. Owens. *The Impact of Foreign Students at US Universities on Graduate Programs in Chemistry, Physics, Mathematics and Computer Science*. Arlington, Va.: SRI International, 1989.

22   Crane, D. *Invisible Colleges*. Chicago: University of Chicago Press, 1972.

23   Davis, C.H., and T.O. Eisemon. "Mainstream and Non Mainstream Scientific Literature in Four Peripheral Asian Communities." *Scientometrics* 15 (1989), nos. 3, 4: 215–239.

24   Decat, M., and T. Mercier. *Etudiants d'Afrique, des Caraïbes et du Pacifique*. Paris: Karthala, 1989.

25   Dedijer, S. "Under Developed Science in Underdeveloped Countries." *Minerva* 2 (1963), no. 1: 61–81.

26   Eisemon, T.O. "The Implantation of Science in Nigeria and Kenya." *Minerva* 12 (1979), no. 4: 504–526.

27   ———. *The Science Profession in the Third World*. New York: Praeger, 1982.

28   Eisemon, T.O. et al. "Transplantation of Science to Anglophone and Francophone Africa." *Science and Public Policy* 2 (1982), no. 4.

29   Frame, D.J. "Problems in the Use of Literature-based S&T Indicators in Developing Countries." In: H. Morita-Lou, ed. *Science and Technology Indicators for Development*. Boulder, Colo.: Westview, 1985, pp. 117–122.

30   Frame, D.J., F. Narin, and M.P. Carpenter. "The Distribution of World Science." *Social Studies of Science* 7 (1977): 501–516.

31   Franklin, M.N. *The Community of Science in Europe*. London: Gower, in association with the Commission of the European Communities, 1988.

32   Gaillard, J. "Les chercheurs des pays en développement." *La Recherche* 18 (1987), no. 189: 860–870.

33   Gaillard, J. "Les chercheurs et l'émergence de communautés scientifiques nationales dans les pays en développement." Ph.D. diss., CNAM/STS, Paris, 1989.

34   ———. "La science du Tiers Monde est-elle visible?" *La Recherche* 20 (1989), no. 210: 636–640.

35   ———. "La communauté scientifique thaïlandaise: un développement rapide mais une reproduction difficile." *Inter-Mondes* 1 (1990), no. 2: 43–57.

36   ———. *Scientists in the Third World*. Lexington: University of Kentucky Press, 1991.

37   Gaillard, J., and S. Ouattar. "Purchase, Use and Maintenance of Scientific Equipment in Developing Countries." *Interciencia* 13 (1988), no. 2: 65–70.

38   Gaillard, J., and R. Waast. "La recherche scientifique en Afrique." *Afrique Contemporaine* 148 (1988): 3–29.

39   Garfield, E. "Mapping Science in the Third World." *Science and Public Policy* (June 1983): 112–127.

40   Glaser, W.A. *The Brain Drain: Emigration and Return*. Oxford: Pergamon Press, 1978.

41   Goudineau, Y. "Etre excellent sans être pur: potentiel technologique et pouvoir technocratique à Singapour." *Cahiers des Sciences Humaines de l'ORSTOM* (Paris) 26 (1990), no. 3: 379–405.

42   GRADE. "Comunidad científica y científicos en el Perú: un estudio de cuatro campos." Lima, 1985. Typescript.

43   Guerrero, R.C. "La idea de comunidad científica: su significado teórico y su contenido ideológico." *Revista Mexicana de Sociología* 42 (1980), no. 3: 1217–1230.

44   Hagstrom, W. *The Scientific Community*. New York: Basic Books, 1965.

45   Hansen, G.E. *Beyond the Neoclassical University: Agricultural Higher Education in the Developing World, an Interpretive Essay*. A.I.D. Program Evaluation Report, no. 20. Washington, D.C.: A.I.D., 1990.

46   ICSPS (International Council for Science Policy Studies). *Science and Technology in Developing Countries: Strategies for the 90s, A Report to UNESCO*. Paris: Unesco, 1992.

47   Jacobs, S. "Scientific Community: Formulations of a Sociological Motif." *British Journal of Sociology* 38 (1987), no. 2: 266–276.

48   Knorr-Cetina, K.D. "Scientific Communities or Transepistemic Arenas of Research? A Critique of Quasi-economic Models of Science." *Social Studies of Science* 12 (1982), no. 1: 101–130.

49   ———. "New Developments in Science Studies: The Ethnographic Challenges." *Cahiers Canadiens de Sociologie* 8 (1983), no. 2: 153–177.

50   Krishna, V.V. "Scientists in Laboratories: A Comparative Study on the Organization of Science and Goal Orientation of Scientists in CSIRO (Australia) and CSIR (India) Institutions." Ph.D. diss. University of Wollongong, Australia, 1987.

51   ———. "The Colonial 'Model' and the Emergence of National Science in India: 1876–1920." Paper presented at the International Colloquium on Science and Empires, Unesco, Paris, 2–6 April 1990.

52   Krishna, V.V., and A. Jain. "Country Report: Scientific Research, Science Policy and Social Studies of Science and Technology in India." Paper presented at the First Workshop on the Emergence of Scientific Communities in the Developing Countries, 22–27 April 1990, Paris: ORSTOM.

53   Kuhn, T.S. *The Structure of Scientific Revolutions*. Chicago: University of Chicago Press, 1962.

54   Latour, B. *Science in Action*. Cambridge, Mass.: Harvard University Press, 1987.

55   Latour, B., and S. Woolgar. *Laboratory Life: The Social Construction of Scientific Facts*. Beverley Hills: Sage, 1979.

56   Lawani, S.M. "Citation Analysis and the Quality of Scientific Productivity." *BioScience* 27 (1977), no. 1: 26–31.

57   Longuenesse, E., ed. *Bâtisseurs et bureaucrates: ingénieurs et sociétés au Maghreb et au Moyen-Orient*. Paris: Maison de l'Orient, 1990.

58   Maliyamkono, T.L., ed. *Policy Developments in Overseas Training*. Tanzania: Black Star Agencies, 1980.

59   Maliyamkono, T.L., A.G. Ishumi, S.J. Wells, and S.E. Migot-Adholla. *Training and Productivity in Eastern Africa*. London/Ibadan/Nairobi: Heinemann Educational Books, 1982.

60   Merton, R.K. *The Sociology of Science*. Chicago: University of Chicago Press, 1973.

61   Moravcsik, M.J. *Science Development – The Building of Science in Less Developed Countries*. 2nd ed. Bloomington: PASITAM, 1976.

62   ———. "Science in the Developing Countries: An Unexplored and Fruitful Area for Research in Science Studies." *Journal of the Society for Social Studies of Science* 3 (1985), no. 3: 2–13.

63   NSF (National Science Foundation). *Foreign Citizens in U.S. Science and Engineering: History, Status and Outlook*. Washington, D.C.: NSF, 1986.

64   Polanco, X., ed. *Naissance et développement de la science-monde*. Paris: Editions La Découverte/Council of Europe/Unesco, 1989.

65   Polanyi, M. *The Republic of Science: Its Political and Economic Theory*. Chicago: Roosevelt University, 1962.

66   Price, D.J. de Solla. *Little Science, Big Science*. New York: Columbia University Press, 1963.

67   Raj, K. "Hermeneutics and Cross-Cultural Communication in Science: The Reception of Western Scientific Ideas in 19th Century India." *Revue de Synthèse*, 4th series, nos. 1, 2 (1986): 107–120.

68   Ramírez, M.A. *Requerimientos de estudios de postgraduado en Costa Rica*. San José: CONICIT, 1987.

69   Rengifo, R.M. "Emergencia de la comunidad científica en Venezuela: la polémica originaria." Paper presented at the First Workshop on the Emer-

gence of Scientific Communities in the Developing Countries, Paris: ORSTOM, 22–27 April 1990.

70   Roche, M. "Aspects sociaux du progrès scientifique dans un pays en voie de développement." *IMPACT, Science et Société* 16 (1966), no. 1: 53–63.
71   Rossi, G. "La science des pauvres." *La Recherche* 30 (1973): 7–15.
72   Salomon, J.-J., and A. Lebeau. *Mirages of Development*. Boulder, Colo.: Lynne Rienner, 1993. Originally published in French as *L'écrivain public et l'ordinateur*. Paris: Hachette, 1988.
73   Schojet, M. "The Condition of Mexican Science." *Minerva* 12 (1979), no. 3: 381–412.
74   Schwartzman, S. "Struggling to Be Born: The Scientific Community in Brazil." *Minerva* 26 (1988), no. 4: 545–580.
75   ———. *A Space for Science: The Development of the Scientific Community in Brazil*. University Park, Penn.: Pennsylvania State University Press, 1992.
76   Shils, E.A. "Scientific Community: Thoughts after Hambourg." Bulletin of the Atomic Scientists 10 (1954): 151–155.
77   Shiva, V., and J. Bandyopadhyay. "The Large and Fragile Community of Scientists in India." *Minerva* 18 (1980), no. 4: 575–594.
78   Velho, L. "Science on the Periphery: A Study of the Agricultural Scientific Community in Brazilian Universities." Ph.D. diss., SPRU, University of Sussex, 1985.
79   ———. "The Meaning of Citation in the Context of a Scientifically Peripheral Country." *Scientometrics* 9 (1986), nos. 1, 2: 71–89.
80   ———. "Sources of Influence on Problem Choice in Brazilian University Agricultural Science." *Social Studies of Science* 20 (1990): 503–517.
81   Vessuri, H. "The Search for a Scientific Community in Venezuela: From Isolation to Applied Research." *Minerva* 22 (1985), no. 2: 196–235.
82   ———. "La cultura científica en el futuro de Venezuela." In: Silva Michelena, ed. *Venezuela hacia el 2000*. Caracas: Editorial Nueva Sociedad, 1987, pp. 299–317.
83   Waast, R. et al. *Proceedings of Two Workshops on the Emergence of Scientific Community in Developing Countries*. Paris: ORSTOM. Forthcoming.
84   Williams, R., and J. Law. "Beyond the Bounds of Credibility." *Fundamenta Scientiae* 1 (1980): 295–315.
85   Yanchinski, S. "The Newly Industrialized Countries." *Biofutur* (July–August 1987).

7

Technology, economics, and late industrialization

Jorge Katz

Public policies to induce changes in the organization of production or in patterns of international trade have now become quite common even in nations that otherwise support free market principles. Such policies normally involve efforts to restructure individual firms and industries, as well as to encourage more flexible forms of automation, new patterns of subcontracting and market organization, etc.

It would be helpful if there were satisfactory theoretical tools for approaching the problems that these policies are trying to tackle, such as the sources of innovation, dynamic comparative advantage, and productivity growth. Unfortunately, the social sciences are some way from offering a comprehensive body of theory on these questions. On the one hand, almost all of the issues raised by technological change, innovation, and production organization are interdisciplinary by nature, i.e. they involve knowledge concerning economic as well as engineering, organizational, institutional, or educational aspects and hence demand a high degree of interaction between specialists of different disciplines. Instead of this interaction, however, each discipline tends to develop a self-contained body of principles and analytical tools with which to look at any one specific question. This fragmentation usually results in only partially satisfactory answers and incomplete descriptions of reality.

On the other hand, and even if we confine ourselves to the more limited area of one particular discipline – say, economics – it is by no means obvious that we can use the same theoretical models to describe the complexities and idiosyncracies of societies of extremely disparate degrees of maturity and economic development. The organization of production at the individual company level, the spread of markets, their functioning and their degree of imperfection, the nature and behaviour of regulatory institutions, etc., vary considerably across nations and it is hard indeed to accept that one single model could be used equally well to understand the process of innovation and technological change of societies that differ substantially as regards major aspects of social organization.

In view of this, it is somewhat surprising that much of current thinking among professional economists still takes place under standard neoclassical assumptions of perfect market functioning, equilibrium, and profit maximization [35], and that different forms of market failure – externalities, "public goods," etc. – have so far received comparatively little attention. Both in the realm of theory and in the design and implementation of public policies, this neglect has certainly had strong adverse implications.

Finally, a similar argument to the previous one – but this time of an inter-temporal nature – can be advanced if we concentrate simply on the economics of technological change, innovation, and production organization in countries of "late industrialization," which we can imagine to be at a somewhat similar level of economic development. A major structural "break" can be identified in the growth process and in the regulatory regime of many developing countries around the mid-1970s. This break came about as a consequence of the dramatic change of circumstances that occurred in the global political economy in the early 1970s, and particularly after the debt crisis of the early 1980s, which forced many of these countries into a complex structural adjustment effort. It is now evident that these efforts have had a major impact on the rate and nature of technological change and innovation, and indeed on the overall development process of many developing nations. As in our previous example – that of the comparison between countries of differing degrees of maturity and economic development – it is by no means obvious that the same model that could help us to understand the development process and the technological and innovative performance of countries of "late industrialization" in the 1960s and 1970s could be used equally well to examine these same aspects in the 1980s.

On the contrary, the propensity to save and the rate of new capital formation, the inflow of foreign manufacturing investment, and, indeed, entrepreneurial dynamism and business behaviour in general all seem to have changed quite dramatically in recent years compared with the immediate post-war period. These changes had a major –

238

and as yet very imperfectly understood – impact upon the economic and social performance of many developing nations.

As a response to the deficiencies of neoclassical growth models, an alternative set of theoretical ideas stemming from Schumpeter began to be developed by a number of academic economists during the 1970s and 1980s. Notions of disequilibrium, imperfect information, "bounded rationality," "evolutionary sequences," etc., have gradually found their way into the technical change and growth literature through the pioneering work of scholars such as R. Nelson, S. Winter, C. Freeman, N. Rosenberg, G. Dosi, and others [13].

It is interesting to note that much of this new strand of "evolutionary economics" comes from English-speaking social scientists who draw their basic inspiration from both a critical appraisal of conventional neoclassical ideas as well as from stylized observations of how contemporary firms, markets, and institutions behave in mature capitalist societies. There are various reasons for believing that, a priori, their theoretical constructions cannot adequately capture the highly idiosyncratic industrial and social organization of countries such as Argentina or Brazil. It is typical of these developing nations that they face a highly unstable and volatile macroeconomic situation that has adversely affected the process of expectations formation, entrepreneurial dynamism, and capital accumulation, encouraging entrepreneurs to prefer opportunistic and rent-seeking activities rather than technological and innovative efforts. Yet the neo-Schumpeterian scholars have not found it necessary to deal with these matters when thinking about industrialized countries. Their models and theories cannot easily be reconciled with the kind of socio-economic environment to be found in many developing countries.

Lights and shadows of conventional neoclassical growth theory

The main body of neoclassical growth theory – and its empirical uses in growth accounting exercises – was presented in the professional literature in the late 1950s and throughout the 1960s. Few authors were as influential as MIT's Robert Solow, who in 1988 received a Nobel Prize in recognition of his path-breaking contributions in this field. He reminds us in his Nobel lecture [33] that what came to be known later as the basic neoclassical growth model started from a certain dissatisfaction with the Harrod-Domar growth model, in which the rate of growth of the economy depends upon three exogenous parameters: the rate of savings, the rate of population growth, and the capital/output ratio. The ultimate question the model is trying to answer is whether or not decentralized market mechanisms could lead the economy to a stable growth path undisturbed by labour shortages, on the one hand, or by high unemployment, on the other.

The Harrod-Domar answer to this question is relatively simple: although in theory there are situations in which the capital stock increases at the same rate as the labour force, allowing for "steady state" expansion, the chances of actually achieving such a growth path are small. Entrepreneur expectations determining savings and investment play a major role in this respect.

In the basic Harrod-Domar presentation, production functions are rigid and there is no room for capital/labour substitution in response to changes in relative factor prices. This is precisely Solow's point of departure when he specifies a model that admits a certain degree of technological flexibility that was absent in the Harrod-Domar formulation. In Solow's model, the rate of technical progress plays a major role as a determinant of the equilibrium growth rate attained by the economy. Technical change, however, is exogenous to the system. It falls like manna from heaven.

In order to deal with questions of technical change and productivity growth at this level of abstraction, the neoclassical model needs to make a number of stringent assumptions concerning the behaviour of economic agents. In particular, it has to postulate a well-defined relationship between short- and long-term scenarios in order to make the idea of dynamic equilibrium possible to sustain. The likelihood of disequilibrium is eliminated from neoclassical growth models through an elegant – but "ultimately unacceptable," to use Solow's words – simplification. Solow himself explains the case as follows:

The idea is to imagine that the economy is populated by a single immortal consumer, or by a number of identical immortal consumers. The immortality itself is not a problem. Each consumer could be replaced by a dynasty each member of which treats his or her successors as extensions of himself or herself. But no shortsightedness can be allowed. He or she is supposed to solve an infinite time utility function. The next step is harder to swallow in conjunction with the first. For this consumer every firm is just a transparent instrumentality, an intermediary, a device for carrying out intertemporal optimization subject only to technological constraints and initial endowments. Thus, any kind of market failure is ruled out from the beginning. . . .

I find none of this convincing. The markets for goods and labour look to me like imperfect pieces of social machinery with important institutional peculiarities. They do not seem to behave at all like transparent and frictionless mechanisms for converting the consumption and leisure desires of households into production and employment decisions. [33, pp. 310–311]

Prior to the mid-1960s, the neoclassical growth model lacked any endogenous theory of technical change. New production techniques, new product designs, etc., arrived stochastically from heaven. Kennedy [20] and Ahmad [1] tried to specify technological change as an endogenous "search" process, i.e. as yet one more economic activity performed by economic agents, and for this purpose imagined the existence of an "innovation possibility frontier," which they defined as an ex ante description of all of the labour- and capital-saving technological innovations available to the firm. With perfect knowledge of such options and free access to the required know-how, the entrepreneur is assumed to choose between "search" options exclusively on the basis of relative factor prices.

The problem with this specification of technological behaviour is that it is entirely devoid of the component of uncertainty and risk that normally underlies the very notion of innovation. As Nordhaus pointed out some years later [28], we are bound to assume that the firm knows in advance the complete set of innovative options, as well as the results of each one of them. Now, if this is so, the obvious question is, Why carry out the search efforts in the first place if there is no uncertainty to resolve?

In spite of these drawbacks, neoclassical growth models provided an interesting set of instruments for "growth accounting" exercises. The measurement of the "residual," i.e. that fraction of the observed rate of total factor productivity growth that is not accounted for by the expansion of capital and labour conventionally measured – and its ex post "explanation" in terms of better education, better machines, structural change – captured the attention of numerous economists in those years. Denison's work is probably the best known of these studies and accurately reflects the spirit of that time [11].

In spite of the fact that neoclassical growth models and growth accounting exercises served to illuminate important aspects of the development process, such as the role of capital/labour substitution or of human capital improvement through education and health, they none the less provide an oversimplified view of how an economy actually operates. On the one hand, they are based on highly unrealistic assumptions concerning (a) the nature and behaviour of firms, (b) the role of market imperfections, (c) the complexities of the institutional structure underlying the operation of the economy, etc. Equilibrium, perfect information, profit maximization, costless and timeless access to technological know-how, a very elementary institutional scenario, etc., appear as central features of the neoclassical growth metaphor [25] and are rather difficult to accept.

It is precisely on account of these assumptions that many academic economists have in recent years felt that they needed to proceed along a different route if they were to understand technological change and innovation. In the following section I review some of these newly emerging theoretical efforts.

Alternative theoretical routes

Over the last decade there has been a revival of heterodox thinking in the field of technical change and innovation. Factors other than the price mechanism, imperfect markets, disequilibrium, behaviour under conditions of imperfect information, "learning sequences," "technological trajectories," and similar concepts are central features of a newly emerging theoretical paradigm.

Some of these new ideas are intellectually rooted in Schumpeter's work, particularly in his *Capitalism, Socialism and Democracy* [29], where he writes:

The first objectionable concept in the model is that of competition. For many years economists only thought in terms of price competition. This idea refers to a scenario of given conditions in which production methods and, even more particularly, the forms of industrial organization are invariable. Nevertheless, in the reality of capitalism and, in contrast to what happens in textbook models, this is not the type of competition that matters the most. Competition via new products, new technologies, new sources of supply, new ways of organizing the production process, etc. are more important. This competition presents a decided advantage in terms of costs or quality over what went before. It doesn't matter if competition in the conventional sense of prices works better or worse. The powerful force that expands production and reduces long term prices comes from another source.

These ideas simultaneously enrich and add to the complexity of the economist's professional tool-box. Imperfect information, uncertainty, and disequilibrium allow for behavioural differences among firms [25], as well as for innovative leads and lags and endogenous changes in market structure. Technological learning can differ from company to company as a function of how much a given firm spends on R&D activities, but also as a function of the quality of its research and engineering staff, or of its luck in the "search" for new technology. It now becomes possible to postulate models of "adaptive" behaviour in which we do not have to assume that the firm has complete ex ante knowledge of all of its future technological possibilities, nor that its only objective function is that of maximizing profits [31]. The door is now open to organizational and behavioural models of the firm of the sort advanced by authors such as March and Cyert [23], Williamson, and others [21].

Once the idea of regular and predictable behaviour inherent in the neoclassical logic is abandoned, the notion of evolutionary performance can be introduced [26]. Current behaviour is strongly influenced by the recent past, and this past includes not just the individual company's history but also that of the market and of the macro and institutional environment in which any given actor operates. A certain "biological" flavour is imparted to these models by the evolutionary mechanism that underlies the dynamics of firm behaviour and of market structure.

A number of academic economists have pursued this promising line of work in recent years (e.g. Boyer [7], Clark and Juma [9], Silberberg et al. [30]). It is important to realize, however, that most of their ideas are inspired by stylized observations of what is at present going on in developed industrial societies in relation to changes in the organization of production at the individual firm level, in market organization, and subcontracting practices, and in the behaviour of regulatory institutions, etc. In each one of these areas industrialized countries are currently undergoing major structural transformations that are gradually being captured by the stylized growth models with which economists operate.

It is by no means obvious, however, that such models could successfully be used to throw new light upon the complex socio-economic and institutional environment of countries that came late to industrialization, such as Argentina, Brazil, Mexico, India, and China. Both during the period of import substitution industrialization in the 1960s and 1970s and also during the 1980s, these countries exhibited patterns of social and production organization quite unlike those prevailing in more mature industrial societies.

Import substitution industrialization in the 1960s and 1970s

Policies to promote import substitution industrialization started in many developing nations in the 1930s and 1940s as a consequence of the breakup of the gold standard. Needless to say, these efforts began under extremely unfavourable conditions as regards lack of skilled manpower and markets (particularly the absence of capital markets that could adequately finance long-term capital formation), institutional fragility, etc.

Such features, combined with a small domestic market and an inward-oriented import substitution strategy that aimed at that point to cater only for local consumers, underlay the creation of a highly idiosyncratic industrial sector whose structure and performance have not been well understood by development economists and social scientists in general. I first review some of the main features of the industrial structure that developed at that point before proceeding with an examination of the impact of the import substitution industrialization process upon domestic technological capabilities, as well as upon the functioning of the national system of innovation.

The following features could be oberved in the industrialization process. First, foreign manufacturing investment rapidly acquired a leading role within the newly emerging production structure. Domestic subsidiaries of large multinational companies brought along with them new product designs, production processes, and organization technologies that acted as a training ground for local human resources. These technological transfers from abroad had both positive and negative consequences for the receiving societies. On the one hand, they significantly affected local production practices by disseminating quality control standards, patterns of subcontracting, models of production organization, etc., largely unknown to local firms. On the other hand, however, their arrival pre-empted the "technological path" industrial firms were to follow thereafter, establishing the new consumption, production, and industrial organization paradigm within which the development process was to take place. In consequence, domestic technological capabilities grew up within the limits imposed by this paradigm.

It is important to note, however, that several authors (e.g. Amsden [2]) have recently argued that Korea has followed an alternative strategy in this respect: foreign manufacturing capital was not allowed to play a major role in the early stages of the industrialization process and was invited to participate only in more recent times, when the country had a competitive domestic industry already in operation.

Secondly, the newly created manufacturing facilities had highly idiosyncratic features as regards size of plant, degree of vertical integration, range of product "mix," etc. Locally established plants were seldom much bigger than, say, one-tenth the size of comparable production units operating in industrialized countries. Because of the immaturity of the local industrial structure, the degree of vertical integration of these companies was much greater than the one prevailing in more developed societies. As a result of the small size of the local market, their output mix was significantly more diverse than in similar firms in industrialized countries.

Lastly, market organization and performance, as well as regulatory institutions, also evolved along highly individual lines, hardly comparable to those exhibited by more developed industrial societies. Oligopolistic and monopolistic situations and a high rate of external protection prevented market forces and competition from adequately performing their disciplinary role. Government failure turned out to be at least as important as a source of difficulties as market failure itself.

The development of domestic technological capabilities took place within the limits imposed by this "inward-looking" process of growth. Industrial firms were forced to supply themselves with a significant amount of "in-house" engineering and technological knowledge on the basis of which to adapt to the local environment both products and production technologies transferred from abroad. These domestic engineering and technological efforts had the purpose of adapting product designs to the preferences of local consumers and production processes to a different set of raw materials, a much smaller scale of operation, and a different pattern of work automation, etc. Thus, local R&D efforts in developing countries started as an *endogenous* answer to signals coming from the particular industrial organization and institutional environment in which the industrialization process took place. These signals were clearly different from those received by engineers and technicians working for comparable firms and industries in more mature industrial societies, and therefore the local technological trajectory was bound to be different from the one followed by somewhat similar companies and industries in industrialized countries.

On the other hand, the other parts of the national innovation system – i.e. universities, public R&D laboratories, etc. – remained rather isolated from the industrialization process and confined themselves to more basic research ventures carried out for the sake of scientific knowledge rather than for the development of production technology. In other words, the national innovation system grew up

245

as a fragmented and heterogeneous network of agencies and institutions that maintained only a very weak connection with the emerging industrialization process. As far as industry is concerned, the lion's share of the national technological search efforts were of the adaptive type, seldom carried out with the purpose of attaining "state of the art" production technology.

The determinants of firms' R&D efforts

Various different micro and macro forces influenced the technological search strategy followed by local firms during the import substitution industrialization period. Consider first those forces that are strictly firm-specific. No two factories in the world are exactly alike and each one tends to develop its own particular bottlenecks, intersectoral imbalances, etc. Troubleshooting activities are normally undertaken by technical-assistance-to-production personnel with a view to keeping the available facilities running smoothly. In the course of carrying out their tasks, troubleshooters generate a steady flow of incremental technical and engineering knowledge concerning product design, process technology, etc., that is normally not "new" at the world level but is certainly "new" for the firm in its particular circumstances [15].

A second set of forces influencing the individual firm's technological search efforts is related to the market's competitive climate. Markets are dynamic institutions whose structure and competitive atmosphere normally change through time, *pari passu* with the entry of new competitors, the introduction of new products, etc. Monopolistic market situations have been shown to induce capacity-stretching technological search efforts [24], i.e. engineering activities intended to extract more output from a given set of machines, whereas imperfect and oligopolistic competition have been shown to induce engineers and technicians to search for quality improvements as well as for product differentiation opportunities [16]. Whereas in the former case process engineering R&D activities have usually been more prominent, in the latter product design efforts tend to be given higher priority.

A third set of forces affecting technological behaviour relates to macroeconomic variables such as the exchange rate, interest rates, the effective degree of protection, wage rates, etc., i.e. macroeconomic "prices" pertaining to the economy as a whole. The level of uncertainty also belongs in this category. All of these forces are macroeconomic in nature and do not affect one particular firm or industry but instead cut across the overall production structure. Available empirical evidence [8] indicates, for example, that engineering efforts in the field of production organization – such as, for example, time and motion studies, plant layout balancing efforts, etc. – were undertaken by metalworking firms operating in Argentina after capital markets were de-regulated in the late 1970s and the rate of interest became highly positive. Companies tried in this way to cut down on idle time and inventories, and used their engineering personnel for the purpose.

Lastly, a fourth set of variables influencing individual firm technological strategy has to do with the company's perception of – and capacity to decode – what is going on at the world's technological frontiers in its particular field of activity. World trade fairs, scientific and technological publications, patent files, information from equipment suppliers, etc., normally act as diffusion channels through which technical and engineering knowledge are disseminated. Plant engineers and technicians are frequently exposed to such information and carry out R&D efforts in order to adapt the knowledge to their particular needs.

I am now in a position to summarize briefly my argument concerning the determinants of the individual firm's technological search strategy during the period of import substitution industrialization: rather than being *exogenous* to the firm – as the neoclassical model assumes – technological change resulted from in-house adaptive R&D and engineering efforts carried out by plant technical personnel. These efforts produced a steady flow of incremental units of technical information concerning product design, production processes, and production planning and organization. In their search for better ways of doing things, engineers responded, on the one hand, to signals that were localized and firm-specific and, on the other, to forces emerging from the competitive atmosphere of the market, the macroeconomic scenario, and the firm's perception as to how the state of the art was changing at the world level in their specific field of activity.

Not all of the variables mentioned above had the same weight and importance throughout the period, nor did they play a similar role in each and every country. In the early stages of the import substitution industrialization process – i.e. in the immediate post-war years – intra-plant technical matters and questions related to the competitive atmosphere in which firms operated seem to have played a major role as determinants of in-house R&D and engineering efforts. These variables, however, became much less significant during the course of the 1970s and 1980s, when the level of uncertainty and the degree of macroeconomic turbulence became much more noticeable in most of the third world [16].

It is also important to note that companies very seldom expanded their R&D commitments beyond the adaptive stage, trying to develop more permanent and "state of the art" technological capabilities [10]. Nor did they search for a more intimate relationship with other agents and institutions of the national innovation system – i.e. universities, public R&D laboratories, etc. – which thus remained isolated from the industrialization process.

Technological search efforts, productivity growth, and dynamic comparative advantages

Engineering and technical capabilities did not develop all at the same time within any given company. Available empirical evidence suggests that efforts to increase in-house technological capabilities passed through several phases associated with the absorption of different types of qualified human resources.

Product development capabilities appeared first in the early postwar period. Since the war cut off supplies from traditional capital goods producers, many small family metalworking enterprises started local production of goods such as lathes, electric motors, harvesters, etc., in the late 1940s and 1950s, particularly in Argentina [16] and Brazil [27]. Quite frequently, local capital goods were outmoded versions of US or European machines that were successfully copied locally through some kind of reverse engineering effort. Local firms usually began domestic production on the basis of second-hand machinery, with a casual plant layout and with very little in the way of production planning and organization. Under such circumstances, product design capabilities tended to develop first, followed later by production and process engineering capabilities, and much further along the line – perhaps as much later as 10 or 15 years [18] – by production planning and organization skills.

Firms seem to have proceeded from the simpler to the more complex technological tasks. These last usually demanded a greater degree of technological sophistication on the part of the local engineering team. In the course of this evolutionary sequence from simple to more complex technological search efforts, firms gradually learned to operate pilot plants, to build prototypes and other forms of experimental equipment. The accumulation of skills naturally took time: in many cases as much as 10 years were needed in order to develop in-house capabilities in product design, process engineering, production planning and organization [16].

Economists have long been interested in the relationship between in-house engineering efforts and total factor productivity growth. As early as the 1940s, Lundberg reported that although no new investment was made in the Horndal Ironworks in Sweden for some 15 years, productivity rose on average at a rate of 2 per cent per annum [32].

Arrow's article [3] on the economic implications of learning by doing presented a theoretical model in which he identifies the endogenous nature of the learning process. The article opened the way for a long series of empirical studies on the "learning curve" that appeared in the industrial economics literature throughout the 1960s. However, more than just accumulated experience is needed; a whole host of "minor" technological improvements – some of them embodied in the existing capital stock, others more disembodied and related to production organization – find their way into the company's daily routine.

A micro study carried out in the early 1960s by Hollander at the DuPont Rayon plants in the USA [15] showed that the bulk of cost-reducing improvements introduced by these firms throughout three decades came from "plant personnel attached to the Technical Assistance to Production groups, which played the most important role in the development of minor technical changes. Such groups were intimately linked with current operations and their function was to keep existing processes 'out of trouble'" (p. 196).

A number of case-studies carried out by economists and engineers in countries of "late industrialization," such as Argentina, Brazil, Mexico, or India, confirm the fact that in developing countries, too, in-house R&D and engineering activities constitute the major explanatory variable in total factor productivity growth. As one of these studies pointed out:

We are now in a position to summarize the various factors that underlie this company's growth performance. Three different sets of growth-inducing forces have been hereby identified:

The first – and quantitatively the more important – is associated with technological changes generated by the firm's engineering team itself. It includes: increases in operational speed and improvements in product quality. On the whole these are 'disembodied' technical changes which were incorporated in the existing – albeit slightly modified – capital equipment. 35% of the observed changes in labour productivity were achieved through the first of the above mentioned set of changes and 30% through the second. This means that close to two thirds of what happened in terms of labour productivity growth between 1941 and 1967 can be accounted for by this group of explanatory forces.

The second set of technological changes affecting labour productivity – also originated in intra-firm engineering efforts – concerns the sphere of production organization and, more particularly, the company's degree of vertical integration and its use of subcontractors.

Finally, the third set of forces bringing about productivity growth includes a number of technological changes originated outside the firm. We are mostly referring to technological changes embodied in the new capital equipment imported second hand from the US. [17, p. 208]

Similar results have been reported by many other economists who studied the technological behaviour of manufacturing firms in developing countries [22, 6, 10, 24, 12].

In addition to the evidence presented so far, two further points can be made in support of the suggestion that adaptive R&D and engineering efforts exerted a major influence upon productivity growth in developing countries. On the one hand – and given the fact that not every firm followed the same technological search strategy or had identical success in terms of productivity growth – we should a priori expect market shares and industrial structure to change as a consequence of inter-firm differences in attitudes to and results of in-house engineering and R&D activities, i.e. as a result of forces *endogenous* to the market.

On the other hand, and considering that there are large inter-industry differences in the rate at which the world's technological frontier moves over time, we find strong grounds a priori to expect that in those cases where a rapid rate of technological learning and productivity growth on the part of the local firm obtained simultaneously with a low rate of expansion of the world's technological frontier, the local firm would be able gradually to catch up with the international technological frontier, achieving growing competitiveness in both domestic and foreign markets.

This was what probably happened in many of the success stories of Brazilian, Mexican, Indian, or Korean firms, which did increasingly well on the export side throughout the 1970s. The exports were not exclusively of technologically sophisticated industrial goods but also involved pure technology in the form of licensing agreements, complete manufacturing plants sold on a turnkey basis to entrepreneurs from other developing countries [22], and infrastructure projects such as roads, pipelines, or airport facilities, etc. [18].

250

Thus in the 1970s, the development of domestic technological capabilities seems to have been associated with a gradual change in dynamic comparative advantages and in the degree of internationalization of locally based companies in many developing countries.

A "catching up" model of this sort underlies much of the professional thinking concerning the case of Japan and, more recently, Korea. Economists have not as yet accepted that the case of many Brazilian, Mexican, Indian, or Argentinian firms that have built up substantial export capabilities in the 1970s could be regarded in a similar way.

We have so far examined some of the technological and organizational consequences of the import substitution industrialization process for many developing nations during the 1960s and 1970s. Obviously there are large inter-country differences: why Korean or, to a lesser extent, Brazilian firms have been more outward-looking and aggressive [10], whereas their Argentine or Mexican counterparts remained significantly less dynamic and did not expand their technological and export commitments as much as the others remains an interesting and still unresolved question for future investigation. Other aspects of the national system of innovation, such as the role of educational or R&D policies and institutions, as well as the impact of implicit and explicit government industrial policies, are surely major explanatory factors in the observed inter-country differences.

There are important institutional, ideological, and political reasons why the national system of innovation has worked better in certain environments than in others, and therefore why the impact of the import substitution industrialization efforts on the development of domestic technological capabilities has been dramatically different across nations.

In spite of the positive achievements mentioned above – i.e. significant gains in productivity and export capabilities, expansion of the local engineering and technological capabilities, etc. – the import substitution industrialization process largely failed to develop a world class manufacturing sector. In fact, a significant number of the newly created firms and industries found it increasingly difficult to compete both locally and internationally, particularly in the late 1970s and early 1980s, when the rapid diffusion of microprocessors and microelectronic technologies opened up the way for an entirely new generation of product designs and production processes that in a matter of just a few years gained wide acceptance in world markets for consumer durables and capital goods. New product designs gradually incorporated digital and numerical control devices, miniaturization,

251

and other features that many producers in developing countries could not incorporate quickly into their locally produced electro-mechanical versions of roughly comparable products.

Concurrently with these changes – and with the debt crisis that took on dramatic proportions in the early 1980s – the macroeconomic scenario facing many developing nations turned out to be highly uncertain and turbulent. The rapid deterioration of fiscal and external accounts and the drastic curtailing of external financing forced many countries to introduce major changes in public policy. Throughout the decade, many developing countries implemented an orthodox and market-oriented policy package that included opening up the economy to foreign competition, de-regulating markets, privatizing public enterprises, etc. Such policy packages – which in many cases had to be enforced through the intervention of the army and with a considerable amount of social repression – gradually induced substantial changes in the structure of the economy as well as in the performance of markets and institutions. The rate and nature of technological change and innovation and the functioning of the national system of innovation are also changing as part of this socio-economic transformation.

The 1980s: Towards a new socio-economic and technological scenario

Bearing in mind the fact that there are important inter-country differences – which I shall briefly examine later when looking at the experience of Argentina, Chile, and Brazil – let us start by summarizing some of the most outstanding features of the contemporary socio-economic scenario:

- The rate of economic growth of many developing countries has slowed quite dramatically in the 1980s compared with the immediate post-war period, and the structure of the GDP has changed significantly in most of them, with manufacturing accounting for a diminishing share of total output. This has certainly been the case throughout Latin America. Even Hong Kong, Singapore, and Taiwan experienced a marginal reduction in their rate of expansion *vis-à-vis* their performance of the 1970s, leaving Korea as the only case among the newly industrialized countries in which the rate of growth was actually higher in the 1980s than in the 1970s.
- The production of consumer durables and capital goods has fallen significantly, especially in Argentina and Chile, and less dramatic-

ally so in Brazil and Mexico. At the same time, resource-based industries have expanded rapidly. The foodstuffs industry in Chile and the production of non-durable consumer goods (footwear, garments, etc.) and industrial commodities (e.g. steel, petrochemicals, aluminium, or pulp and paper) in Argentina and Brazil have taken the lead as dynamic sectors within manufacturing industry.

– The organization of work and the social division of labour are gradually experiencing significant changes as new patterns of subcontracting, of flexible automation, etc., are introduced by domestic companies. This is particularly noticeable in areas such as automobile manufacturing, "made-to-order" capital goods, footwear and garment production, etc.

– Manufacturing exports are also undergoing a major transformation. Raw material processing industries now account for a much larger share of exports than in the past, as the cases of Chile, Argentina, and Brazil show. This has been accompanied by a drastic reduction in exports of relatively more sophisticated products with high value added, such as electrical and non-electrical machinery, transport equipment, electromechanical instruments, and capital goods in general. This decline in the mechanical engineering sector is particularly noticeable in Argentina and Brazil. Thanks to special vertical integration arrangements with the US vehicle industry, Mexican car producers have been able to expand production and exports over the same period, in spite of the deteriorating performance of the Mexican economy in general and of its capital goods industry in particular.

– As a result of take-overs, mergers, and closures, the degree of business concentration has increased substantially throughout manufacturing industry in almost all of the countries mentioned above. In many developing countries – notably in Argentina, Chile, and Brazil – a relatively small group of large domestic corporations has acquired strong control over the local economy in a rapid process of horizontal expansion and economic concentration.

– The share of multinationals in manufacturing production has fallen in many developing countries as major firms have decided to move out altogether in recent years or have considerably reduced their commitment to growth, for instance in pharmaceutical or vehicle production in Argentina, Mexico, and Chile. In fact, aggregate foreign investment flows are now proceeding from the South to the North, i.e. mainly from developing to industrialized countries (the United States and Britain in particular).

- Industrial relations and, indeed, the overall functioning of the labour market have also experienced drastic changes. The bargaining strength of trade unions has significantly diminished relative to the early 1970s. Massive social repression and physical intervention by the armed forces have been instrumental in this respect in many Latin American countries, notably Argentina, Chile, and to a lesser extent Brazil.
- Regulatory institutions are also undergoing a major transformation. A new regulatory regime is emerging, while the earlier emphasis on import substitution steadily declines. A lower level of external protection, de-regulation of markets, a gradual transfer of public production activities to the private sector, capitalization of the stock of debt through the acquisition of public utilities and enterprises, etc., are some components of a new regulatory package to which many developing countries are nowadays gradually accommodating themselves in the face of external pressure.

These various aspects add up to a major structural change and not just to a marginal adjustment. The role and *modus operandi* of manufacturing industry, the organization of production at the individual plant level, the pattern of foreign trade, the degree of business concentration, the participation of foreign capital, the functioning of the labour market, trade union bargaining strength, a new breed of regulatory institutions – all seem to be part of a far-reaching socio-economic transformation whose final form and consequences are still far from clear.

What is the impact of this transformation likely to be upon the national system of innovation and the development of domestic technological capabilities? Is there any reason a priori to believe that scientific and technological institutions and micro patterns of behaviour relating to technical change and innovation are going to change as a consequence of this socio-economic restructuring process?

The available empirical evidence in this respect is as yet imperfect. We lack basic knowledge about major issues, and further research is certainly needed if we want to achieve better understanding of what is happening at present. Nevertheless, in a very preliminary way, we can identify some interesting new trends that we now briefly examine for Argentina, Chile, and Brazil.

Argentina
Savings and investment have both fallen quite sharply relative to the 1960s and 1970s. A much higher degree of macroeconomic uncer-

tainty and volatility account for the fact that capital flights out of Argentina increased dramatically during the course of the 1980s. The lower propensity to invest locally is reflected in the low level of imports of machinery and equipment that has prevailed throughout recent years and still obtains today. Direct foreign investment has also fallen sharply during the 1980s, indicating the lack of interest and low expectations foreign firms now have concerning the future of the Argentine economy. This macro scenario appears to be associated with an expansion of opportunistic and rent-seeking activities on the part of the local business community and with a general decline in entrepreneurial dynamism that permeates the whole production structure and particularly affects the industrial sector.

Within this general climate, three basic points reflecting recent patterns of technological behaviour in Argentina stand out. First, as manufacturing output is smaller today than two decades ago, and as the structure of industry has changed quite considerably with the contraction of engineering and capital goods industries and the concomitant expansion of raw materials processing sectors, the country's industrial R&D and engineering efforts have significantly contracted in recent years. In the mid-1970s, Argentina produced some 350,000 cars, 25,000 machine tools, and 60,000 tractors per annum. The equivalent output figures now are only 120,000, 6,000, and 5,000 respectively, and the factories have changed dramatically as regards production organization, import content, subcontracting practices, etc. The metalworking sector alone has reduced its payroll by more than 200,000 workers compared with the early 1970s. By contrast, Argentina increased its production of petrochemical products from 865,000 to 1,794,000 tons, its steel output from 2.25 to 3.67 million tons, and its sunflower oil production from 630,000 to 2 million tons. Very little new employment has been generated by these expanding sectors, which now constitute the backbone of the country's manufacturing exports.

Metalworking firms and capital goods producers have cut back their product design activities, as well as their production planning and organization efforts, considering them to be an indirect cost of production largely unjustified at their present low level of operation. As a result, and concomitantly with the contraction in the volume of output, metalworking firms have proceeded to reduce their engineering departments and the use of technologists and engineers.

On the other hand, raw materials processing industries involved in the production of industrial commodities such as steel, petrochem-

255

icals, aluminium, pulp and paper, and edible oil have expanded rapidly in recent years on the basis of new and highly capital-intensive facilities demanding minimal domestic R&D and engineering efforts. Most of these new firms produce low value-added standardized products that are sold in highly competitive, undifferentiated, international markets in which Argentina is a marginal "price taker."

Secondly, a slow and gradual diffusion of computer-based technologies seems to be taking place among medium-size and small family enterprises. Industries such as shoe manufacturing, garments, made-to-order machines and equipment (industrial boilers, canning and bottling plants, etc.), where family businesses are an important source of supply, seem to be increasingly adopting computer-aided design and manufacturing technologies and new ideas concerning production organization and subcontracting.

Lastly, another important dimension of Argentina's contemporary technological situation concerns the country's primary sector. After a rather long period – nearly 30 years – of technological stagnation, agriculture started a process of rapid technological transformation in the 1960s with the introduction of agricultural machinery and production organization technologies. The pace of technological change quickened thanks to the massive diffusion of maize hybrids, agrochemicals and, more recently, various kinds of biotechnological developments.

Chile

Between 1974 and 1983, manufacturing output fell by nearly a quarter, from 25.5 per cent to 19.9 per cent of the GDP. Some 5,000 industrial plants closed, with a loss of nearly 150,000 jobs [34]. De-industrialization has been particularly severe in metalworking and textiles. By contrast, the foodstuffs industry has substantially increased its share of manufacturing production. Employment, investment, and exports in the sector have all expanded rapidly. A major part of this expansion relates to fresh fruits and vegetables: grapes, apples, asparagus, etc. – all products in which Chile has managed to capture important markets in industrialized countries.

The spread of agrochemicals and hybrids as well as of new production organization techniques has been significant. In areas such as packaging, freezing, and transportation, Chilean exporters have become markedly more sophisticated than in the past. New institutions and different social structures within production – including much

256

higher proportions of women in employment, the emergence of a new vintage of very dynamic rural entrepreneurs, and the development of a fairly concentrated intermediary sector handling transportation and distribution – have appeared in association with the expansion of the Chilean foodstuffs industry. Moreover, the modification and local production of agricultural equipment are also showing signs of substantial advance, in what could be interpreted as a case of "learning by doing" and adaptive R&D efforts in the mechanical engineering sector linked to the expansion of agriculture.

In contrast to the experience of Argentina, Chile is one of the few Latin American examples – Colombia is another one – in which public fiscal accounts as well as the country's external sector seem to be close to equilibrium and do not constitute a major source of macroeconomic turbulence and declining entrepreneurial dynamism. In fact, and basically due to political and ideological circumstances, the flow of external financing never constituted a major constraint on the country's growth process in the way that it has in many other developing countries. Chile is probably the example that best supports the modern neoclassical advice of "get your macro accounts right and let the market mechanism carry out your resource allocation task," though we should not forget that a long period of repressive military intervention, massive unemployment, and plentiful external financing preceded this rather peculiar liberal experiment.

Brazil
Throughout the 1960s and 1970s, Brazil was one of the fastest growing economies in the world. Between 1965 and 1980, the country's overall rate of expansion was 8.8 per cent per annum – in fact, not very different from Korea or Hong Kong, which most observers consider extremely successful cases – while the rate of growth of manufacturing production reached about 10 per cent per year. This process of expansion slowed down significantly during the course of the 1980s. Between 1980 and 1988, the GDP managed to grow at only 2.9 per cent per annum, and manufacturing activities at an even lower rate, 2.2 per cent. Although not as dramatically as in Argentina or Chile, the industrial sector's contribution to the GDP has declined over the last decade. In fact, between 1981 and 1984, industry experienced three consecutive years of worsening performance, particularly in areas such as electrical and non-electrical machinery, transport equipment, and instruments, i.e. the core of the metalworking and

257

capital goods sector. After very rapid expansion – 25 per cent per annum – during the 1970s, the Brazilian capital goods sector contracted significantly – by 12 per cent per year – during the 1980s.

Albeit less dramatically than in Argentina, the Brazilian metalworking industry is also showing signs of decline. The adverse technological consequences of this process are probably similar to those previously mentioned for Argentina. Among them are the contraction of investment and the cutting back of R&D and engineering activities. An interesting possible difference is the fact that Brazil has managed to establish a relatively more sophisticated segment of metalworking firms involved in the production of aircraft, military equipment, etc., where product design and process engineering R&D efforts and capabilities have been maintained, largely on the basis of public sector subsidies that might very well be cut off in the immediate future because of current macroeconomic difficulties. Yet another interesting difference between Brazil and Argentina concerns the computer industry, where Brazil had a consistent – and also controversial – policy of market reservation throughout the 1970s. While supporters of this policy tend to emphasize its positive impact upon investment and R&D activities, as well as in the development of domestic technological capabilities, critics have pointed to the high cost of locally produced small computers and the adverse effect this has had on the rate of diffusion of computer-controlled techniques throughout the economy.

Brazil has also managed to develop a considerable production infrastructure and strong export capabilities in non-durable consumer goods such as shoes, orange juice, instant coffee, frozen meat products, etc., as well as in industrial "commodities" such as cast iron, steel, pulp and paper, etc. R&D and engineering activities are less significant – though by no means negligible – in these more traditional segments of industry. In particular, market organization know-how is highly relevant in many of these cases, as the lion's share of the action is in the hands of medium-size and small family enterprises supported by a network of trading firms that handle the international side of the operation. As in Chile, packaging, transport, and marketing technologies are crucial and constitute an area where Brazilian trading companies have in recent years developed valuable proprietary technology.

On the whole, it is fair to argue that Brazil has so far maintained a more consistent policy of domestic market reservation than Argentina or Chile. It still has a strong and well-qualified civil service retaining much of the spirit of the "old" import substitution industrialization period, and in spite of external pressure and official talk about de-regulation and free market operation, still manages – through non-tariff barriers – to maintain a more closed local economy than other Latin American countries. Whether that has been a good or a bad thing for technological change and innovation and for the development of domestic technological capabilities, and whether or not Brazil will be able to maintain such strong regulatory positions in the future, are major, still unresolved issues.

We have so far examined three quite different situations within Latin America. It is obvious that the dynamics of the present socio-economic restructuring process varies enormously across nations and that no simple generalization can be offered as a substitute for detailed research. Nevertheless, several points stand out from the foregoing discussion.

First, savings and investment as well as the rate of new capital formation have decreased in many of the "late industrializers" in the 1980s compared with the immediate post-war period. With few exceptions – such as the vehicle industry – foreign direct manufacturing investment has also fallen relatively to the 1960s and 1970s. Entrepreneurship seems to be at a particularly low ebb, and entrepreneurs tend to have turned more to opportunistic and rent-seeking activities than to technological and innovative efforts. Macroeconomic stability, fiscal equilibrium, and capacity for growth are still difficult to obtain in many developing countries, even in those that achieved an impressive growth performance in the 1960s and 1970s, as was the case of Brazil, Mexico, or Argentina. Chile and, to a lesser extent, Colombia belong in a different category, where a more stable macroeconomic environment, external financing, and fiscal equilibrium have managed to sustain a reasonably high level of entrepreneurial dynamism among local businessmen.

Secondly, the contraction of engineering and capital goods industries has clearly induced a fall in domestic R&D and engineering activities, as well as in the use of highly qualified staff. Low value-added industries have gained ground within the newly emerging pattern of international trade. In this respect, the experience of Argentina does not seem to be very different from that of Brazil, in spite of the fact that the production of arms, aircraft, and small computers has probably allowed Brazil to develop – and so far maintain – a somewhat larger stock of domestic engineering skills and technological capabilities than the one Argentina has been able to preserve.

Thirdly, both the primary sector and different branches of the foodstuffs industry seem to be involved in a rapid process of technical change. The diffusion of hybrids, agrochemicals, biotechnological processes, etc., seems to be slowly taking place among rural producers. On the other hand, downstream service sectors such as packaging, freezing, and transportation also seem to be experiencing a number of important institutional and technological transformations, adapting themselves to new opportunities in the international market-place. The Chilean example is probably the most interesting of these cases within the Latin American region.

Fourthly, raw materials processing industries such as petrochemicals, steel, pulp and paper, and edible oil have expanded significantly in recent years. Exports have grown rapidly in these sectors, but, with few exceptions – for example, Usiminas in Brazil [10] – most of these firms did not themselves engage in state-of-the-art R&D and technological activities. "Specialty" chemicals or steels – where a much higher degree of technological sophistication and value added is involved – have not so far received much attention. Such "downstream" industries might, however, become very important in the future once these countries have managed to develop a strong raw materials processing sector.

Fifthly, medium-size and small family enterprises engaged in the production of shoes, garments, made-to-order pieces of capital equipment, etc., are gradually introducing – probably more so in Brazil than in Argentina – computer-based technologies and new forms of market and production organization. It would be wrong to assume that such phenomena constitute a generalized and sweeping process, but it would also be a serious mistake a priori to believe that small and medium-size local firms are entirely left behind by current developments in micro-electronics and informatics.

Lastly, the degree of technological heterogeneity seems to be growing within the industrial structure as a dramatic process of "creative destruction" takes place on a massive scale. A much higher degree of economic concentration and a noticeable slide in welfare standards also seem to be part of the contemporary Latin American scene.

It is thus quite clear that a major socio-economic transformation is under way in the region and that it is going to have a deep impact upon the rate and nature of technological change and innovation, as well as upon the growth process of many of these nations. It is also quite clear that this transformation is far from complete and that we still do not fully grasp its principal consequences in very many spheres besides those for the national system of innovation. Structural unemployment, an extremely adverse impact upon income distribution, equity and welfare standards, increasing and more complex forms of social conflict, etc., appear to be possible consequences of the ongoing socio-economic restructuring process, which should certainly be examined because they affect fundamental aspects of social equity and political governability. It is, however, outside the scope of the present paper to explore such aspects in any detail.

Concluding remarks

The above discussion naturally leads us into policy issues. For the sake of brevity, I shall touch upon only three major aspects that, in my opinion, deserve serious consideration on the part of policy-oriented researchers. The first relates to macro-policy management and to the role of the state in the process of capital accumulation. Many developing countries are now dealing with a complex situation in which extreme forms of external and fiscal disequilibrium demand a very rigorous macroeconomic policy management if negative expectations are to be reversed and the process of capital accumulation and growth revitalized. It seems unlikely that without a significant flow of external financing, the present vicious circle of stagnation and diminishing entrepreneurial dynamism could be easily overcome in the near future. The industrialized countries probably need to revise their views in this regard if they seriously want to help developing countries to enter into a new sustainable growth path.

Secondly, beyond sound macro-policy management, i.e. of "getting the macro prices right," developing countries require explicit industrial and technological policies capable of dealing with market failure in areas such as education, technology generation and diffusion, etc. Downward linkages towards high value-added industries in fields where strong, resource-based sectors have already been created in recent years – i.e. petrochemicals, pulp and paper, steel, etc. – have to be developed, and there are strong reasons a priori to believe that the "invisible hand" will encounter significant difficulties in inducing a socially optimal pattern of development in this direction.

Thirdly, questions of social equity and political governability should be examined closely. The process of socio-economic restructuring is far from being fair to the poorest one-third (or even more) of the local population, and those are the people who are now "paying for" the current structural transformation. Public expenditure in areas such as health, education, and social security have contracted sharply in recent years, to the severe detriment of welfare standards. Science and technology policies could be called upon in these areas in order to improve their performance and thus reduce the damaging impact of current trends.

References

1   Ahmad, S. "On the Theory of Induced Innovation." *Economic Journal* 76 (1966).

2   Amsden, A. *Asia's New Giant: South Korea and Late Industrialization*. New York: Oxford University Press, 1989.

3   Arrow, K.J. "The Economic Implications of Learning by Doing." *Review of Economic Studies* 29 (1962).

4   Arthur, B. "Competing Technologies: An Overview." In: Dosi et al., eds. *See* ref. 13.

5   Balassa, B. *Policy Choices in the Newly Industrializing Countries*. World Bank Working Papers, no. 432. Washington, D.C.: World Bank, 1990.

6   Berlinski, J. "Una planta argentina de motores diesel." In: Katz, ed. *See* ref. 16.

7   Boyer, M. "Formalizing Growth Regimes." In: Dosi et al., eds. *See* ref. 13.

8   Castaño, A., J. Katz, and F. Navajas. "Una empresa argentina productora de máquinas herramientas." In: Katz, ed. *See* ref. 16.

9   Clark, N., and C. Juma. "Evolutionary Theories in Economic Thought." In: Dosi et al., eds. *See* ref. 13.

10   Dahlman, C. "From Technological Dependence to Technological Development: The Case of the Usiminas Steel Plant in Brazil." In: Katz, ed. *See* ref. 17.

11   Denison, E. *The Sources of Economic Growth in the US and the Alternatives Before Us*. New York: Committee for Economic Development, 1962.

12   Desai, A. *Technology Absorption in Indian Industry*. New Delhi: Wiley Eastern, 1988.

13   Dosi, G., C. Freeman, R.R. Nelson, G. Silverberg, and L. Soete, eds. *Technical Change and Economic Theory*. London: Frances Pinter, 1988.

14   Freeman, C. *The Economics of Industrial Innovation*. London: Frances Pinter, 1982.

15   Hollander, S. *The Sources of Efficiency Growth: A Study of DuPont Rayon Plants*. Cambridge, Mass.: MIT Press, 1966.

16   Katz, J., ed. *Desarrollo y crisis de la capacidad tecnológica Latinoamericana*. Buenos Aires: BID/CEPAL/IDRC/PNUD, 1986.

17   ———. *Technology Generation in Latin American Manufacturing Industries*. London: Macmillan, 1987.

18   Katz, J., and E. Ablin. "De la industria incipiente a la exportación de tecnología: la experiencia argentina en la venta internacional de plantas industriales y obras de ingeniería." In: E. Ablin, J. Katz, F. Gatto, and B. Kosakoff, eds. *Internacionalización de empresas y tecnología de origen argentino*. Buenos Aires, CEPAL, 1985.

19   Katz, J., and B. Kosakoff. *El proceso de industrialización en la Argentina: Evolución, retroceso y prospectiva*. Buenos Aires: CEPAL/CEAL, 1989.

20   Kennedy, C. "Induced Bias in Innovation and the Theory of Distribution." *Economic Journal* 74 (September, 1964), no. 4.

21   Koutsoyiannis, A. *Modern Microeconomics*. New York: John Wiley & Sons, 1975.

22   Lall, S. *The New Multinationals: The Spread of Third World Enterprises*. Paris: J. Wiley and Sons, 1983.

23   March, J., and R. Cyert. *A Behavioural Theory of the Firm*. Englewood Cliffs, N.J.: Prentice Hall, 1963.

24   Maxwell, P. "Adequate Technological Strategy in an Imperfect Economic Context: A Case Study of the Evolution of the Acindar Steel Plant in Rosario, Argentina." In: Katz, ed. *See* ref. 17.

25   Nelson, R. "Research on Productivity Growth and Productivity Differentials: Dead Ends and New Departures." *Journal of Economic Literature* 19 (1981).

26   Nelson, R., and S. Winter. *An Evolutionary Theory of Economic Change*. Cambridge, Mass.: The Belknap Press, 1982.

27   Nogueira da Cruz. "Una planta brasilera de equipos para el procesamiento de cereales." In: Katz, ed. *See* ref. 16.

28   Nordhaus, W. "Some Skeptical Thoughts on the Theory of Induced Innovation." *Quarterly Journal of Economics* (December 1973).

29   Schumpeter, J. *Capitalism, Socialism and Democracy*. New York: Harper & Row, 1942.

30   Silberberg, G., G. Dosi, and L. Orsinengo. "Innovation, Diversity and Diffusion: A Self-organizing Model." *Economic Journal* 94 (December 1988).

31   Simon, H. "On the Behavioural and Rational Foundations of Economic Dynamics." *Journal of Economic Behaviour and Organization* 5 (1984).

32   Solow, R. *Capital Theory and the Rate of Return*. Amsterdam: North Holland Publishing Co., 1963.

33   ———. "Growth Theory and After." *American Economic Review* 78 (June 1988).

34   Vignolo, C., M. Castillo, A. Animat, R. Donoso, J. Tampier, S. Merino, and J. Weinstein, eds. *La industria chilena: Cuatro visiones sectoriales*. Santiago de Chile: Centro de Estudios de Desarrollo, 1988.

35   Williamson, J. *Latin American Adjustment: How Much Has Happened*. Washington, D.C.: Institute of International Economics, 1990.

8

Technological capabilities

Sanjaya Lall

This chapter is a review of the nature of technological activity in developing countries and the case for government interventions to strengthen technological and, hence, industrial development. Much of the traditional literature, theoretical and empirical, has neglected the need for, and production of, technological activity in developing countries. Simple neoclassical writing simply assumes the problem away. In the highly simplified models used in trade theory, for instance, technology is taken to be freely available to all countries and, within countries, to all firms. Countries simply settle on an appropriate level of capital/labour intensity in accordance with their factor price ratios, determined by their relative endowments of physical capital and labour. Firms in a given industry are all on the same production function and select their techniques, again, with reference to the relative factor price ratio, shifting costlessly along the function as this ratio changes. To the extent that technological lags are admitted, developing countries are assumed to receive all relevant improvements from developed country innovators: there is no problem in assimilating the transferred technology in the developing country; there are no adaptations required, since alternatives are available for all factor prices; all firms remain equally efficient; firm-specific learning or technical effort are unnecessary and irrelevant; and so on [53].

These traditional approaches to technology also assume that innovation (movements of the production function rather than along it)

Slightly revised version of the article "Technological Capabilities and Industrialization" published in *World Development*, Vol. 20, No. 2 (1992), pp. 165–86.

is a completely distinct activity from gaining mastery of technology or adapting it to different conditions (because the only admissible differences between countries in theory are capital/labour ratios, adaptations are necessarily restricted to movements along the function). Innovative activity is an investment in something unrelated to production. In theoretical modelling, such investment is guided by a known innovation possibility frontier, with marginal returns equalized with other returns [52]. The model assumes that major innovations all occur in the advanced industrialized countries; developing countries select and costlessly apply those innovations that are useful or appropriate. As the general level of capital accumulation (and skills) rises, more capital-intensive (or complex) technologies become economic – these are also bought from the international technology shelf.

The general thrust of conventional approaches is to minimize not just the role of technological activity in developing countries, but also the need for policies to support, protect, and induce such activity [62]. What are now termed "neoclassical approaches" to development (associated with Balassa, Krueger, and others) tend to confine themselves to prescriptions like "get prices right," "reduce or eliminate protection," or "free international flows of capital and technology," and cut back on government intervention in industrial activity. Where more moderate neoclassicals admit the need for interventions in industry, they favour neutral (or "functional") rather than selective interventions (i.e. those that support the functioning of markets, like education or R&D, rather than those that promote some industries or technologies over others). These approaches disregard the peculiar nature and costs of technological learning in specific activities, the externalities it generates, and the complementarities it enjoys, which may lead to market failures and may call for a more selective approach to policy than conventional theory admits [47]. Yet selective interventions can be justified within the neoclassical framework if such sources of market failure are taken into account.

In contrast to the analyses just mentioned, a number of "unconventional" approaches to the issues of technology in developing countries have appeared in the past decade or so. These have assigned a central role to indigenous technological effort in mastering new technologies, adapting them to local conditions, improving upon them, diffusing them within the economy, and exploiting them overseas by manufactured export growth and diversification, and by exporting technologies themselves. They can be framed in neoclassical terms but their emphasis is often on reasons why markets are not efficient. This chapter provides a brief review of these approaches. It draws out the industrial policy implications that arise from the specific characteristics of technological development and illustrates their relevance with reference to the experience of the newly industrializing countries (NICs) of East Asia and other, less spectacularly successful, countries.

Firm-level technological capabilities (FTC)

The micro-level analysis of technology in developing countries has drawn inspiration from the "evolutionary theories" developed by Nelson and Winter [55], and explained in Nelson [52, 53] and Dosi [15]. The starting point of these theories is that firms cannot be taken to operate on a common production function. Technological knowledge is not shared equally among firms, nor is it easily imitated by or transferred across firms. Transfer necessarily requires learning because technologies are tacit, and their underlying principles are not always clearly understood. Therefore, simply to gain mastery of a new technology requires skills, effort, and investment by the receiving firm, and the extent of mastery achieved is uncertain and necessarily varies by firm according to these inputs. Furthermore, firms have more knowledge of their "own" technology, less about similar technologies of other firms, and very little about dissimilar alternatives, even in the same industry. They operate, in other words, not on a production function but at a point, and their technical progress, building upon their own efforts, experience, and skills, is (to varying degrees) "localized" around that point [3]. The extent to which firm-level differences in technological effort and mastery occur may vary by industry, by size of firm or market, by level of development, or by trade/industrial strategies pursued.

There is little doubt that as a description of reality, in developed or less-developed countries, the evolutionary approach is far more plausible than the production function approach. As Dosi [15] puts it, evolutionary theories can explain the "permanent existence of asymmetries among firms, in terms of their process technologies and quality of output" (p. 1155). Scale economies and vintage differences in capital goods explain part of this asymmetry, but they "are also the effect of different innovative capabilities, that is, different degrees of technological accumulation and different efficiencies in the innovative search process" (p. 1156). Once firm-level technological change is understood as a continuous process to absorb or create technical knowledge, determined partly by external inputs and partly by past accumulation of skills and knowledge, it is evident that "innovation" can be defined much more broadly to cover all types of search and improvement effort. From the firm's point of view, there is little difference in essence between efforts to improve technological mastery, to adapt technology to new conditions, to improve it slightly or to improve it very significantly – though in terms of detailed strategies, degrees of risk, and potential rewards these efforts will certainly be different.

There are various ways to categorize firm-level technological capabilities (FTC). Drawing upon Katz [36, 37], Dahlman et al. [12] and Lall [44], table 1 shows an illustrative matrix of the major technical functions involved. The columns set out the major FTCs by function, the rows by degree of complexity or difficulty, as measured by the sort of activity from which the capability arises. The categorization is necessarily indicative, since it may be difficult to judge a priori whether a particular function is simple or complex [76]. Nor is it meant to show a necessary sequence of learning. Though the very nature of technological learning (i.e. accumulated experience of problem-solving, aided by external inputs or formal research effort) would seem to dictate that mastery would proceed from simpler to more difficult activities, different firms and different technologies adopt different sequences. This would depend on various factors, described below.

The functions set out in table 1 may not be exhaustive, and not all of them have to be performed for every industrial venture. Even where they are performed, moreover, not all need be undertaken by the firm itself – several specialized services can be bought in from (domestic or foreign) contractors, consultants, or other manufacturing firms. Yet there is a basic core of functions in each major category that have to be internalized by the firm to ensure successful commercial operation. If a firm is unable by itself to decide on its investment plans or selection of equipment processes, or to reach minimum levels of operating efficiency, quality control, equipment maintenance, or cost improvement, or to adapt its product designs to changing market conditions or to establish effective linkages with reliable suppliers, it is unlikely to be able to compete effectively in open markets. Moreover, the basic core must grow over time as the firm undertakes more complex tasks. The ability to identify a firm's scope for efficient specialization in technological activities, to

267

Table 1 **Illustrative matrix of technological capabilities**

Degree of Complexity		Investment		Production			Linkages within economy
		Pre-investment	Project execution	Process engineering	Product engineering	Industrial engineering	
Basic	Simple, routine (Experience-based)	Prefeasibility and feasibility studies; site selection; scheduling of investment	Civil construction; ancillary services; equipment erection; commissioning	Debugging; balancing; quality control preventive maintenance; assimilation of process technology	Assimilation of product design; minor adaptation to market needs	Work flow; scheduling; time-motion studies; inventory control	Local procurement of goods and services; information exchange with suppliers
Intermediate	Adaptive, duplicative (Search-based)	Search for technology source; negotiation of contracts; bargaining suitable terms; info. systems	Equipment procurement; detailed engineering; training and recruitment of skilled personnel	Equipment stretching; process adaptation and cost saving; licensing new technology	Product quality improvement; licensing and assimilating new imported product technology	Monitoring productivity; improved coordination	Technology transfer of local suppliers; coordinated design; S&T links
Advanced	Innovative, risky (Research-based)		Basic process design; equipment design and supply	In-house process innovation; basic research	In-house product innovation; basic research		Turnkey capability; cooperative R&D; licensing own technology to others

268

extend and deepen these with experience and effort, and to draw selectively on others to complement its own capabilities is the hall-mark of a "technologically mature" firm. Before full "maturity" is achieved, firms will vary in their mastery of the various functions involved. While this is true of any economy, it is likely that the typical firm in developing countries, with deficiencies in skills and limited experience of manufacturing, will use the same technology less efficiently than its counterpart in developed countries. Scattered evidence confirms that this is in fact the case, and that such differences also exist between more and less advanced developing countries [61].

Investment capabilities are the skills needed before a new facility is commissioned or existing plant is expanded: to identify needs, prepare and obtain the necessary technology, then design, construct, equip, and staff the facility. They determine the capital costs of the project, the appropriateness of the scale, product mix, technology, and equipment selected, and the understanding gained by the operating firm of the basic technologies involved (which, in turn, affect the efficiency with which it later operates the facility). *Production capabilities* range from basic skills like quality control, operation, and maintenance to more advanced ones like adaptation, improvement, or equipment "stretching" to the most demanding ones of research, design, and innovation. They cover both process and product technologies, as well as the monitoring and control functions included under industrial engineering. The skills involved determine not only how well given technologies are operated and improved, but also how well in-house efforts are utilized to absorb technologies bought or imitated from other firms (on the significance of R&D for assimilating external innovations see ref. 9). *Linkage capabilities* are the skills needed to transmit information, skills, and technology to, and receive them from, component or raw material suppliers, subcontractors, consultants, service firms, and technology institutions. Such linkages affect not only the productive efficiency of the enterprise (allowing it to specialize more fully) but also the diffusion of technology through the economy and the deepening of the industrial structure, both essential to industrial development. The significance of extra market linkages in promoting productivity increase is well recognized in the literature on developed countries (survey in ref. 8).

The emerging empirical literature of FTC in developing countries has touched on various aspects of the development of FTC (apart from the references above, see refs. 11, 82, 26, 80, 32, 63, 19). These need not be reviewed at any length here, but it is worth noting the

269

main influences on the demand for, and supply of, FTC. On the "demand" for efforts to build FTC, the most important factors are threefold. First, there is an inherent need for the development of new skills and information simply to get a new technology into production. This operates regardless of policy regime and provides the elemental drive for firms to invest in capability building; the form that capability building takes depends on the nature of the technology (process or batch, simple or complex, large to small scale).

Second, apart from this inherent pressure for capability acquisition, external factors strongly influence the process. As with any investment decision, the macroeconomic environment, competitive pressures, and the trade regime all affect the perceived returns to FTC development efforts. A stable, high growth environment conduces to higher investments in FTC. So does competition, with international competition probably the most potent inducement to skill and technology upgrading. However, competition is a double-edged sword, and, given the necessary costs of learning, can stifle capability building in newcomers when certain market failures exist. This type of "infant industry" argument is taken up in the next section. Trade orientation also affects the content and pace of FTC development. The evidence (and the present author's comparison of technological development in similar industries in India and Korea: see ref. 44; also 38, 1) suggest that inward-oriented regimes foster learning to "make do" with local materials, "stretch" available equipment for downscale plants, while export-oriented regimes foster efforts to reduce production costs, raise quality, introduce new products for world markets, and often reduce dependence on (expensive) imported technology.

Third, technological change itself, which proceeds continuously in almost all industries in the developed world, stimulates developing country firms to try to keep up. Exposure to competition mediates this incentive, and highly protected firms can delay their upgrading for long periods. Nevertheless, the existence and potential availability of more efficient technologies can create their own incentives to invest in FTC.

On the "supply" side, the ability of firms to produce new capabilities depends on: the size of firm (where technologies are complex and call for large-scale production, large amounts of skilled manpower, or intense technological effort, and particularly where capital markets are deficient); access to skills from the market; organizational and managerial skills in the firm and its ability to change structures to

270

absorb new methods and technologies [37, 33]; access to external technical information and support (from foreign technology sources, local firms and consultants, and the technology infrastructure of laboratories, testing facilities, standards institutions, and so on); and access to appropriate "embodied" technology, in the form of capital goods, from the best available sources, domestic or foreign.

In sum, FTC development is the outcome of investments undertaken by the firm in response to external and internal stimuli and in interaction with other economic agents, both private and public and local and foreign. Thus, there are factors that are firm-specific (leading to micro-level differences in FTC development and to "idiosyncratic" results) and those that are common to given countries (depending on their policy regimes, skill endowments, and institutional structures). It is these common factors to which we now turn.

National technological capabilities

Let us now consider national technological capabilities in developing countries. National capabilities are not simply the sum of thousands of individual firm-level capabilities developed in isolation. Because of externalities and interlinkages, there is likely to be synergy between individual firm-level capabilities. Despite individual idiosyncracies, there is a common element of response of firms to the policy, market, and institutional framework. It makes sense, in other words, to conceive of national differences in technological capabilities. Clearly, countries – developing or developed – differ in their ability to utilize or innovate technologies, and this difference manifests itself in their productivity, growth, or trade performance. There is little by way of theory that brings together all the factors that may influence these variables (but see refs. 21, 22, 24, 25, 57, 59). The analysis of national technological capabilities is nevertheless important because of the current dominance of some partial explanations of industrial success, which may lead to misleading policy conclusions [46, 47]. In particular, it is necessary to look again at approaches that, as mentioned in the introduction, trace success to "getting prices right" and non-interventionist strategies, treating them as both necessary and sufficient conditions. These approaches are based on particular readings of technological capability and the efficiency of markets in developing countries.

The OECD explains long-term differences in the performance of advanced industrial economies thus:

271

Over the longer term, economic growth arises from the interplay of *incentives* and *capabilities*. The capabilities define the best that can be achieved; while the incentives guide the use of the capabilities and, indeed stimulate their expansion, renewal or disappearance. In the advanced economies, the capabilities refer primarily to the supplies of human capital, of savings and of the existing capital stock, as well as to the technical and organizational skills required for their use; the incentives originate largely in product markets and are then more or less reflected in markets for factor supplies – thereby determining the efficiency with which capabilities are used. Both incentives and capabilities operate within an institutional framework: institutions set rules of the game, as well as directly intervening in the play; they act to alter capabilities and change incentives; and they can modify behaviour by changing attitudes and expectations. [57, p. 18]

This three-pronged approach, involving the interplay of capabilities, incentives, and institutions, is a useful way of organizing the numerous factors that influence national technological capabilities in developing countries [46].

Capabilities
At the country level, capabilities can be grouped under three broad headings: physical investment, human capital, and technological effort. These three are strongly interlinked in ways that make it difficult to identify their separate contributions to national performance [52], but they do not always go together. If physical capital is accumulated without the skills or technology needed to operate it efficiently, national technological capabilities will not develop adequately; or if formal skills are created but not combined with technological effort, efficiency will not increase dynamically (see ref. 67 for a theoretical analysis); and so on. Physical investment is in some sense a "basic" capability, in that plant and equipment are clearly necessary for industry to exist, but it is the efficiency with which capital is utilized that is of greater interest. The ability to muster the financial resources and the embodied technology that make up physical investment (and the need for an efficient financial system to support this) need not be spelled out at any length here.

The term "human capital" is used broadly here to include not just the skills generated by formal education and training but also those created by on-the-job training and experience of technological activity, and the legacy of inherited skills, attitudes, and abilities that aid industrial development. Literacy and primary education are essential for all forms of efficient industrialization, and may be largely suf-

272

ficient for early industrial efforts utilizing simple technologies [48]. However, as more sophisticated technologies are adopted, the need for more advanced, specialized skills on the part of both workforce and managers emerges [75]. Moreover, the gap between the workforce and engineers has to be reduced to facilitate skill transfer [50]. The quality of formal education, especially of technical training, and the relevance of the curriculum to changing technical needs are clearly very important. To the extent that public or private training facilities do not meet the need for such skills, firms have to invest in their own training facilities, but will do so only if mobility is low and their investments yield appropriate benefits [40]; low mobility thus has this benefit but is offset by the restraint it places on the diffusion of knowledge. Ergas [21] and the OECD [57] outline the very different systems dealing with these problems in the United States, Germany, and Japan, each with its own strengths and weaknesses.

The final capability relates to national technological effort. Trained labour and physical capital are fully productive only when combined with efforts by productive enterprises to assimilate and improve upon the relevant technology. As discussed earlier, such effort comprises a broad spectrum of production, design, and research work with firms, backed up by a technological infrastructure that provides information, standards, basic scientific knowledge, and various facilities too large to be owned by private firms. It is impossible to measure properly such technological effort, but rough proxies are available in the form of technical manpower available for technical tasks, or expenditures on formal R&D (input measures), or innovations, patents, and other indicators of technological success (output measures). The interpretation of all such measures is fraught with difficulties [8], since not all effort is equally efficiently made, and no measure captures fully routine engineering work devoted to minor innovation or mastery. Nevertheless, it is evident that different countries devote different levels of effort to technology (see refs. 21, 57, 13, 59 on developed countries, and 78, 35, 46 on developing countries), and even a crude measure is of some use.

Apart from domestic technological effort, the extent and nature of a country's reliance on foreign technology is also directly relevant to national technological capabilities. All countries need to import technology, but different modes of import have different impacts on local technological development. In semi-industrialized countries, for instance, a heavy reliance on foreign direct investment (FDI) may become a substitute for domestic effort at the "advanced" levels shown in table 1 above, because FDI is an efficient means to transfer the results of innovation rather than the innovative process itself. The alternative strategy, following the example of Japan, of building a strong domestic technological base, may therefore entail a selective curtailment of FDI entry, at least at certain stages of the development process (see below).

Incentives

While both physical and human capital are necessary for industrial development, they will not be utilized effectively if the structure of incentives for investment and production is inappropriate. Incentives, arising from market forces, institutional functioning, and government policies, affect the pace of accumulation of capital and skills; the types of capital purchased and the kinds of skills learnt; and the extent to which existing endowments are exploited in production. In most developing countries, the role of policies assumes great importance, in positive as well as negative ways: positive because structural and market failures call for remedial action, negative because interventions can be excessive or misjudged, and even justifiable interventions can be poorly administered.

Three broad sets of incentives affect the development of national technological capabilities:

1. MACROECONOMIC INCENTIVES. Under this heading, I include signals that emanate from GNP growth (rate and stability), price changes, interest rates, exchange rates, credit and foreign exchange availability, and similar economic variables, as well as political stability or exogenous shocks (e.g. terms of trade). The impact of growth, stability, sensible balance of payments, monetary or fiscal policies, favourable external circumstances, etc., on investment and capability building are obvious and need not be discussed in detail here.

2. INCENTIVES FROM COMPETITION. Competition is, as discussed earlier, the most basic of incentives affecting capability development. Domestic competition is influenced by the size of the industrial sector, its level of development and diversification, and government policies on firm entry, exit, expansion, prices, ownership, small-scale industry, and so on. Most developing countries impose constraints on internal competition to prevent excessive entry (and so fragmentation) in protected markets, to preserve employment,

274

to promote small firms or public enterprises, to hold down prices, to force industry to locate in backward areas, or to prevent the growth of large-size firms or the concentration of economic power. Some industrial regulation is clearly necessary in every economy, but high levels of intervention can frustrate or dissipate the development of healthy capabilities and prop up non-viable enterprises that should die out (see ref. 88 for a brief review of the most common types of competition-retarding policies).

International competition – from imports, entry of foreign investors, or export activity – can be an even greater stimulant to healthy technological development than domestic competition, in small or large countries (size of economy does not affect whether or not enterprises in the country are exposed to such competition). Yet governments place many barriers to such competition, often in a sweeping, irrational, and prolonged way that retards technological development, efficiency, export growth, and structural change. The recent development literature has analysed the costs of inward-oriented trade strategies at great length (for a useful review, see ref. 87). Most of the conventional arguments are not couched in terms of the impact of trade strategies on technological capabilities, but the implicit assumptions made about technological capability development are relevant to the issue.

The debate over intervention in trade flows is of long standing (review in ref. 5). While acknowledging the benefits of market competition, economic theory accepts that interventions in the incentive framework of free trade, in the form of infant industry protection or promotion, are needed to overcome many (but not all) market failures affecting resource allocation [82, 83, 62, 46]. It is important to be clear about the correct case for such intervention. Some arguments for protection are misplaced: if the source of market failure lies outside the firm (e.g. lack of skills, infrastructure, institutions), intervention to protect the firm will do nothing to ensure that costs come down over time. However, to the extent that failures arise from the firm's own lack of investment in capability building, due to externalities (loss of skills or technology or interdependencies between firms [62]), risk aversion, or lack of information (due to missing information markets or "learning to learn" phenomena [71]), intervention may have a justifiable role to play in restoring efficient resource allocation.

The intervention may not necessarily take the form of import protection. Theory suggests that subsidies are preferable because

275

they involve lower consumption costs than import restrictions. But protection is easier (and cheaper) for the government to administer, and historical evidence suggests that tariffs have been used by every developed country in critical stages of industrialization [80]. While protection has often been misused, as the trade strategy debate shows, it has also accompanied entry into difficult and complex activities with high learning costs. In fact, the existence of such costs in developing countries (with imperfect capital and information markets and strong linkages and externalities) suggests that protection is a necessary condition for development beyond technologically simple activities. However, it may not usually be sufficient, because market failures in factor markets and institutions (see below) can hold back full gains in efficiency.

Such interventions have to be selective, requiring that policy makers identify specific sectors, activities, or even firms for promotion over others to exploit their superior growth potential, linkages, or externalities. There are two basic requirements for such intervention to be effective. First, since protection itself reduces incentives to invest in FTC, it should not be too widespread, indiscriminate, or prolonged, and should be offset by other incentives for increased efficiency. The best combination may be the selective and temporary protection of domestic markets, together with strong incentives for export activity and domestic competition. Second, policy makers should be able to identify suitable activities for protection and have the authority to correct mistakes and modify choices over time (i.e. shut down inefficient operations). This requires considerable informational and organizational resources, as well as political strength, on the part of the government. Some countries can provide such resources, but many cannot; I return to this below.

3. INCENTIVES FROM FACTOR MARKETS. Theory suggests that well-functioning, flexible factor markets and correct relative factor prices are necessary to achieve efficient production and resource allocation. Efficiency in capital markets requires that long-term financing be available, especially for risky projects involving new technologies, and that price signals achieve proper inter-firm and inter-industry resource allocation. Efficient labour markets should be responsive to changing needs, not hampered by restrictive practices, and be equipped with requisite skills. Similarly, efficient technology markets should provide adequate flows of information

to enterprises as well as of "public goods" such as standards, testing facilities, and basic research. In general, incentives should be sufficient to ensure that private firms do not under-invest in their own technological development. Where market failures occur and firms invest less than is socially desirable, governments must be able to step in to enable firms to internalize markets (e.g. provide self-financing or subsidize training of workers) and to remedy the failures directly by providing finance (loans, venture capital financing, R&D subsidies, and so on) to firms or activities where social returns exceed private returns. Such interventions are often regarded as functional rather than selective, and so are considered with greater favour by those who mistrust selectivity ("picking winners") by governments. However, the distinction is often spurious. Interventions in finance, education, research, information, or retraining are generally selective above a certain (fairly low) level: for instance, after providing for general levels of secondary education, the training of university level engineers may need to be guided towards specific industrial needs. Given resource limitations, selectivity in industrial support is inevitable. But there is a stronger case for selectivity in factor market interventions: some activities have greater linkages and externalities than others. As Grossman [30] argues, "When market activity is too low relative to an efficient outcome, it is because the active and potentially-active firms fail to appropriate all the benefits from some aspect of their operation. Corrective government policy should be targeted to the particular activity that generates positive spillovers, and not merely encourage firms to produce more output" (p. 118).

Institutions
The development of capabilities and the play of incentives express themselves only through specific market and non-market institutions. If markets throw up the necessary institutions naturally, there is no need to consider them separately. If they do not, however, the development of a proper institutional framework becomes an area of concern. Since development is almost definable by the deficiency of institutions, clearly the subject requires consideration. Of the vast array of institutions that affect economic life, I note only those that are external to firms and that most directly affect industrial capabilities. In addition to the legal framework supporting industrial activity and property rights, these are: industrial institutions (those that promote inter-firm linkages in production, technology, or training, or

provide support to smaller enterprises, or help firms to restructure and upgrade); training institutions (where firms under-invest in training or fail to provide the right kind or quality of training); and technology institutions (on the US, see refs. 72, 73, 54; on Japan, 28, 51, 59; and 21, 57, on developed countries in general).

National technological capabilities: Some evidence from developing countries

This section applies the above framework to a selection of eight industrializing countries: the four East Asian NICs (Korea, Taiwan, Hong Kong, and Singapore), India, the two dominant Latin American industrial economies (Brazil and Mexico), and one second-tier NIC, Thailand. These give a fair coverage of the different types of countries that have achieved a measure of success with industrial development. There is also some consensus about their strategies and achievements, which makes a classification possible to incorporate relevant elements that cannot be easily quantified.

Despite their obvious importance, institutions are not considered here because it is practically impossible to compare institutional structures and performance across countries.

Table 2 sets out some relevant data on industrial structure and performance and two sets of determinants of national technological capabilities on which figures could be obtained: education and science and technology [46]. The top section of the table is intended to provide background information and illustrates some features of the sample countries. The four East Asian NICs are the most dynamic and efficient (in terms of international competitiveness) of the group. There are, however, significant differences between their industrial structure, export specialization, and reliance on overseas investment (these are taken up below). Of the larger countries, Brazil has the biggest industrial sector, with an advanced technology in many areas of heavy industry; however, it has large areas of uncompetitiveness [10], a high foreign presence in modern industry, and a large public sector. Mexico is similar in many ways, but has a smaller capital goods capability, a higher foreign presence, and a lower manufactured export base. India's industrial sector is very diverse but riddled with inefficiency and technological obsolescence; it has suffered low rates of growth of exports and value added (until very recently), but has the distinction of having a very low level of reliance on foreign investment and technology imports in other forms [44, 43, chap. 10].

278

Finally, Thailand is a relative newcomer, with a shallow industrial base but very dynamic export growth based on the relocation of labour-intensive activities away from Japan and the older NICs.

The pattern is well known: indeed, such diversity of industrial performance, as typified by the relative success of the East Asian NICs (and the emergence of "new NICs" in the region), has prompted much theorizing on the virtues of liberal trade strategies [87]. Our framework suggests that simple incentive-based explanations may be partial and misleading, but let us look at the available evidence.

Incentives

Macroeconomic management has, with one hiccough in Korea in 1979–1980, been excellent for the four East Asian NICs and Thailand, moderately good for India, and poor for the two Latin American NICs. Their trade strategies are well known: consistently highly export-oriented (i.e. with incentives that were neutral between domestic and export markets, or biased in favour of the latter) over a long period for the East Asian NICs, with little or no protection in Hong Kong and Singapore but with selective, variable, and often high protection for several industries in Korea and Taiwan; more inward-oriented for Brazil and Mexico, with large areas of high effective protection, but with export incentives to partially offset the bias; highly and consistently inward-oriented for India; and increasingly export-oriented for Thailand, but still with remnants of protected import substitution. At the trade strategy level, therefore, export-oriented strategies seem to be positively correlated with industrial success, supporting the arguments of the liberal school that competition in international markets stimulates efficient specialization and healthy FTC development; in addition, it is suggested, export orientation provides free inflows of information from world markets, gives greater and more stable access to foreign technology and equipment, and is associated with lower rent-seeking behaviour [4, 56].

However, these simple categorizations of "export orientation" may be misleading depictions of strategies that are much more complex in their impact on national technological capabilities. There are several varieties of export orientation [18]. Hong Kong is at one extreme, with fully *laissez-faire* economic policies combined with stable administration, a strong presence of British trading and financial enterprises (with considerable spillover benefits), a concentration of textile-related skills and technology (from Shanghai), and a long tradition of entrepôt trade (which created a variety of contacts and

279

Table 2 **Indicators of national technology capability in selected NICs**

	South Korea	Taiwan	Hong Kong	Singapore	India	Brazil	Mexico	Thailand
A. Structure and performance								
1. Mfg. value added $b. (1985)	24.5	22.2	6.7	4.3	35.6	58.1	43.6	7.7
Mfg. growth 1965–80/1980–86	18.7/9.8	16.4/12.9	17.0/7.0	13.3/2.2	4.3/8.2	9.6/8.2	7.4/0.0	10.9/5.2
2. Mfd. exports (1986) $b. (1986)	31.9	35.9	32.6	14.7	7.2	9.1	4.9	3.9
Growth of merchandise exports: 1965–80/1980–86	27.3/13.1	19.0/12.7	9.5/10.7	4.7/6.1	3.7/3.8	9.4/4.3	7.7/7.7	8.5/9.2
3. Gross domestic investment as % GDP (1986)	29	19	23	40	23	21	21	21
4. Capital goods prod. as % of total mfg. (1985)	23	24	21	49	26	24	14	13
5. Capital goods imports $b. (1985)	10.6	5.6	7.1	8.1	3.7	2.2	6.1	2.7
(as % MVA)	(43.3)	(25.2)	(106.0)	(188.4)	(10.4)	(3.8)	(14.0)	(35.1)
6. Stock of foreign direct investment $b. (1984–86)	2.8	8.5	6.0–8.0	9.4	1.5	28.8	19.3	4.0/5.0
7. FDI stock as % GDP	2.8	8.1	20–26	53.8	0.7	9.6	13.6	10.5–13.1
B. Education								
1. (a) Education expenditure as % household consumption (1980–85)	6	n/a	5	12	4	5	5	6
(b) Public expenditure % GNP (year)	4.9 (1985)	5.1 (1986)	2.7 (1978)	2.9 (1980)	3.7 (1985)	2.9 (1984)	2.6 (1985)	3.9 (1984)
2. Central government expenditure on education % total government expenditure (1986)	18.1	20.4	n/a	21.6	2.1	3.0	11.5	19.5

3. % Age group enrolled (1985)								
–primary	96	100	105	115	92	104	115	97
–secondary	94	91	69	71	35	35	55	30
–tertiary education	32	13	13	12	9	11	16	20
4. Vocational ed. enrol. (1984) nos. ('000s)	815	405	32	9	398	1,481	854	288.0
as % population working age	3.06	3.24	0.86	0.5	0.07	1.83	2.0	0.96
5. No. of tertiary level students								
–in S/E fields ('000s)	585	207	36	22	1,443	535	563	360
% population	1.39	1.06	0.67	0.89	0.21	0.40	0.70	0.70
(year)	(1987)	(1984)	(1984)	(1984)	(1980)	(1983)	(1986)	(1985)
–in engineering ('000s)	228	129	21	15	397	165	282	n/a
% population	0.54	0.68	0.41	0.61	0.06	0.13	0.35	
C. Science and technology								
1. Patents granted: total (1986)	3,741	10,615	n/a	598	2,500	3,843	2,005	n/a
of which % local	69	56	n/a	8	20	9	9	n/a
2. R&D % GNP	2.3	1.1	n/a	0.5	0.9	0.7	0.6	0.3
(year)	(1987)	(1986)		(1984)	(1984)	(1982)	(1984)	(1985)
3. R&D in productive sector % GNP	1.5	0.7	n/a	0.2	0.2	0.2	0.2	n/a
4. R&D financed by productive enterprises % GNP	1.9	0.6	n/a	0.2	0.1	0.1	0.01	0.04
5. Scientists/engineers in R&D per million population	1,283	1,426	n/a	960	132	256	217	150
6. All scientists/engineers								
(a) Total nos. ('000s)	361.3	n/a	145.5	38.3	1,000–2,000	1,362.2	565.6	20.3
(b) Per million population	8,706	n/a	26,459	15,304	1,282–2,564	11,475	10,720	472
(year)	(1986)		(1986)	(1980)	(1985)	(1980)	(1970)	(1975)

Sources: refs. 2, 23, 64–66, 78, 79, 86–88.

skills).. Singapore offers no protection, but intervenes heavily in several ways, in guiding investment, setting up public enterprises (these account for 10 per cent of value added in manufacturing), directing wages, encouraging savings, and so on [41]. It permits only very selective immigration (of skilled personnel) and is generally highly involved in guiding the economy's development, especially by inducing foreign investors to upgrade the skill and capital intensities of the projects they undertake. As a result, the industrial structures of the two island economies differ quite sharply [41]. Hong Kong has remained specialized in light consumer goods, essentially assembling imported components, while moving up the quality scale – its industry does not have great technological depth or high vertical linkages [7], and competitive pressures are forcing it to relocate in cheap-labour areas (chiefly China) rather than to deepen domestic industrial activity. Singapore has a much "heavier" industrial structure, with strong emphasis on producer goods and very high requirements of technical skills.

Korea and Taiwan have been much more interventionist, with the former traditionally far more so than the latter [42, 81]. Until the 1980s, the Korean government protected and promoted selected (strategic) industries highly, sometimes set up public enterprises (like its highly efficient Pohang steel plant), directed investment at the sectoral and, often, the firm, level, promoted exports by several direct measures, intervened in technology transfer agreements and technology development (as in petrochemicals, see ref. 20), restructured industries, and enforced labour training [62, 86, 1, 83]. Even today, despite considerable liberalization, a strong element of "guidance" remains in Korea. Taiwan also protected emerging industries, guided expansion along particular lines, and had a very active technology development policy [81, 34]. However, the Korean strategy was more specifically directed at creating and supporting giant firms (the *chaebol*) that could internalize many inefficient markets, though at the risk of a high level of government direction and the rigidities associated with size. Taiwanese strategy concentrated on providing support to small and medium-size firms, providing great flexibility but holding back large, risky investments in technology by the firms themselves. It was perhaps a safer, more "incremental," strategy, while the Korean one was more risky but permitted larger leaps into high-tech activities. In the production of semiconductor (DRAM) chips, for instance, Korean *chaebol* were able to cross-subsidize and enter into production and export in a major way with little explicit government support [39, 49]; an electronics research institute set up to launch semiconductor technology was quickly bypassed as the *chaebol* went directly into production with massive facilities. The Taiwanese government, on the other hand, had to adopt a far more interventionist strategy because of its earlier "hands off" stand on promoting firm size. Its DRAM production facility was set up by a public sector firm, and the government had to coordinate related technology import, design, manufacture, and marketing by several private firms. In effect, "the government is doing in Taiwan what forward and backward integration does for companies" [68, p. 67].

The large countries were also very interventionist in their industrial and technology policies. Brazil promoted several large public research organizations, and its giant public enterprises invested in R&D. It intervened in technology imports to support the development of local capabilities in the selected industries (the best-known case being minicomputers). Despite its heavy investments and major successes in some specifically targeted areas (aircraft, minicomputers, special steels, armaments), however, Brazilian strategy in technology development was to a large extent ineffective in achieving competitiveness for large parts of industry [10]. Mexico also pursued policies to build up domestic industry behind import protection, but did not adapt Brazilian-style interventions to develop specific technologies; it also lagged in the development of local capital goods. As a result, Mexican technological prowess is generally considered to be behind Brazil's.

India's industrial strategy has remained highly interventionist within its import substitution approach. The Indian government was suspicious of private enterprise in general, and large private firms and foreign investors in particular; and barriers to entry, exit, growth, and diversification were rife. It set up a large network of science and technology institutions, but these were divorced from manufacturing enterprises and excessively bureaucratic. The administration of its policies was slow, complex, and prone to corruption.

Capabilities

Let us start with human capital, measured via data on *education*. Based on 1958–1959 data, Harbison and Myers [31] developed a composite index of human resource development in a large international sample of countries. At that time, Argentina emerged with the highest rank in the developing world, followed by Korea and Taiwan. Then (of our sample) came India, Mexico, and Brazil (others were not included). In 1965, enrolment in secondary schools (as a percentage of the relevant age group) was distinctly higher in East Asia (Korea 35%, Taiwan 38%, Hong Kong 29%, Singapore 45%) than in other countries (Brazil 16%, Mexico 17%, India 27%, or Thailand 14%). Enrolment in tertiary education was also ahead (6%, 7%, 5%, 10% respectively in East Asia, 2% and 4% in Latin America, 2% in India and Thailand).

By 1985, the East Asian lead in secondary education had been maintained or widened, while that in tertiary education had been narrowed or eroded – with the exception of Korea. Mexico and Thailand had made particularly large gains in tertiary education. (It should be noted that, according to Unesco data, Hong Kong and Singapore have large proportions of students in higher education overseas, 32% and 25% respectively, so the figures in table 2 are underestimates.) India has the smallest stock. In Latin America, Mexico is ahead of Brazil. Thailand is expanding very rapidly from a low base.

Figures for enrolment in education by themselves may be misleading. The true impact on technological capability development also depends on the drop-out rate, the technical orientation of the students, and the quality of teaching. Drop-out rates are exceptionally low in East Asian NICs [38, 58]. The technical orientation of education is highest in Singapore (60% of tertiary students are in science and technology subjects), followed by Mexico (48%), Hong Kong (46%), Korea (42%), Brazil (36%), India (27%), and Thailand (21%). There is no information on Taiwan, but we can safely assume the figure to be high.

More important is the proportion of each country's population enrolled in science and engineering. This broad measure of technological capacity is led by Korea (1.39), followed by Taiwan (1.06), Singapore (0.89), Mexico (0.7), Hong Kong (0.67), Brazil (0.4), and India (0.21). Allowing for students abroad (and taking the proportion in science and engineering to be the same as at home), the figures for Singapore and Hong Kong rise to 1.01 and 0.81. Korea's rises to 1.41, while those of others (Taiwan data are missing) are not affected. Taking engineering on its own, Taiwan leads the sample, followed by Singapore and Korea. These three NICs have figures some 10 times higher than India's, and four or five times higher than Brazil's.

The only relevant indicators of the quality of education are of the performance of primary and secondary school students on the *Inter-*

national Education Review's test scores in science and mathematics. In one test, administered in 19 (mostly developed) countries, with only Korea and India included from our sample (quoted in ref. 85), Korea came second only to Japan in nearly all tests, and in one it beat Japan. It consistently outperformed countries like the United States, Britain, Germany, Sweden, Austria, and so on. Its primary school pupils did 2.5 times better than India; its secondary school pupils 3.8 times better. In another test, reported in OTA [59], two other sample countries, Hong Kong and Thailand, were included in a sample of 14, again mostly developed, countries. Twelfth graders (17–18 year olds) were tested in geometry and algebra in the mid-1980s. The top performer in both was Hong Kong, followed by Japan. The United States came twelfth in geometry and thirteenth in algebra. Thailand came last in both tests. These tests should, however, be treated with caution, because they may not be robust indicators of educational standards across the board.

The technical competence of an industrial workforce is improved by education imparted by various formal training systems and by in-firm training. While the precise nature of the benefits of vocational as opposed to general training, and pre-employment as opposed to post-employment training, is still the subject of debate [17], it is indisputable that the speed of technical change in modern industry necessitates increasing inputs of training and retraining. Data are most readily available on vocational training (from Unesco); these are shown in table 2, in total and in relation to the size of the population. Korea and Taiwan are far in the lead (over 3% of the population of working age), exceeding relative levels in Latin America (about 2%) and other East Asian NICs. Singapore is also relatively low (0.5%), but this is misleading because of the large size of its employee training programmes run on a cooperative basis by government and industry. Hong Kong has a relatively poor showing (0.86%), behind that of Thailand, reflecting the specialized and technologically undemanding nature of its industrial structure. India has very small enrolments, suggesting widespread skill deficiencies.

In-firm training figures are not widely available, but McMahon [48] singles out Korea as an exceptional case, in that "since 1960 South Korea has insisted that companies spend at least 5 to 6% of their total budget on education and training programs, involving the private sector in the education process in a meaningful way" (p. 19). It is doubtful whether any other country in the sample has a training effort comparable with this: presumably, it has provided the basis for efficient production in Korea's rapid drive into new, demanding industries.

The impressions that emerge from these data are:

- The East Asian NICs have the largest stock of human capital in a broad sense (formal education at secondary and tertiary levels). They are followed by Mexico, then Brazil and Thailand, with India clearly at the bottom.
- In terms of technical education and vocational training, Korea and Taiwan are clear leaders (with Korea pulling ahead at general high level, and Taiwan in engineering, education), with Singapore close behind. Hong Kong comes next, followed by Mexico, then Brazil or Thailand (depending on the measure), with India again lagging well behind.
- In terms of the quality of education, patchy evidence suggests that the East Asian NICs, with their strong cultural emphasis on education, are ahead of the others.
- In firm-level training, Korea is likely to be the leader. Singapore leads in employee training provided externally.

These impressions conform broadly to the patterns of "revealed national technological capabilities" discussed earlier. While the most successful countries have the largest investments in human capital formation, preceding and accompanying their industrial growth, Korea and Taiwan are in a different class from Hong Kong and Singapore. The larger relative technical-skill endowments of the former two explain their greater ability to tackle more complex, demanding industrial technologies. Hong Kong is distinctly behind Singapore, which conforms to the observed differences in their industrial structures and technological prowess. Interestingly, Singapore's heavy reliance on foreign investors in its high-tech industries does not relieve it of the need to provide educated and trained technical manpower; multinational corporations are able to set up such industries there only because of the availability of appropriate manpower (and Singapore is widely regarded as having one of the world's best systems for employee training).

Mexico seems to have a better trained workforce than Brazil by every measure. Its apparent lag in national technological capabilities must then be attributed to specific industrial and technological policies, which have failed to develop technological capabilities (at least in selected areas) as forcefully as Brazil. India's substantial lag in human resources may appear surprising, because of the general aura it has of a country with an oversupply of technical and educated manpower. There is certainly a large absolute supply (although of highly variable quality), and graduate unemployment and emigration are real problems. In relation to the size of the economy, however, the stock is poor, and what there is seems to be concentrated in the larger establishments. The apparent oversupply is more a reflection of the economy's poor performance than anything else: wrong policies have held back even the absorption and effective utilization of its meagre human resources.

Science and technology

The most common measure of national technological effort is total R&D spending in relation to GNP. By this measure, sample data (not available for Hong Kong) show that Korea, with 2.3% in 1987, is now well ahead of the others (more than double that of Taiwan, its nearest rival) and planning to reach 5% by 2000. Taiwan and India are close to each other, around 1%, followed by Brazil and Mexico, Singapore and Thailand.

Total R&D expenditures may be less relevant a measure of industrial technical effort than R&D performed or financed by productive enterprises. Total R&D includes large elements of non-industrial R&D, or industrial R&D performed in government laboratories, or performed in productive enterprises but financed by others. Each has different implications for industry, in terms of effectiveness, control, and relevance. It is usually a safe assumption that R&D effectiveness is higher the more it is performed and financed by productive enterprises (Griliches [29] finds, for instance, that privately financed R&D in the United States yields much higher returns than R&D financed by the Federal government and performed by the same enterprises). On this criterion, table 2 shows again that Korea is far in the lead, with Taiwan some distance, and other countries much further, behind. The bulk of Korean private R&D is performed by its giant *chaebol*, themselves the products of earlier policies to select, protect, and subsidize large firms to lead the industrialization drive. In this sense, even the private R&D of Korea is traceable to selective intervention: to create *chaebol*, direct them into heavy and complex activities, and force them to compete internationally.

Patent data are also available but are notoriously difficult to compare meaningfully. Nevertheless, the figures on the proportion of patents taken out by residents (which may include foreigners) are suggestive [23]. Korea and Taiwan (69% and 56%) are far ahead of

India (20%), Brazil and Mexico (9% each), or Singapore (8%). The commercial value of these patents may be questionable, but it is instructive in this context to refer to Fagerberg's [25] growth accounting exercise, which used patents taken out internationally as a measure of innovative activity, and included Asian NICs (Hong Kong, Korea, Taiwan) as well as Latin American NICs (Argentina, Brazil, Mexico) as sub-samples.

Fagerberg's calculations showed that both groups of NICs grew faster than the "frontier" countries (United States, Switzerland, Germany, Japan, Sweden), East Asia 6% faster and Latin America 1.9% faster. The difference between the two subgroups was primarily due to their innovative efforts. For Asian NICs, this contributed 2.9% of their relative growth performance, for Latin America −0.1%. Such exercises suffer from well-known limitations and interpretation problems, but the general results are plausible and in line with other sorts of evidence: innovative effort is important for growth even among NICs, and East Asia performs far better than Latin America.

The employment of scientists and engineers in R&D in relation to population is another common measure of technological effort. The figures in table 2 show Taiwan ahead of others (1,426 per million population in 1986, higher than France's 1,365 in 1984). Korea is a close second with 1,283, followed by Singapore with 960. There is then a large gap, with Brazil and Mexico having 256 and 217 respectively. Thailand has 150 and India 132. The quality of R&D scientists and engineers may differ by country, and their economic value may depend on the type of R&D they are engaged in, but there is no reason to believe that, as far as NICs are concerned, these factors would reduce the apparent lead of East Asia. If anything, they would strengthen it.

A similar measure of the total "potential stock" is the number of scientists and engineers. The data (taken from Unesco, which collects the figures by questionnaires) are sometimes dubious (especially for Hong Kong, where they appear to be overestimates), but they show the two island NICs of Asia with the highest stocks, followed by Brazil, Mexico, and Korea. India comes out ahead of Thailand on this measure, but well behind the others.

The technological data broadly support the trends revealed by the figures on education. The Asian NICs, in particular Korea and Taiwan, have invested not only in educating and training their populations, but also in technological innovation. This investment was primarily oriented to the commercial needs of productive enterprises, and has drawn upon a large pool of scientists and engineers. Combined with a highly skilled workforce, these investments yielded the competitiveness and dynamism that revealed themselves in growth and export performance. Export orientation played a permissive and stimulative role, and as such was necessary – but it was not sufficient.

Technology imports
All sample countries import large amounts of technology, but their patterns of import differ greatly. In part this is due to differing rules and controls on buying know-how and services abroad: the international technology market is subject to a spectrum of failures caused by asymmetric information, opportunism, missing markets, and so on, and different governments have adopted different measures to overcome such failures and help national enterprises to purchase technology on fair terms. In part, however, it is due to a more fundamental difference, national technological strategy. This concerns the relative roles of foreign and local enterprise in building indigenous capabilities. There are striking variations across the leading semi-industrial countries in the extent to which they have drawn on foreign direct investment (FDI) to provide technology and skills.

FDI can, in appropriate conditions, be a very efficient means of transferring a package of capital, skills, technology, brand names, and access to established international networks. It can also provide beneficial spillovers to local skill creation and, by demonstration and competition, to local firms. Where local skills and capabilities are inadequate, FDI can sometimes be the only means to upgrade technologies and enter high-tech activities. However, the very fact that FDI is such an efficient transmitter of packaged technology based on innovative activity performed in advanced countries has serious implications. With few exceptions, the developing country affiliate receives the results of innovation, not the innovative process itself: it is not efficient for the enterprise concerned to invest in the skill and linkage creation in a new location. The affiliate, in consequence, develops efficient capabilities up to a certain level, but not beyond: in the literature this is called the "truncation" of technology transfer. Such truncation can diminish not only the affiliate's own technological development, but also its linkages with the host country's technological and production infrastructure, and so beneficial externalities. Moreover, a strong foreign presence with advanced technology can prevent local competitors from investing in deepening their own capabilities (as opposed to becoming dependent on imported technology or, where the technology is not available at reasonable prices, withdrawing from the activity altogether).

For these reasons, countries with technological potential may find it beneficial to restrict FDI and import technology in "unpackaged" forms (including foreign minority-owned joint ventures). The choice of mode of technology imports is thus not neutral – some are more beneficial than others for certain strategies and at certain stages of development. The sample countries cover the whole range of FDI strategies. Rows A6 and A7 of table 2 set out data on stocks of foreign investment in each country and on FDI as a percentage of GDP in the relevant year as a measure of the relative significance of FDI. It shows, at one extreme, low levels of reliance on FDI by India and Korea, and, at the other, very high levels by Singapore and Hong Kong, and fairly high levels, among large countries, by Mexico, Thailand, and Brazil. The interesting cases are those of Korea and Singapore, both successful NICs that have opted for opposing strategies on foreign capital.

South Korea has developed arguably the most advanced and competitive base of technological capabilities in the developing world, drawing on foreign technology mainly in non-equity forms (i.e. by capital goods imports, licensing and minority foreign ventures [84, 83]). In order to nurture this massive effort, it followed the Japanese example of some decades earlier – protection against imports and selective exclusion of foreign investment, accompanied by the upgrading of skills, huge investments in R&D, and the sponsoring of the giant *chaebol* to internalize various markets and so cope with the rigours of international competition. The strategy may be characterized as one of "protecting domestic technological learning" at a stage of development when externalities and uncertainties abound, information linkages are imperfect, and basic capabilities are in their infancy. This stage is similar in many respects to the micro-level process of developing a new innovation by a developed country firm, when (as Grossman [30] argues) "the strongest case for government intervention may arise . . . [because this would] involve substantial research outlays and costly learning-by-doing [and] private firms often are unable to capture more than a fraction of the benefits they create for consumers and for other firms in the industry" (p. 119).

The Korean strategy went well beyond supporting R&D, to restricting imports and direct investment, because technological development by an industrializing developing country is different in a critical sense from a firm innovating a new technology: the developing country faces an external environment where several competitors have already undergone the learning process and have developed the necessary institutional structures. The need for intervention in developing countries is concomitantly greater. Korea demonstrates that protection of the learning process can be highly effective when complex, large-scale, fast-moving technologies are involved [83]. Singapore, by contrast, relied entirely on technology generated elsewhere, but intervened (selectively) to induce investors to move up the technological scale and (functionally) to provide a well-trained workforce. The strategy worked well for Singapore – but whether it can be emulated by larger economies, and whether it will lead to a broad base for sustained industrial development (*à la* Japan or Korea) is open to question. The Latin American economies have come somewhere in between. Brazil has set up large public enterprises and restricted foreign entry in certain sectors to protect indigenous learning, Mexico also doing so on a much smaller scale. The heavy reliance of these countries on multinational firms for a great deal of advanced technology may well have pre-empted indigenous capability development in the sectors concerned. India has had a very different experience, excluding multinationals in much of manufacturing, but also suffering technological lags and inefficiency as a result of its trade and industrial policies and poor human capital endowments.

Conclusions and implications

The analysis presented above on the determinants of national technological capabilities provides a broad, suggestive framework rather than a precise set of causal connections. It has been suggested in this chapter that the development of capabilities is the outcome of a complex interaction of incentive structures (mediated by government interventions to overcome market failures) with human resources, technological effort, and institutional factors (each also strongly affected by market failures and so needing corrective interventions). Partial explanations of the development of national technological capabilities, which concentrate exclusively on market-driven incentives, on the one hand, or on capability-building measures, on the other, are apt to be misleading for analytical and policy purposes. It is the interplay of all these factors in particular country settings that determines at the firm level how well producers learn the skills and master the information needed to cope with industrial technologies and, at the

291

national level, how well countries employ their factor endowments, raise those endowments over time, and grow dynamically in the context of rapidly changing technologies.

In view of the current prevalence of non-interventionist views on economic development strategy, it is important to be clear about the implications of the framework of national technological capabilities presented here. One set of determinants cannot by itself produce dynamic, broad-based, sustained industrial development. Just getting proper incentives in place will be better, *ceteris paribus*, than giving the wrong signals, but just "getting prices right" may lead to specialization in activities with static comparative advantage if the skills, technology, or institutions are not present to permit efficient diversification. Similarly, generating skills by itself would achieve little if incentives for efficient industrial activity were lacking. Given skills and incentives, performance would still differ (as it does between developed countries), depending on the ability of institutions and government policies to overcome market failures and protect activities with genuine dynamic potential. The existence of market failures considerably modifies what are regarded as neoclassical prescriptions for development, even within the strict rules of neoclassical analysis.

Government policy affects all three components of technological development. Let us reiterate, starting with incentives. A consensus is emerging on the trade and industry policies that promote healthy national technological capabilities development. These are largely taken to be market-oriented policies that promote competition, specialization by comparative advantage, and free flows of technology and capital internationally. However, it is recognized that there can be serious failures in the provision of correct signals from free markets. The existing configuration of prices and costs may not be a reliable guide to resource allocation (including investments in capability building) where there are externalities, complementarities, uncertain learning gains, or capital market failures [70, 71]. There may then be little theoretical or empirical justification for some fashionable policy prescriptions, such as free trade, or giving low and uniform effective protection to different activities. There may be a valid case for intervening in free trade on infant industry grounds. There may also be a valid case for selectivity: some activities may well need much higher protection (and capability-building support) than others, depending on their technical requirements, externalities, and the cost and risk involved in developing the necessary capabilities. By the same reasoning, there may be justifiable reasons for promoting "strategic" industries (because of extensive linkages) or selected individual firms (to realize economies of size and scope by internalizing deficient markets) [83].

As far as capabilities are concerned, there is perhaps more agreement on the need for policy interventions to promote physical and human capital development and technological effort. However, the interventions needed may be selective as well as functional if education and technology strategies are to be geared to realizing specific forms of dynamic comparative advantage. At early stages, industrial development needs basic human capital (literacy and numeracy, with some vocational skills); the period needed to absorb simple industrial technologies is short and needs little protection or external support. At this stage, relatively non-selective educational interventions may be appropriate. As development proceeds, more difficult technologies are used and the need for more sophisticated and specialized education/training grows. To the extent that the education "market" lacks information on these specialized needs, or under-invests in providing facilities of the right kind and quality, there arises the need for selective intervention. Moreover, since there is a serious risk of private under-investment in training at the firm level when labour is mobile, human capital development requires measures to induce more investment to support employee training, by firms individually or cooperatively, or by governments where private agents consistently under-invest. These measures may be functional, applied to all activities, or they may be selective, targeting emerging sectors.

The need for specific technological effort to acquire technological capabilities also rises with industrial development. Easy capabilities may be acquired by brief training combined with learning by doing (i.e. repetition without technical search, investment, or experimentation). More difficult capabilities necessarily require more training and technological effort to master, with concomitant risk and uncertainty. As technologies grow more complex, the development of capabilities runs into problems of appropriability, externalities, lumpiness, and requirements of very specialized skills [74]: policies may be needed to overcome these problems in firm-level effort. The policies must also cover the development of institutions external to firms, to provide information, standards, basic research, and other similar "public goods" relevant to capability development [30]. As development proceeds, moreover, institutional interventions may grow more selective as the initial basic needs are met and markets function more efficiently.

Technological development always needs imports of technology from advanced countries. However, the extent of dependence on imported technology, and the form that technology imports take, affect national technological capabilities development. A passive reliance on foreign skills, knowledge, and technology may lead to national technological capabilities stagnation at a low level, while selective inputs of foreign technology into an active domestic process of technology development can lead to dynamic national technological capabilities growth. Imports of technology must therefore be directed to forms that feed into local efforts rather than suppress them. Adverse effects can arise from a massive foreign presence in the form of multinational corporations that keep their main R&D functions overseas. They can, however, also arise from licensing or use of foreign consultants in ways that do not transfer "know why" to local agents, and that transfer all the benefits of learning abroad. Licensing can be deep or shallow, a stimulus to local learning or a drain on it: national technological capabilities development requires appropriate information selection and negotiation. Thus specific interventions are needed to promote national technological capabilities development, and these will have selective as well as functional aspects.

The above is not meant to suggest that there is a single optimal path to industrial development for all developing countries. The experience of NICs shows clearly that there are many roads to success. Some differences in viable strategies are given by the "state of nature": viz. size, resource endowment, or location. Small countries are not, other things being equal, handicapped by their size, but the sorts of industries they can set up and the technological options they can pursue differ from those for large countries. But there are other differences in possible strategies that depend more on the strategic choices of policy makers than on the "state of nature." The extent and pace of industrial deepening, for example, is a strategic variable for the policy maker: this determines, in turn, the pace and content of human resource development, incentives needed via protection or credit allocation, requirements for technical support or infrastructure, and so on. A country (like Hong Kong) that is content to specialize in light industry needs to invest heavily in (generic) human capital, infrastructure, and some (selective) support for likely export activities, but it needs to intervene less (and less selectively) in other ways than one that aims for heavy industry of particular types. Similarly, the desired extent of national ownership or depth of indigenous technological capability (the two may be closely linked) determines the need for efforts on local skill creation and investments in R&D.

Each of the NICs represents a different model of industrial development because of its choice among strategic variables: the promotion of selected industries or of selected enterprises, fostering of particular types of industrial structures, reliance on domestic as opposed to foreign ownership of industry, and development of an indigenous base of technology and skills. These choices dictate, in turn, different degrees and combinations of selective and functional interventions. It is an open question which set of choices constitutes an ideal long-term development strategy. What is evident is that many strategies are viable, that each is based on a different combination of incentives, capabilities, and institutions, and that each carries its own set of concomitant interventions.

The choice of a less selective set of interventions (*à la* Hong Kong) reduces the risks of backing expensive losers, but it has its own demands and drawbacks. To achieve something approximating the industrial success of Hong Kong, a government would need to intervene initially to build up a comparable base of skills, entrepreneurship, trading know-how, and infrastructure. To enable competitive new activities to emerge without selective promotion, furthermore, the government would have to intervene over time to create new skills, technologies, and institutions. If the objective is to establish a deep and diverse industrial structure (as it should be in larger economies), such functional measures would have to be very extensive indeed. It may even be the case that dynamic industrial development with non-selective interventions would place greater demands on administrative capabilities (to mount functional interventions) rather than less. If such capabilities were lacking, the process of development may be slower or more lopsided than with a package that included careful selective interventions. In any case, it is not clear that, in the absence of selective interventions (in factor or product markets), such a country would be able to diversify into more complex, demanding industries with heavy learning costs. Certainly industrialization experience does not suggest that it would.

In the final analysis, therefore, a large role remains for government policies in promoting each of the three determinants of technological development. But

governments face information and incentive problems no less than does the private market. . . . Good policy requires identifying them [market failures], asking which can be directly attacked by making markets work more effectively (and in particular, reducing government imposed barriers to the effective working of markets) and which cannot. We need to identify which market failures can be ameliorated through nonmarket institutions (with

295

perhaps the government taking an instrumental role in establishing these nonmarket institutions). We need to recognize both the limits and strengths of markets, as well as the strengths, and limits, of government interventions aimed at correcting market failures. [71, p. 202]

The experience of developing countries is replete with instances of misguided intervention. It has been suggested here that many of these failed interventions were neither economic nor truly selective. The relatively few cases of successful selective intervention that exist suggest that interventions are necessary in the presence of wide-spread market failures. Consequently, improved methods of intervening are worth striving for. Much depends on the competence, honesty, and political strength of the policy makers: where governments are so weak or corruptible that selective interventions inevitably lead to the "hijacking" of policy by entrenched interests, it may be better to suffer market failure than pervasive "government failure" [6]. In such cases, however, it is not evident that non-intervention would lead to industrial success. It should be feasible to strengthen the administrative capabilities and power of governments by providing better information and building in measures to safeguard sensible economic policies and to limit interventions in scope to prevent the worst abuses. But this takes us well beyond the scope of the present discussion, into the realms of political economy proper, where again fears of "government failure" may have been overdone [69].

References

1   Amsden, A. *Asia's New Giant: South Korea and Late Industrialization*. New York: Oxford University Press, 1989.
2   Asian Development Bank. *Foreign Direct Investment in Asia and the Pacific*. Manila: Asian Development Bank, 1988.
3   Atkinson, A.B., and J.E. Stiglitz. "A New View of Technological Change." *Economic Journal* 79 (1969), no. 4: 573–578.
4   Balassa, B. and Associates. *Development Strategies in Semi-Industrial Economies*. Baltimore: Johns Hopkins Press, 1982.
5   Bhagwati, J.N. "Is Free Trade Passé After All?" *Weltwirtschaftliches Archiv* 125 (1989), no. 1: 17–44.
6   Biggs, T., and B. Levy. "Strategic Interventions and the Political Economy of Industrial Policy in Developing Countries." In: D. Perkins and M. Roemer, eds. *Economic Systems Reform in Developing Countries*. Cambridge, Mass.: Harvard University Press, 1990.
7   Chen, E.K.Y. "The Changing Role of the Asian NICs in the Asian Pacific Region towards the Year 2000." In: M. Shinohara and F. Lo, eds. *Global Adjustment and the Future of the Asian-Pacific Economy*. Tokyo: Asian and Pacific Development Centre, 1989, pp. 207–231.

8   Cohen, W.M., and R.C. Levin. "Empirical Studies of Innovation and Market Structure." In: R. Schmalensee and R.D. Willig, eds. *Handbook of Industrial Organization*, vol. 2. Amsterdam: North-Holland, 1989, pp. 1060–1107.

9   Cohen, W.M., and D.A. Levinthal. "Innovation and Learning: The Two Faces of R&D." *Economic Journal* 99 (1989), no. 4: 569–596.

10   Dahlman, C.J., and C. Frischtak. "National Systems Supporting Technical Advance in Industry: The Brazilian Experience." Washington, D.C.: World Bank, 1990. Typescript.

11   Dahlman, C.J., and L.E. Westphal. "Technological Effort in Industrial Development – An Interpretative Survey of Recent Research." In: F. Stewart and J. James, eds. *The Economics of New Technology in Developing Countries*. London: Frances Pinter, 1982, pp. 105–137.

12   Dahlman, C.J., B. Ross-Larson, and L.B. Westphal. "Managing Technological Development: Lessons from Newly Industrializing Countries." *World Development* 15 (1987), no. 6: 759–775.

13   Dertouzos, M.J., R.K. Lester, and R.M. Solow. *Made in America: Regaining the Productive Edge*. Cambridge, Mass.: MIT Press, 1989.

14   Desai, A.V., ed. *Technology Absorption in Indian Industry*. New Delhi: Wiley Eastern, 1988.

15   Dosi, G. "Sources, Procedures and Microeconomic Effects of Innovation." *Journal of Economic Literature* 26 (1988), no. 3: 1120–1171.

16   Dosi, G., C. Freeman, R.R. Nelson, G. Silverberg, and L. Soete, eds. *Technical Change and Economic Theory*. London: Frances Pinter, 1988.

17   Dougherty, C. *The Cost Effectiveness of National Training Systems in Developing Countries*. World Bank, PPR Working Papers, WPS 171. Washington, D.C.: World Bank, 1989.

18   Edwards, S. *Openness, Outward Orientation, Trade Liberalization and Economic Performance in Developing Countries*. National Bureau of Economic Research Working Paper no. 2908. Cambridge, Mass.: NBER, 1989.

19   Enos, J. *Learning How: The Creation of Technological Capability in Developing Countries*. Geneva: International Labour Office. Forthcoming.

20   Enos, J., and W.H. Park. *The Adoption and Diffusion of Imported Technology: The Case of Korea*. London: Croom Helm, 1987.

21   Ergas, H. *Why Do Some Countries Innovate More than Others?* CEPS Paper no. 5. Brussels: Centre for European Policy Studies, 1984.

22   ———. *Does Technology Policy Matter?* CEPS Paper no. 29. Brussels: Centre for European Policy Studies, 1986.

23   Evenson, R.E. "Intellectual Property Rights, R&D, Inventions, Technology Purchase, and Piracy in Economic Development." In: R.E. Evenson and G. Ranis, eds. *Science and Technology: Lessons for Policy*. Boulder, Colo.: Westview Press, 1990, pp. 325–356.

24   Fagerberg, J. "A Technology Gap Approach to Why Growth Rates Differ." *Research Policy* 16 (1987), no. 1: 87–99.

25   ———. "Why Growth Rates Differ." In: Dosi et al., eds., pp. 432–457. *See* ref. 16.

26   Fransman, M. *Technology and Economic Development*. Brighton: Wheatsheaf Books, 1986.

27   Fransman, M., and K. King, eds. *Indigenous Technological Capability in the Third World*. London: Macmillan, 1984.

28   Freeman, C. "Japan: A New National System of Innovation." In: Dosi et al., eds., pp. 330–348. *See* ref. 16.

29   Griliches, Z. "Productivity, R&D and Basic Research at the Firm Level in the 1970's." *American Economic Review* 76 (1986), no. 1: 141–154.

30   Grossman, G. "Promoting New Industrial Activities: A Survey of Recent Arguments and Evidence." *OECD Economic Studies* 14 (Spring 1990): 87–125.

31   Harbison, F.H., and C.S. Myers. *Education, Manpower and Economic Growth*. New York: McGraw-Hill, 1964.

32   Herbert-Copley, B. "Technical Change in Latin American Manufacturing Firms: Review and Synthesis." *World Development* 18 (1990), no. 11: 1457–1469.

33   Hoffman, K. *Technological Advance and Organizational Innovation in the Engineering Industry*. Industry and Energy Department Working Paper, Industry Series Paper no. 4. Washington, D.C.: World Bank, 1989.

34   Hou Chi-ming. "Relevance of the Taiwan Model of Development." *Industry of Free China* 71 (1989), no. 2: 9–32.

35   IDB (Inter-American Development Bank). *Economic and Social Progress in Latin America: 1988 Report*. Washington, D.C.: IDB, 1988.

36   Katz, J. "Domestic Technological Innovation and Dynamic Comparative Advantage: Further Reflections on a Comparative Case Study Program." *Journal of Development Economics* 16 (1984), no. 1: 13–38.

37   Katz, J., ed. *Technology Generation in Latin American Manufacturing Industries*. London: Macmillan, 1987.

38   Kim, L. "Korea's Acquisition of Technological Capability: Macro and Micro Factors." Seoul: Korea University Business Management Research Centre, 1988. Mimeo.

39   Kim, L., J. Lee, and J. Lee. "Korea's Entry into the Computer Industry and Its Acquisition of Technological Capability." *Technovation* 6 (1987), no. 2: 277–293.

40   King, K. "Science, Technology and Education in the Development of Indigenous Technological Capability." In: Fransman and King, eds., pp. 31–63. *See* ref. 27.

41   Krause, L.B. "Hong Kong and Singapore: Twins or Kissing Cousins?" *Economic Development and Cultural Change* 36 (1988), no. 3: S45–66.

42   Kuznets, P.W. "An East Asian Model of Economic Development: Japan, Taiwan and South Korea." *Economic Development and Cultural Change* 36 (1988), no. 3: S11–43.

43   Lall, S. *Multinationals, Technology and Exports*. London: Macmillan, 1985.

44   ———. *Learning to Industrialize: The Acquisition of Technological Capability by India*. London: Macmillan, 1987.

45   ———. "Human Resource Development and Industrialization, with Special Reference to Africa." *Journal of Development Planning* 19 (Summer 1989), 129–148.

46   ———. *Building Industrial Competitiveness in Developing Countries*. Paris: OECD, Development Centre, 1990.

47   ———. "Explaining Industrial Success in the Developing World." In: V.N. Balasubramanyam and S. Lall, eds. *Current Issues in Development Economics*. London: Macmillan, 1991, pp. 118–155.

48   McMahon, W.W. "Education and Industrialization." Background paper for *World Development Report*. Washington, D.C.: World Bank, 1987. Mimeo.

49   Mody, A. "Institutions and Dynamic Comparative Advantage: Electronics Industry in South Korea and Taiwan." Washington, D.C.: World Bank, 1989. Mimeo.

50   ———. "Firm Strategies for Costly Engineering Learning." *Management Science* 35 (1989), no. 4: 496–512.

51   Nagaoka, S. *Overview of Japanese Industrial Technology Development*. Industry and Energy Department Working Paper no. 6. Washington, D.C.: World Bank, 1989.

52   Nelson, R.R. "Research on Productivity Growth and Productivity Differences: Dead Ends or New Departures." *Journal of Economic Literature* 19 (1981), no. 3: 1029–1064.

53   ———. "Innovation and Economic Developments: Theoretical Retrospect and Prospect." In: Katz, ed., pp. 78–93. *See* ref. 37.

54   ———. "Institutions Supporting Technical Change in the U.S." In: Dosi et al., eds., pp. 312–329. *See* ref. 16.

55   Nelson, R.R., and S.J. Winter. *An Evolutionary Theory of Economic Change*. Cambridge, Mass.: Harvard University Press, 1982.

56   Nishimuzu, M., and S. Robinson. "Trade Policies and Productivity Changes in Semi-Industrialized Countries." *Journal of Development Economics* 16 (1984), no. 1: 177–206.

57   OECD. *Structural Adjustment and Economic Performance*. Paris: OECD, 1987.

58   Oshima, H.T. "Human Resources in East Asia's Secular Growth." *Economic Development and Cultural Change* 36 (1988), no. 3: S103–122.

59   OTA. *Making Things Better: Competing in Manufacturing*. New York: Oxford University Press, 1990.

60   Pack, H. *Productivity, Technology and Economic Development.* New York: Oxford University Press, 1987.

61   ———. "Industrialization and Trade." In: H.B. Chenery and T.N. Srinivasan, eds. *Handbook of Development Economics*, vol. 1. Amsterdam: North Holland, 1988, pp. 334–380.

62   Pack, H., and L.E. Westphal. "Industrial Strategy and Technological Change: Theory versus Reality." *Journal of Development Economics* 22 (1986), no. 1: 87–128.

63   Rath, A. "Science, Technology, and Policy in the Periphery: A Perspective from the Centre." *World Development* 18 (1990), no. 11: 1429–1443.

64   Republic of China. *Education Statistics of the Republic of China 1984*. Taipei: Government of Taiwan, 1985.

65   ———. *Science and Technology Data Book*. Taipei: Government of Taiwan, 1987.

66   ———. *Statistical Yearbook of the Republic of China 1988*. Taipei: Government of Taiwan, 1989.

67   Romer, P.M. "Capital, Labor, and Productivity." *Brookings Papers on Economic Activity: Microeconomics, 1990* (1990): 337–367.

68   Saghafi, M.M., and D.-S. Davidson. "The New Age of Global Competition in the Semiconductor Industry: Enter the Dragon." *Columbia Journal of World Business* 24 (1990), no. 1: 60–70.

69   Shapiro, H., and L. Taylor. "The State and Industrial Strategy." *World Development* 18 (1990), no. 6: 861–878.

70   Stiglitz, J.E. "Learning to Learn, Localized Learning and Technological Progress." In: P. Dasgupta and P. Stoneman, eds. *Economic Policy and Technological Development*. Cambridge: Cambridge University Press, 1987, pp. 125–155.

71   ———. "Markets, Market Failures and Development." *American Economic Review, Papers & Proceedings* 79 (1989), no. 2: 197–203.

72   Tassey, G. "Infratechnologies and the Role of the Government." *Technological Forecasting and Social Change* 21 (1982), no. 2: 163–180.

73   ———. "The Role of the National Bureau of Standards in Supporting Industrial Innovation." *IEEE Transactions on Engineering Management* 33 (1986), no. 3: 162–171.

74   Teece, D.J. "Economic Welfare and the Allocation of (Private) Resources to Innovation." Berkeley, Calif.: Walter A. Haas School of Business, 1989. Mimeo.

75   Teitel, S. "The Skill and Information Requirements of Industrial Technologies: On the Use of Engineers as a Proxy." In: M. Syrquin and S. Teitel, eds. *Trade, Stability, Technology and Equity in Latin America*. New York: Academic Press, 1982, pp. 333–348.

76   ———. "Technology Creation in Semi-Industrial Economies." *Journal of Development Economics* 16 (1984), no. 1: 39–61.

77   ———. "Science and Technology Indicators, Country Size and Economic Development: An International Comparison." *World Development* 15 (1987), no. 9: 1225–1235.

78   UNESCAP (United Nations Economic and Social Council for Asia and the Pacific). *Statistical Yearbook for Asia and the Pacific 1986–1987*. Bangkok: UNESCAP, 1988.

79   Unesco. *Statistical Yearbook, 1988*. Paris: Unesco, 1989.

80   Vernon, R. *Technological Development: The Historical Experience*. Seminar Paper no. 39. Washington, D.C.: Economic Development Institute, World Bank, 1989.

81   Wade, R. "The Role of Government in Overcoming Market Failure: Taiwan, the Republic of Korea and Japan." In: H. Hughes, ed. *Achieving Industrialization in East Asia*. Cambridge: Cambridge University Press, 1988, pp. 129–163.

82   Westphal, L.E. "Fostering Technological Mastery by Means of Selective Infant-Industry Protection." In: M. Syrquin and S. Teitel, eds. *Trade, Stability, Technology, and Equity in Latin America*. New York: Academic Press, 1982, pp. 255–279.

83   ———. "Industrial Policy in an Export-Propelled Economy: Lessons from South Korea's Experience." *Journal of Economic Perspectives* 4 (1990), no. 3: 41–59.

84   Westphal, L.E., Y.W. Rhee, and G. Pursell. "Foreign Influences on Korea's Industrial Development." *Oxford Bulletin of Economics and Statistics* 41 (1979), no. 4: 359–388.

85   World Bank. *Korea: Sector Survey of Science and Education*. Report no. 3775-KO. Washington, D.C.: World Bank, 1981.

86   ———. *Korea: Managing the Industrial Transition*. 2 vols. Washington, D.C.: World Bank, 1987.

87   ———. *World Development Report 1987*. Washington, D.C.: World Bank, 1987.

88   ———. *World Development Report 1989*. Washington, D.C.: World Bank, 1989.

9

The environmental challenge

Ignacy Sachs

The "environmental revolution" [69] happened in the 1960s. Ecology, hitherto a quasi-esoteric discipline belonging to the realm of sciences of life, captured the attention of the general public. It even became the scientistic, if not scientific, foundation of the Green ideology [7], combined with the discontent generated by the deterioration in the quality of life ("les dégâts du progrès") and the rebirth of a religious feeling towards Nature in reaction to a world that appeared increasingly artificial.

This time the old Malthusian spectre of exhaustion of food supplies (and, by extension, of other natural resources) as a result of the population explosion was combined with the realization that the capacity of Nature to act as a sink was also limited. According to *The Limits to Growth* [61], the most influential book written with this viewpoint at the request of the Club of Rome, humankind is heading towards disaster: unless it departs promptly and sharply from its present growth-oriented path, the only cruel choice left within a few decades will be between death by starvation or death by excess pollution [24, 12].

This new wave of pessimism came at a moment when technological optimism was widespread, the competition between the two major socio-political systems – capitalism tempered by the Welfare State and "true socialism" – seemed to be judged in terms of their capacity to sustain high economic growth, and decolonization created a hope-

This chapter was written in 1991, before the United Nations Conference on Environment and Development held in Rio in June 1992.

ful mood regarding the emancipation and modernization of the newly independent countries.

How can this paradox be explained?

Ecological awareness and Green movements originated in the richest part of our planet as a reaction to the excesses of boundless optimism projected centuries ahead without any serious consideration of the natural as well as social limits to growth [45] and its ecological costs. Herman Kahn's writings, at least as popular as the reports of the Club of Rome, epitomize this attitude, and his projections covered a two-century span [50]. Berry [6] extended them over 10,000 years, dismissing the fear of exhaustion of resources by suggesting that the problem would be overcome through the colonization of other planets! Technological optimism was also an article of faith among Marxists [54, 77].

Yet, the everyday experience of people living in the industrialized regions was quite different: urbanization and the phenomenal growth of industries brought in many inconveniences: highly damaging pollution and even disasters (Minamata, Seveso, Three Mile Island, and Chernobyl acted as eye-openers), unhealthy working conditions, a shortage of public housing, overcrowding of mass transportation systems, proliferation of private cars, and, above all, the inability to overcome the problems of poverty, social exclusion, and spatial segregation in spite of the unprecedented growth of GNP. Misdirected and misappropriated economic growth did not result in an improvement in the quality of life for significant segments of the industrialized societies, even though their material standards of living, measured by GNP per capita, went up.

This brings us to the interface between environment and development. We know today that the phenomenal growth of material production since the Industrial Revolution involved a predatory and so far mostly unaccounted incorporation of the capital of Nature, degrading the life-supporting systems (air, water, soils, forests). The very conditions of human life on our planet are menaced not only by the prospect of a nuclear holocaust, but also by the global warming of the atmosphere mainly due to the overuse of fossil fuels and massive destruction of forests. Furthermore, careless dumping of waste constitutes a powerful factor of environmental disruption.

On the other hand, three UN-sponsored decades of development, largely rhetorical, did little to overcome the gap between the minority of affluent countries and people, and the rest. Per capita consumption ratios in the North range from 2.9 that of the South for cereals,

5.7 for meat, 8.1 for milk, 19.9 for iron and steel, 20.3 for chemical products, 20.6 for metals, and 23.6 for cars. The per capita consumption of liquid fuels in the North is 9.8 times that in the South, and that of electricity is 13.4 times higher in the North than in the South. The respective shares in global emissions of CO_2 per capita are in the region of 8:1.

Under these circumstances, it can be argued that past development, concentrated mostly in the North, has put such pressure on the carrying capacity of our planet that there is no room for newcomers. Given the resource-intensive and environmentally disruptive path followed by the industrialized countries – and that by Taiwan and Korea, often presented as a model for the third world – the planet would collapse if these models were to be extended to the rest of the world, i.e. if all the world's poor were to become rich in the sense that the affluent minority now gives to this term.

The first debate on environment and development

The landmark UN Conference on the Human Environment in Stockholm in 1972 tried to steer a middle path between two extreme and still influential views: the narrowly economic and the unconditionally ecological.

The partisans of the "growth first" approach claimed that all the other dimensions of development either would be taken care of automatically by the "trickle down effect" of rapid growth or else could be attended to in better conditions once the country concerned had achieved a much higher per capita GNP. This approach is still present in the discussions on how to deal with "global change": on the basis of a controversial cost benefit analysis, Nordhaus [70], to quote the most extreme example, makes a plea for postponing measures aimed at reducing the "greenhouse effects" until the danger really knocks at our doors.

At the other extreme were the partisans of the zero rate of growth. Some applied this concept only to the population. Others extended it to both population and material growth, claiming that real development should concentrate on qualitative rather than quantitative aspects (for an up-to-date formulation of this argument insisting on the need to reconstruct the natural capital rather than to expand man-made capital as an investment priority, see Daly [17]). In their most extreme formulations, the partisans of an end to growth demanded the "de-industrialization" of the rich countries and the non-industrialization of the poor ones; the latter could serve in the meantime as a recreational and cultural reserve for the rest of the world [20].

The population controversy

The demographic argument played an important role in the debate in the late 1960s and early 1970s, although this argument should be qualified in at least three respects.

First, contrary to a widespread belief, curbing the numbers of "non-consumers" will not greatly reduce the pressures on resources and the environment. This point was rightly made by Barry Commoner [13] and is today recognized even by the Ehrlichs in their most recent book [21]. The environmental impact is a function of the population, its affluence (GNP per head), and the technology employed:

$$\text{environmental impact} = \text{population} \times \text{affluence} \times \text{technology}$$

If we take as a proxy for affluence and technology the per capita consumption of commercial energy, "a baby born in the United States represents twice the destructive impact on Earth's ecosystems and the services they provide as one born in Sweden, three times one born in Italy, thirteen times one born in Brazil, 35 times one born in India, 140 times one in Bangladesh or Kenya, and 280 times one in Chad, Rwanda, Haiti or Nepal" [21, p. 134]).

Seen from the resource consumption angle, the population problem is essentially one of the rich people (wherever they are) and countries. Moreover, the Commoner equation clearly shows that the environmental impact may be reduced by acting on the other two variables. Thus, lifestyles, consumption patterns, and technologies in the North and in the affluent enclaves in the South should be our first concern, without underestimating the difficulty of achieving meaningful results with respect to voluntary self-limitation of the growth of material consumption on the part of the affluent minority.

Secondly, policies directed at birth control in third world countries, desirable as they may be to slow down the *rate* of population growth, are likely to prove deceptive if they do not come as part of a social development package, including the education of women, effective public health policies resulting in reduced infant mortality, access to subsidized, rationed, or distributed food for those who cannot afford to buy the minimum ration, and some protection in old age.

305

The studies of Kerala by Raj et al. [76] and of Sri Lanka by Panikhar et al. [73] documented the possibility of important social advances in very poor regions. The same conclusion can be drawn from the experience of China. More generally, it is possible to argue that developing countries need not repeat the historical sequence followed by the industrialized countries, where the welfare concern appeared at a late stage of development. The time sequence can be inverted provided that adequate human resources (paramedical personnel, primary school teachers, etc.) are trained, service delivery techniques with a high labour content are chosen, and research is directed towards modern yet inexpensive, preventive, and therapeutic techniques and practices, as shown by Unicef (for a theoretical discussion, see Sachs [81]).

By contrast, the experience of India shows that enforcing (sometimes in the literal sense of this word) birth control practices does not lead very far so long as the broader contextual conditions outlined here are not present. And while the impact of urbanization on reducing fertility cannot be denied, the attendant social costs of massive migrations of rural refugees to urban shanty towns are very high indeed [38].

Thirdly, the spatial maldistribution of the world population poses at least as serious a problem as the rates of demographic growth. This observation applies equally to rural and urban areas. Some rural regions have a population that clearly exceeds their carrying capacity. Others, on the contrary, do not have the minimum density required for meaningful social health and educational policies. Less than half the world's rural population have access to basic health care. Half the rural women over 15 years old are illiterate. In most developing countries, those who live in the countryside typically earn 25 to 50 per cent less than those in the towns. Three-quarters of the poor people in the South live in ecologically fragile zones. In order to survive, they overexploit the natural resources to which they have only a very limited access. The number of environmental refugees is estimated at 14 million people.

The situation is particularly dramatic in sub-Saharan Africa. Mortality of children under five still stands at 178 deaths for every 1,000 live births. Almost two-thirds of the population lack safe water, 18 million suffer from sleeping sickness, and malaria kills hundreds of thousands of children each year [103].

The most dramatic social and environmental challenge in terms of quality of life for billions of people is, however, the urban explosion.

The third world cities continue to expand as a result of the massive rural exodus. They are attractive as "lotteries of life," allowing upward mobility for the lucky few or, perhaps, their children, and they are also places where things still happen ("bread and circuses," but also schools, hospitals, and jobs for some). According to UN estimates, the urban population in the South will grow from 1 to 2 billion between 1980 and the year 2000 and double again in 25 years to reach 4 billion inhabitants in 2025. How many among them will be condemned to live in shanty towns, enduring the double plight of pollution by poverty and of pollution generated by other peoples' affluence, which they may help produce while sharing very little of its bounties? According to Hardoy et al. [42], 600 million urban residents are exposed to serious health risks on account of deficient water supply, sanitation, drainage, and removal of household waste.

Without being equally dramatic, the situation in many cities of the North – and certainly of eastern Europe – is far from satisfactory from both the social and environmental viewpoints. Urban infrastructures have grown obsolete. Huge investments are required to modernize and expand them, or even for straightforward repair. Pockets of destitution remain. Social exclusion and spatial segregation have not been overcome. In several American cities, urban centres abandoned by affluent people have been transformed into socially and economically distressed ghettos inhabited by social minorities. Social exclusion is increasingly present in European towns, resulting in racial, religious, and ethnic conflicts.

Therefore, there is the need to put very high on the environmental agenda the issues of the habitability of urban agglomerations, of new rural-urban configurations, and, also, of organized migrations from areas whose density of population clearly exceeds their carrying capacity to places that can still absorb the incomers. The first two of these have an important science and technology component, while the third is eminently political and ethical since it presupposes a willingness to receive alien people on one's territory. It cannot be cast in objective terms. This is all the more so in that the evaluation of the carrying capacity is already a subjective and highly controversial matter. In a study prepared for the Canadian Conserver Society, Goldsmith [33] claimed that Canada is already overpopulated! By contrast, pleading for "more immigrants, please" so as to reach a population of 40 million, his critics argued that if the 10 per cent of Canada's territory that is habitable were as densely populated as the Netherlands, Canada would have over 400 million people!

While the concept of carrying capacity is useful in so far as it reminds us of the existence of outer limits, it cannot be quantified once for all, as both the *pattern of demand* for goods produced and the *technological capability* to produce more while destroying less are likely to change over time. Tricart and Killian [102] have used the same argument to question the concept of "agricultural vocation" of different soils widely used in cartography. The only objective approach is to list the physical constraints that new technologies may or may not overcome.

The harmonization game

The middle path suggested by the UN Conference on the Human Environment in Stockholm in 1972 consisted in reaffirming the need for further growth with equity while incorporating explicitly a concern for the environment as a dimension of development conceived as a positive-sum game with Nature. Hence the challenge of applying simultaneously to development thinking the following three criteria:

- *equity* in the formulation of social goals of development, as an ethical imperative expressing the synchronic solidarity with all the present travellers on the Spaceship Earth;
- *ecological prudence* as an ethical postulate of solidarity with the future travellers and, also, as a means to improve the present-day quality of life;
- *economic efficiency* instrumental in making good use of the manpower and material resources from the macrosocial point of view, i.e. by taking into consideration the hitherto externalized social and ecological costs.

As this last criterion does not necessarily coincide with the microeconomic profitability at the enterprise level, it follows that ecodevelopment strategies – a shorthand for socially equitable, environmentally viable, and economically efficient strategies [82, 83] – cannot be implemented in a pure market economy. They call for a set of regulations on behalf of the state within the broad framework of "mixed economies." Tinbergen and Hueting [100] rightly point out that market prices send wrong signals for sustainable economic success that mask environmental destruction. "If collective side-effects (externalities) are substantial and important, the classical doctrine of the blessings of free trade simply becomes irrelevant as a guideline for economic policy" [39].

Neoliberals interpret the collapse of the command economies in eastern Europe as a proof *a contrario* of the excellence of the unrestricted free-market model. However, when government fails, will the market do better? Barry Lester's well-argued reply [55] to this question shows that it need not be so, even in terms of productive efficiency, not to mention that the free marketeers relegate equity considerations to second place in the development paradigm [43], while equity and efficiency should be considered as complementary, not conflicting, goals [98]. Toye [101] is right when he postulates a case by case pragmatic analysis that is more costly, to decide whether failure should be blamed on the state or the market.

The variables of the harmonization game are situated at both the demand and supply levels, as well as in the location of productive activities.

DEMAND. The most decisive variable here, but at the same time politically the most difficult to manage, is the *consumption pattern* reflecting the development style. Resource saving through demand management implies one of the following solutions:
- resource saving through greater discipline on the part of consumers, retrofitting of the existing housing stock to improve its energy efficiency, time scheduling of activities to reduce peak hours, and, above all, better organization of the production and distribution cycle. In so far as resources saved in this way and through better maintenance of equipment and infrastructures may be considered as a "development reserve" [84], they constitute an important source of "non-investment growth" [51];
- a reduction in the consumption standards as postulated by the advocates of "voluntary simplicity" and self-restraint [48];
- acceptance of more or less far-reaching substitutions between material and non-material consumption: fewer goods and more services or, in a more radical version, less time spent in market-oriented economic activities and more time allocated to non-economic activities and/or small-scale environmentally benign material production for self-consumption [46, 36, 84];
- shift from individual cars to mass transportation systems or bicycles or else new kinds of environmentally benign vehicles such as small electric cars;
- reduce the demand for intra-urban transportation by redesigning the cities (instead of traditional zoning, put housing, work, trade, and leisure within walking distance);
- reduce the demand for long distance transportation by better integrating local, regional, and national economies, greater selectivity in

external trade (without falling into the autarky trap), and, in so far as it is feasible, substituting communication for professional (but not tourist) travel.

While the main obstacles will lie, as already mentioned, in the political sphere, much will depend also on the availability of attractive technical solutions, but not "technical fixes" isolated from the cultural, ethical, institutional, and political contexts.

SUPPLY. It is here, at the intersection between Nature and society, that technology plays a leading role. Nature provides energy, space, and resources, i.e. those elements of the natural environment that, thanks to the knowledge accumulated, can be transformed into some "use value" deemed as such by the society. The concept of "resource" is therefore essentially cultural and historical.

Society sets the values and the societal goals, builds the institutions, and produces the knowledge – both traditional and scientific (*techne* and *episteme*) – used to design the goods corresponding to societal needs and aspirations, to identify the resources, to invent the product and process technologies and the necessary equipment. It also supplies the workforce.

The production process combines in a given site resources and energy with work and previously produced equipment to generate a flow of "goods" that go to the market (or reach the consumer through other institutional mechanisms) and of "bads" that are dumped back in Nature, this time acting as a sink.

It immediately follows from this schematic description that *technology constitutes potentially a privileged locus to harmonize the three concerns of social equity, ecological prudence, and economic efficiency*. This can be achieved by a variety of means:

– Promoting energy and resource saving through product and process design, as well as upgrading the environmentally sound traditional techniques.

– Finding novel ways of using the specific resources of each ecosystem, with special emphasis on renewable resources, while recognizing that the conditions of their renewability must be respected; a forest that is felled without ensuring its regeneration or replanting is a mine of timber, not a renewable resource. Furthermore, assessing the value of biological resources cannot be restricted to the value of products that are commercially harvested ("productive use value") or collected for self-consumption ("consumption use value"). It also calls for considering the indirect values of ecosystem functions, such as watershed protection, regulation of climate,

310

and production of soil ("non-consumptive use value"), the intangible values of keeping the options for future by preserving the biodiversity ("existence and option values") [59].

– Minimizing the "bads" by resorting to low-waste technologies.
– Recycling and reusing non-renewable resources (aluminium becomes a renewable resource in so far as it can be reused several times).
– Using the natural ecosystem as a paradigm for man-made production systems; taking a horizontal view of development in order to explore the potential complementarities and synergies, in sharp contrast with the prevailing compartmentalization and narrow specialization; closing whenever possible the loops, using the waste from one production module as an input in the next module of the system, as illustrated by the traditional Chinese dyke-pond systems [78] and all other integrated food-energy production systems, with different levels of technical sophistication [87].

By contrast, "careless" technologies prove environmentally disruptive and socially costly. It is only natural that left to itself an enterprise tends to externalize its ecological and social costs in order to maximize the internalized profits, up to the point when the environmental disruption or the social discontent become a hindrance. But this stage is reached only after having done considerable and often irreversible damage, locally and globally. The anthropogenic modifications of the biosphere have reached a worrying scale. Ruffolo [79] suggestively contrasts the increasing might (*potenza*) of our technologies with our utterly deficient political power (*potere*) to control them (see also ref. 49).

LOCATION OF PRODUCTIVE ACTIVITIES. This is the third strategic variable of the harmonization game. The environmental impact of productive activities will greatly depend on the climatic and topographic features of the site and the density and nature of the human activities in the proximity. The ecodevelopment approach calls for ecosystem-specific, culture-specific, and site-specific solutions. In the last instance, global problems can be solved only through a coordinated set of local solutions. The future does not belong, however, to an archipelago of self-contained local development units. Institutional arrangements are called for to better articulate the local, national, and transnational spaces of development, tilting the balance in favour of bottom-up approaches to overcome the inherited bias towards centralization and the cities.

It is important to emphasize that, far from being an attempt to re-

311

turn to ancestral practices, respectful of Nature by necessity in order to survive but situated at a very low level of productivity, the approach that emerged from the UN Stockholm Conference sought a modern development in harmony with Nature, recreating the old peasant rationality at a completely different level of the spiral of knowledge. It suggested searching for knowledge-intensive, energy- and resource-saving, environmentally sound and socially responsive development paths. The utilization of local knowledge is of enormous importance in this endeavour, the task being to extract from it the original ideas it might contain and to study them by applying the resources of modern science. According to Amilcar Herrera, "the most important local contribution would probably be, more than in concrete specific technologies, in new approaches to the solution of old problems, that might stimulate scientific research into hitherto unexplored directions" [44, p. 28].

Slow progress towards ecologically and environmentally friendly development

Twenty years separate the UN Stockholm Conference from the UN Conference on Environment and Development in Rio de Janeiro in June 1992. Yet, compared with the expectations raised, little progress was achieved during these two decades in terms of international action directed at a more rational management of the biosphere. The United Nations Environment Programme (UNEP), the body that emerged from the Stockholm meeting, never had the resources commensurate with the immensity of the task entrusted to it.

Lack of progress in international environmental cooperation prompted the United Nations to set up a high level environment and development commission presided by Ms. Brundtland, prime minister of Norway. Its report, *Our Common Future* [104], did not add much to what was known on the subject, but it had the merit of giving a new impetus to the political discussion on the urgency to promote what is now called "sustainable development." However, the institutional breakthrough at the international level – the Montreal Convention on the protection of the ozone layer – was essentially due to the fears motivated by the adverse effects of human activities on the world's climate, about which scientists produced new evidence.

By contrast, greater progress was achieved in the institutionalization of environmental concern at the national level. Practically all countries now have ministries of the environment. Several promulgated advanced laws. The Brazilian constitution (1988) has an ex-

312

cellent chapter on the environment; Peru has consolidated the legislation on environmental protection and management in an extensive code. Of course, the problem of enforcement of such laws remains. Institutional creativity may even serve as a screen to disguise lack of will to change the status quo. But at least a framework has been built to start action when the political conditions are favourable.

Conceptually, some progress has been made. We shall review it under four headings: the analytical tool-box, the debate on sustainability, the emergence of a new paradigm in ecology, and global change.

The planners' and managers' tool-box

By analogy with technology assessment, a comprehensive environmental impact assessment has been instituted and is today required by law in several countries for projects such as big dams, river diversion, mines, large industrial complexes, siting of potentially dangerous factories (chemicals, nuclear facilities, etc.).

In practice, such exercises are often performed in ways that do not guarantee effective protection to the populations or to the long-term interests of the country. This happens in particular where the investor is supposed to produce the environmental impact statement, but no adequate mechanisms have been set up to control it effectively. While citizen associations are formally consulted, they do not have access to the necessary expertise to analyse in depth the investor's proposals.

Another difficulty arises as regards negotiating compensation for the populations affected or even displaced. The common practice of paying individual compensation lends itself to many abuses. On the other hand, finding meaningful collective solutions proves much more difficult. The imbalance of power between the actors interested in carrying out the project and the defenceless populations is often dramatic. It is not by refining the analytical tools but by perfecting the negotiating and contractual process and by offering adequate institutional protection to the weaker party (advocacy planning) that progress may be achieved [64].

In spite of these shortcomings, the environmental impact statements constitute already an antidote against spatial interventionism and the radical proposals to transform Nature that emerged after the Second World War both in the USSR (diverting southward the flow of Siberian rivers) and in the United States (an artificial sea in the Amazon region proposed by the Hudson Institute).

The growing interest in the environment coincided with the decline

313

of planning and the rise of neoliberal economics. Under these circumstances, considerable effort was displayed to find ways of including the environmental (but strangely enough not the social) externalities within the conventional economic calculus. A new discipline emerged under the name of "ecological economics" [14]. While the journal published under this name contains many interesting contributions, ecological economics is flawed by the underlying assumption that ultimately the decision-making must rest on the economic calculus.

A radical criticism of ecological economics leads, however, to the uncomfortable (but alas lucid) position that planning and decision-making are an art, not a science [37].

In policy terms, the attempt to internalize the environmental costs led to the formulation of the "polluter pays" principle elaborated in great detail by the OECD [71]. Although applicable and practical within certain limits, this principle has several limitations.

What should the polluter pay for: The right to continue to pollute? A compensation to the victims of pollution? The cost of shifting to clean technologies? Should he choose the solution that costs least? Does it make sense to establish a market to trade emission rights locally or even at a global scale? The latter solution, carried to its extreme, might for instance lead a polluting industry in the North to buy large estates in a tropical country in order to stop the methane-releasing paddy production there, the social cost of the operation being left outside the calculation!

Another set of questions concerns the polluter's ability to pass on the cost to the consumer, which depends on the imperfections of the market. Much of the theory underlying the "polluter pays" principle assumes a perfect market, which very rarely exists. Paradoxically, mainstream economists have tended to argue that the environment can be successfully managed within a pure market economy, although the evidence does not bear this out.

Another area in which considerable work has been done is that of "environmental accounting." Two contrasting positions have emerged. One postulates drawing up "accounts" using an array of physical indicators to reflect the changes occurring in the "natural capital": depletion of non-renewable resources, soil erosion, deforestation, etc. This kind of accounting should provide a safeguard against predatory methods of resource use. The other maintains that the depletion of the "natural capital" could be evaluated in monetary terms and therefore could be subtracted from the GNP (for a more general discussion see Ahmad et al. [2]). This, however, leaves aside

314

the non-tangible and non-monetary use, existence, and opportunity values put forward by the conservationists [59].

The debate on "sustainability" and the technology issue
The term "sustainable development" suffers from an ambiguity: Is sustainability to be understood merely in ecological terms? Does it refer to all the facets of development: ethical, social, economic, etc.? How does it relate to economic growth?

As far as philosophies are concerned, the two camps mentioned at the beginning of this chapter maintain their positions: the "Malthusians" sharply attacked the Brundtland Report for its adoption of the goal of sustainable growth, which, in their view, is an oxymoron [17], while the other continues to put much faith in technological progress.

Whether unlimited growth (as distinct from purely qualitative development) is possible depends on the precise meaning given to the two terms. Extensive growth, using more material resources and producing more waste, i.e. increasing the "material throughput," cannot be envisaged. But intensive growth, meaning by this producing more for the same quantity of inputs and releasing less waste per unit of output, is not at all incompatible with the existing ecological constraints. This is what the partisans of "another kind of growth" have in mind. They add yet another clause: growth should be not only environmentally sustainable but socially meaningful, i.e. directed to meeting goals set by people and not through marketing [15, 105]. Presumably, the concept of qualitative development includes the intensive growth as defined above.

While the policy conclusions of the "Malthusians" are open to discussion, their reconceptualization of the field of economics is of great importance. The pioneering work of Georgescu-Roegen [28] was instrumental in reintroducing into the realm of economics the physical processes underlying production, a dimension practically ignored by all the economic schools after the physiocrats. This paradigmatic breakthrough was followed by a careful elaboration of the process of production as a throughput of energy and resources and the explicit introduction of the "natural capital" in the production functions [17].

Much effort was devoted to both the conceptual discussion and practical work on "environmentally friendly technologies." They cover a wide spectrum ranging from small-scale "soft technologies" and upgrading of traditional know-how to major efforts to produce large-scale modern low-waste technologies, as well as anti-pollution equipment.

Special reference should be made to the discussion of agricultural techniques. Can we really speak of sustainable agriculture when it requires growing inputs of fertilizers and pesticides? The concept of "regenerative agriculture" pioneered by Robert Rodale tries to promote agricultural practices that are capable of regenerating the soils without massive additions of industrial inputs. It does not, however, go as far as the "organic agriculture," whose advocates often have an extremely restrictive vision of what is "natural" and, therefore, acceptable.

Lately, as a result of the investigation by a committee chaired by John Pesek, the broad concept of "alternative agriculture" has been recognized by the National Research Council of the United States [67]. Alternative agriculture is defined as any system of food and fibre production that systematically pursues the following goals:

- more thorough incorporation of natural processes such as nutrient cycles, nitrogen fixation, and pest-predator relationships into the agricultural production process;
- reduction in the use of off-farm inputs with the greatest potential to harm the environment or the health of farmers and consumers;
- greater productive use of the biological and genetic potential of plant and animal species;
- improvement of the match between cropping patterns and the productive potential and physical limitations of agricultural lands to ensure long-term sustainability of current production levels;
- profitable and efficient production, with emphasis on improved farm management and conservation of soil, water, energy, and biological resources (see also Dahlberg [16]).

The different schools of thought participating in the debate differ about the scope for applying "soft technologies," sometimes narrowly interpreted as a subset of "intermediate technologies," as well as with respect to the relative importance of low-waste and depolluting technologies. The latter subject involves the question of how much effort should go into preventive actions instead of continuing business as usual – that is, producing "goods" and "bads" and then increasing the national wealth by additional production of equipment to suppress or mitigate the "bads"!

In so far as the spatial concentration of production is a major source of environmental disruption, the opportunities created by flexible specialization, modern small-scale production, and decentralized industrialization are bound to become an important locus of harmonization of economic efficiency and ecological prudence. A thorough revision of the concepts of economies of scale and concentration inherited from the previous stage of industrialization are called for in the light of the recent trends in technical progress (micro-electronics, computers, communication, flexible specialization) [75, 4, 47].

Since instant retooling of the productive apparatus was naturally impossible, the management of technological pluralism [88] and "blending of technologies" became a major policy concern.

Another policy variable is the durability of products. One has to balance the resource-conserving aspects of longer life cycles of products against the need to ensure a reasonable rate of technical change [10, 29]. Developing countries cannot cope with the present trend towards accelerated obsolescence (a perverted form of the Schumpeterian "creative destructiveness"). On the other hand, they must introduce selectively up-to-date technologies in order to achieve competitiveness on international markets. At any rate, better maintenance of infrastructures and equipment offers an excellent opportunity to create jobs financed through the resource saving thus achieved [85]. Discarding the throw-away society is a common objective for the North and the South [109].

To conclude, it is necessary to emphasize once more the ambiguity of the concept of sustainability. Addressing himself to the roots of the problem, Rajni Kothari writes:

In the absence of an ethical imperative, environmentalism has been reduced to a technological fix, and as with all technological fixes, solutions are seen to lie once more in the hands of manager technocrats. Economic growth, propelled by intensive technology and fuelled by an excessive exploitation of nature, was once viewed as a major factor in environmental degradation; it has suddenly been given the central role in solving the environmental crisis. The market economy is given an even more significant role in organizing nature and society. The environmentalist label and the sustainability slogan have become deceptive jargons that are used as convenient covers for conducting business as usual. [53]

Against this rhetoric, Kothari suggests a different meaning of sustainability rooted in ethics and going hand in hand with the search for an alternative mode of development. The essence of his thinking is that a conflict exists between two meanings of "sustainable development": sustainability as a narrow economic ideal referring to maintaining privileges and compromising the future and Nature for the benefit of a minority, as opposed to the ethical ideal of sustainability of life on Earth.

He identifies four primary criteria for sustainable development: a holistic view of development; equity based on the autonomy and self-reliance of diverse entities instead of a structure of dependence founded on aid and transfer of technology with a view to "catching up"; an emphasis on participation; and an accent on the importance of local conditions and the value of diversity. "Our common future cannot lie in an affluence that is ecologically suicidal, and socially and economically exclusive. It can, and must, lie in a curtailment of wants," as Gandhi constantly reminded his countrymen and others.

It would be utopian to believe that these views would be easily accepted by the affluent minority living in the North and the Northern enclaves in the South. It reflects a fundamental difference of opinion between the North and the South.

A new ecology
The rate of progress in integrating the environmental dimension into the social sciences of development in general and economics in particular has been disappointing. Mainstream economic thought resists the change of paradigm that would deprive the "dismal science" more than ever of its pretence of being a hard science. The mechanistic models of growth and the theories of equilibrium are still strongly entrenched. The neoliberal wave is distancing the state at a moment when environmental concern should, on the contrary, lead to a redefinition of the roles of the state, of the markets, and of civil society, seeking their synergy in the management of both the biosphere and society.

Will ecology succeed better in modifying its fundamental underlying paradigm? A pioneering book by Botkin [7] undermines the notion that Nature undisturbed is constant and stable, a myth that led to many catastrophic mistakes in the management of resources. Instead of a balance of Nature, we are in the presence of discordant harmonies created by simultaneous movements of many tones, a combination of processes flowing at the same time and along various scales. The result is "not a simple melody but a symphony at sometimes harsh and at sometimes pleasing" (p. 25). The ecologists borrowed the physical concept of stability from mechanics and accepted the Lotka-Volterra equations on the basis of authority. "Although environmentalism seemed to be a radical movement, the ideas on which it was based represented a resurgence of pre-scientific myths about nature blended with early-twentieth-century studies that provided short-term and static images of nature undisturbed" (pp. 42–43).

The new paradigm proposed by Botkin insists on the great mutual influence of life and environment at a global level. Together they form a planetary-scale system – the biosphere – that sustains and contains life. The total mass of living things is a tiny fraction of the mass of the Earth; if mixed, the concentration of living things would be two-tenths of one part in one billion. Yet, even the geologists are beginning to view life as an integrated part of geological processes.

At the global level, three schools of thought exist about the balance of Nature.

- The first views the biosphere as being in a steady state, extrapolating the nineteenth-century theories of the equilibrium of undisturbed Nature at the local level, related to the metaphor of divine order.
- The second describes the biosphere as a self-regulating entity in which life acts as the Earth's thermostat, a mix or a blend of Nature/machine with organic metaphors; this is the so-called "Gaia hypothesis" [56].
- The third, to which Botkin belongs, rejects the description of the Earth as a mystical organism, proposing a new perspective that blends the older organic metaphor with the new technological metaphor and insists on perpetual change in the biosphere.

This reinterpretation of ecology as natural history is couched in coevolutionary terms between the four dynamic parts of the biosphere – rocks, oceans, air, and life – each with its own ranges of movements and rates of change. "Biological evolution has led to global changes in the environment which, in turn, have led to new opportunities for biological evolution. In this way, a longterm process of change has occurred throughout history of life on the Earth, which is an unfolding, one-way story" [7, p. 148]. Thus, the production of a "biospheric biography" is in order.

This theoretical perspective has far-reaching practical consequences. We must learn to manage the biosphere and the Earth's resources in terms of uncertainty, change, risk, and complexity. Botkin's conclusions point in the same direction as the recent theories of complexity and chaos [65, 31]. They should not be interpreted, however, as a renunciation of scientific analysis and engineering action. "We can engineer nature at nature's rates and in nature's ways: we must be wary when we engineer nature at an unnatural rate and in novel ways" [7, p. 190].

The response to the man-made problems for the environment should not consist in giving up modern technology or in clinging to the belief that everything natural is desirable and good. Botkin concludes: "having altered nature with our technology, we must depend on technology to see us through to solutions" (p. 191). Happily, he qualifies this statement by saying that we must learn how to live with the discordant harmonies of the biosphere so that they function not only to promote the continuation of life but also to benefit our esthetics, morality, philosophies, and natural needs.

Global change

Already in the early 1970s, it was clear that for the first time in history, human intervention was reaching a scale capable of producing significant and irreversible modifications to the working of the biosphere. That evidence transformed itself into an alarm as advances in climatic research confirmed the potentially deleterious consequences of the "greenhouse effect," anticipated a century ago by Svente Arrhenius.

When will the catastrophe occur, for which several more or less plausible scenarios are explored by the media? Which countries will be the most affected? Are all the foreseeable climatic changes negative? Opinions about these questions differ, and none of the existing climatic models can reliably predict the pace and rate of climatic changes [52]. Yet, the presumptions have proved sufficiently strong to mobilize the international community for the first time to undertake preventive action on a significant scale.

Conferences of scientists and politicians succeed each other at an accelerating pace. Important legal precedents have been set. On the one hand, the international community has recognized the need to jointly manage a significant portion of the "international commons" – the atmosphere. On the other, it has agreed to phase out the production of some products releasing greenhouse gases (CFCs) and to enshrine this decision in the international convention on the protection of the ozone layer. Negotiations about global conventions on climate, forests, and biodiversity are under way.

But these initial successes should not be overestimated. Fundamental differences remain between the North and the South about the hierarchy of problems. Is global change to be put above the immediate needs of survival of the poor majority of Spaceship Earth's passengers? Is the recognition of the globality of a problem to be interpreted as a pointer for equal treatment of all the countries whatever their degree of development? How should the costs of adaptation be apportioned? Gallopin, Gutman, and Winograd [27] point out that the recommendations addressed to the South to restrain future energy consumption "sound like a fat man coming out from a fine restaurant and advising a beggar to fast because that is what he is thinking to do after indulging in such a good meal."

The methodology used to estimate the net emissions of greenhouse gases, as well as the data used, have proved highly controversial. The Indian environmentalist Anil Agarwal [1] challenged outright the work of the World Resources Institute (WRI) in Washington [108], which had calculated net emissions of gases by assuming that the same proportion could be applied to all countries to take account of Nature's capacity for the sequestration of gases, and then subtracting the result from gross emissions. Agarwal considers that the "emission rights" due to the natural capacity for self-purification should also be equally distributed among all the inhabitants of the planet. This point is well taken, and it completely upsets the calculations of the WRI, which underestimate the relative share of the industrialized countries in the global warming of the atmosphere.

An even stronger ethical point has been raised by Agarwal. Pollution arising from the need to survive and pollution arising from affluence cannot be treated on an equal footing. Are we going to reduce the livestock population in India or reduce the paddy fields throughout Asia just because cows and paddy fields release a lot of methane? Or should we instead concentrate first on reducing the consumption of fossil fuels by the hundreds of millions of cars that circulate in Northern cities and highways?

As for the primary data, Brazilian scientists and authorities have challenged the WRI's estimates of deforestation in the Amazon region. The figures quoted by the two sides vary by a factor from one to four.

Instead of summarizing the conflicting views about the imminence and the extension of the likely damage produced by global warming, I shall turn to some underlying epistemological and policy questions [52].

A distinction must be made between the realm of "scientific questions" raised by global warming and that of "societal questions." The latter should be evaluated by citizens, not by scientists. As far as the former are concerned, in the forecasting of the climate's future, the concept of "the average" may be misleading; often what matters more are the extremes. Meteorology deals with movements and transportation of energy and water in the atmosphere. Hundreds of thousands of atmospheric "cells" must be taken into account, each one subject to the laws of gravity and of fluid mechanics, and each having many interactions with one another. Modelling this complexity poses very serious problems, and efforts so far leave many uncertainties. "Bold forecasts assisted nowadays by numerical modelling using supercomputers are nevertheless fragile and may even lead to error if one does not consider very carefully the different time-scales which intervene in the processes under study, as well as the degree of confidence that can be attributed to the modelling of different processes" [52, p. 72]. Significantly, Kandel shares Botkin's opinion about the myth of natural equilibria. He uses the concept of dynamic equilibria, which takes into account the evolution of the biosphere.

Even more important than the problems raised by Kandel, scenarios of the consequences of global warming depend on three levels of modelling:
- an economic and industrial model that predicts future rates of emission in the atmosphere of carbon dioxide, methane, and CFCs;
- a bio-geo-chemical model to predict the evolution of the concentration of these gases in the atmosphere, taking into account the rates of emission and the exchange processes between the atmosphere, the oceans, soils, and the biosphere;
- a climatic model to predict how the climate, with its atmospheric and marine components, will change.

Another difficult question concerns feedbacks. The increase in carbon dioxide in the atmosphere, analysed mainly in terms of its impact on global warming, at the same time encourages the growth of vegetation. If properly used, the increased biomass production may be a good, not a bad. The ability to put this biomass to good use depends on another feedback: that of human intelligence [52, p. 77].

The real question is to know whether we really want to steer our planet. If we can really know what the consequences will be of this or that policy, if we can really change the policy on the basis of this knowledge, we can – of course within the limits fixed by the laws of nature – choose our destiny. The scientific knowledge and the political mechanism which can save us from an undesirable climatic change are the same as will allow us deliberately to modify the climate. The future of the climate would in this way be inextricably linked to the future of humankind." [52, pp. 122–123]

Signposts for the future

The conflicting positions are clear by now, and little will be gained at this stage by pursuing the conceptual discussion of sustainable development. Priority should be given instead to designing transition strategies towards the virtuous green path, taking into consideration the diverse configurations in the North and in the South [86]. How do we get there? At what economic cost (or gain)? When?

Such strategies must allow several decades to develop non-linear trajectories with changing priorities over time, to produce a new generation of environmentally friendly technologies, and progressively retool the productive system. A 40-year period seems reasonable.

Given the gap that separates the North and the South in terms of wealth, technical capability, lifestyles, and compelling social problems, globality should not be used as a pretext to impose a unique strategy and equal obligations on both groups of countries. Quite the contrary: each country must find the ecosystem, cultural, and site-specific responses to global problems (thinking globally, acting locally). The main burden of the transition must be assumed by the North. The richer and more advanced scientifically and technically a country, the greater its flexibility – all the more so in that many aspects of the transition may yet prove to be less costly in financial and social terms than maintaining business as usual.

Science and technology appear as a major, but by no means unique, variable capable of speeding up or delaying the transition. If properly handled, the transition towards the virtuous green path offers many opportunities for innovative use of resources. Since it is impossible to review them all, I shall concentrate on four examples chosen because of their importance for a meaningful transition strategy and their implications for science and technology.

A one Kw per capita society

A low-energy profile in the North but also in the South, in particular a sharp reduction of fossil energy consumption, is probably the single most important objective. As Amory Lovins aptly remarked in his pioneering book *Soft Energy Paths* [57], people do not want electricity or oil but comfortable rooms, light, vehicular motion, food, and other real things.

In their important study on *Energy for a Sustainable World*, Goldemberg et al. [32] argue that the systematic introduction of already known efficient techniques of end-use of energy would bring about a considerable decrease of per capita energy consumption in industrialized countries and, at the same time, make it possible to reach the present standards of Western comfort in the South with a very small increase of per capita consumption of energy: one Kw per capita will prove sufficient. In their scenario for the years 1980–2020, they predict a doubling of GNP with a 50 per cent cut in per capita energy use in the developed countries (from 6.8 to 3.5 toe). As for the developing countries, their per capita energy consumption would increase from 1.1 to 1.4 toe. Overall world consumption of energy throughout the 40 years would grow by only 9 per cent. We can speak of a zero rate of energy growth.

The authors do not, however, consider the possible gains from modifying the pattern of demand, a subject already discussed in this chapter that does not lend itself easily to quantitative estimates but nevertheless deserves careful consideration.

Opinions vary about the future of the replacement of fossil fuels by non-conventional energies. A recent monograph of the Worldwatch Institute presents a very optimistic outlook for them, based on the anticipation of a sharp reduction in the costs of wind, photovoltaic, and solar thermal electricity [23]. According to their scenario for 2030, world energy use will increase from 9,300 Mtoe in 1981 to 10,490 Mtoe in 2030. The use of oil will be cut by half, from 3,098 to 1,500 Mtoe, that of coal by a factor of nine (from 2,231 to 240 Mtoe). Natural gas remains stationary (1,707 and 1,750 Mtoe). Nuclear energy (now representing 451 Mtoe) is phased out, while renewables jump from 1,813 to 7,000 Mtoe. The total emission of carbon would in this way be more than halved, from 5,764 to 2,590 Mtoe.

Other studies are much less optimistic, but all of them tend to agree that, where there is the political will, affordable technologies to reduce carbon emissions are now available. A 1991 OTA study considers that the United States can decrease its emission of carbon dioxide by 35 per cent below 1987 levels within the next 25 years. In the short term, most actions for decreasing emissions would focus on reducing total energy demand. These actions might include implementation of performance standards, tax incentive programmes, low-cost loans, carbon-emission or energy taxes, labelling and efficiency ratings, energy audits and research, development and demonstration activities.

According to figures produced by the Worldwatch Institute [8], improving energy efficiency has a cost of 2 to 4 cents/Kwh with a carbon reduction of 100 per cent, an estimated pollution cost of zero cents/Kwh, and a carbon avoidance cost (compared with existing coal-fired power plants) of $0–$16/ton. All other alternatives to fossil fuels have much higher carbon avoidance costs.

What place should be reserved for biomass energy? The largest experiment yet carried out is the controversial Brazilian "Pro-alcool" programme, which launched massive production of sugar-cane eth-anol, used first as an additive to petrol (22 parts per 100) without any modification of the car engines, then as the only fuel for specially adapted cars. There are now several million such alcohol-powered cars in Brazil. In spite of gloomy predictions made by major car manufacturers, technically the experiment runs smoothly.

The drawback of Pro-alcool is its poor economic results. It was in-troduced as a crash programme reminiscent of wartime measures, and the objective was achieved without much regard to the cost; the state paid out lavish subsidies under pressure from the sugar-cane lobby. Arguably, too, it would have been wiser to restrict the alcohol-powered vehicles to city-based service fleets of vans, taxis, etc., rather than to distribute the new fuel all over the country.

Attacked by the oil lobby, Pro-alcool seemed condemned at the time of the Gulf war, but it has since been revived with an emphasis on cogeneration of electricity. In fact, Pro-alcool could become a more economic proposition if it concentrated on sugar cane, which at present is poorly utilized; it could, for example, fuel the distilleries. Energy and financial savings at the field level could be achieved through biological pest control and replacing fertilizers by direct ni-trogen fixation. Brazil is at the forefront of the research in this area [18]. Considerable progress could also be achieved by improving the fermentation processes and broadening the range of uses of the many by-products. For example, bagasse is a good animal feed (some alco-hol refineries maintain large herds of cattle), and they can be also transformed into cardboard or paper or formed into briquettes. It can also be used for the cogeneration of heat for the refinery and of elec-trical energy. Thus, from a single-purpose production process we move into an integrated sugar-cane-based agro-industrial system, closing the loops whenever possible and adding new production mod-ules. The overall economic efficiency of such a system is much greater than that of the sum of single-purpose productions. Furthermore, sugar cane agro-industrial systems need not be managed as one large unit: it is possible to design socially responsive systems based on cooperatives and clusters of small-scale industries.

Another way of improving the efficiency of alcohol use in Brazil would be by spreading smaller production units (mini-distilleries or even micro-distilleries) throughout the country, working for local purposes and thus reducing the prohibitive distribution costs.

Of course, other bio-fuels may also be envisaged. A vegetable-oil-based additive to diesel would solve many of Brazil's problems. In this area, Europe is more active than Brazil. European regions are involved in several biomass fuel experiments with the support of the EEC. The first pilot factory producing a diesel additive from rape-oil is being built in France. Sweden is more ambitious: the Committee for Research on Natural Resources proposes that extensive efforts be made to build up a competitive phytochemical industry by the year 2000, mainly based on forest raw materials and working through decentralized small-scale production units [58].

Hall [40] has demonstrated that even burning wood instead of fossil fuels makes sense in terms of slowing down global warming. Contrary to a superficial view shared by many Greens, one should use as much forest biomass as possible on three conditions: burning should not be used as a way of clearing the ground; the forest biomass should be used only where it can be regenerated or replanted; preference should be given to lasting uses of biomass (wood transformed into houses or furniture becomes a sink of carbon).

The prospects for large-scale biomass-energy production is particularly attractive for countries with large areas of suitable soils and favourable climatic conditions, such as Brazil or Argentina. In countries with an unfavourable land-man ratio, such as China or India, biomass fuel is also a priority, although the emphasis shifts to the use of agricultural, animal, and human waste. The long and not always successful take-off of biogas in the latter two countries should not divert them from redesigning more efficient biogas programmes.

A modern plant (biomass) civilization for the tropical countries
Bio-energy is only one among many products that can be derived from biomass. Following Jyoti Parikh [74], one can speak of a "5-F model" for alternative uses of biomass: as fuel, fertilizer, food, animal feed, and industrial feedstock. The phrase "civilisation du végétal" was coined by Pierre Gourou to describe the traditional civilizations of the Far East: in the Chinese cultural area, for example, bamboo has multiple uses.

With the recent progress of biotechnologies, we can speak now of not only the possibility but the extreme urgency to build a new form of civilization based on the sustainable use of renewable resources [99], at least in the tropical countries, whose climate and ecological conditions are favourable to a high primary productivity of biomass grown on fields, in forests, and in water. Swaminathan's injunction recalls Gilberto Freire's pioneering effort in setting up a permanent seminar on tropicology in Recife. In this way, tropicalization of science and technology has at least been put on the agenda.

Biotechnology has a double potential function: to increase the productivity of biomass and to open up the range of food, energy, and industrial products derived from biomass. Up to now, little progress has been achieved as far as the latter application is concerned, but prospects seem bright.

The main obstacle resides in the lack of access of developing countries in general, and of small rural producers in particular, to the biotechniques necessary for this second "green revolution." On this point, the situation has worsened considerably since the first green revolution, which was already heavily biased towards the interests of large and medium-scale producers (see Glaeser [30]; and for a recent evaluation in India Hanumantha Rao [41]). The present mood is one of extending the private intellectual property rights on an ever wider range of biotechnologies and even of new products obtained through their application.

In so far as the World Bank [106] state-of-knowledge report on biotechnologies applied to agriculture insists on private intellectual property rights and on the comparative advantage of larger producers, its evaluation of the prospect of a second green revolution for the small producers remains very cautious.

By contrast, without ignoring the difficulties of the problem, the project on biotechnology and development organized by the University of Amsterdam [9] explores systematically the package of biotechnologies for small producers. Important efforts in this direction are also being made in India [92]. Biotechnology is expected to increase soil fertility and lower the dependence on fertilizers and chemical pesticides. It is also expected to increase crop yields by incorporating resistance to drought, pests, and disease and enhance the protein, starch, or oil content of crops, and disseminate through micro-propagation the desired fruit-trees and rapidly growing varieties of trees and bushes for fuel wood.

Success will depend to a great extent on the ability to organize publicly sponsored research and extension systems. Producing and disseminating biotechnology packages for small producers constitutes a high priority in development-oriented science and technology policy.

As for biomass-based industrialization, if properly handled, it offers a unique opportunity for environmental, social, and economic

327

gains leading to a new rural-urban configuration through "diffuse industrialization," reducing in this way the flow of refugees from the countryside to the large cities. This idea is at the heart of development strategies in China [22].

The main social advantage resides in the creation of employment and in the reduction of infrastructural costs for the expansion of large cities. Manufacturing industries now create few jobs, essential as they may be for the transformation of developing economies. From the employment point of view, what really matters is the multiplier effect: the higher the workers' wage, the more they will spend on goods and services. But by resorting to biomass in place of oil as a feed stock for chemical industries, one sets in motion a second upstream multiplier, because biomass production is much more labour-intensive than oil production.

As for the environmental advantages, "green plastics" are likely to be more environmentally friendly than their oil-based counterparts, even though one should not automatically conclude that biomass production and the products derived from it are by definition ecologically benign. Moreover, once the biomass-based industry has grown into an important segment of the national economy, careful management of the life-support systems – water, soils, forests – will be internalized in the working of the economic system.

Finally, the choice of the biomass species grown or collected for the purpose of food, energy, and industrial production depends on a very careful examination of the potentialities of each ecosystem, taking into consideration the agro-climatic conditions, the natural capital of bio-diversity, and the social and cultural contexts.

Development for the Amazon region

The approach just suggested should be applied to all major eco-regions. As an example, I shall take the tropical rain forest ecosystem of the Amazon region, known for its climatic importance and ecological fragility [19, 94–96].

It is necessary first of all to discard all the scientifically false information circulated by the media (Amazonia as the lung of the world is just one example). In particular, one should remember that the practical potential for deforestation or reforestation to modify the greenhouse effect is ultimately limited. The amount of carbon in the atmosphere is roughly comparable to the amount contained in the biosphere, and the amount within soils is 1.5 times as much as either. By contrast, 15 times as much carbon as in the atmosphere is stored in the ground as fossilized carbon and peat and 75 times as much in the oceans.

While reversal of deforestation and re-afforestation could be cost-effective means of reducing net carbon dioxide emissions, they must compete against alternative land-use demands and take into account the fact that, in order to remain effective over the long term, the carbon must be sequestrated and the process renewed as the trees mature [3].

The Amazon ecosystem should be protected in the interest of its inhabitants and of all Brazilians as a potential source of wealth, a climate stabilizer, and a repository of biodiversity. But the long-term future of the Amazon region cannot consist in transforming it into a huge forest reserve. Nevertheless, development of the Amazon region can be made compatible with banning new land clearing. The original forest has already been destroyed on 300,000 square kilometres, enough to keep at least a couple of generations busy with rational rehabilitation and use of these "capoeiras" without moving the economic frontier further.

The immediate consequence of this approach would be to define a spatial strategy that establishes an archipelago of more or less intensive "development reserves" in the green ocean so as to slow the pressure on the primeval forest, thereby protecting the remnants of the indigenous population and the biodiversity. A related issue is one of slowing down the growth of Manaus and Belem, two mega-cities in the making (over 60 per cent of the Amazonian population is already urbanized; as Bertha Becker says, the Amazon region was born urbanized). At the same time, it is necessary to ensure the minimum critical size for human settlements sufficient to provide social services and cultural amenities.

The so-called "extractive reserves" constitute an immediate solution for the existing population of destitute *seringueiros* but do not offer a blueprint for a long-term strategy for the Amazon region. One *seringueiro* needs 500 hectares to earn a miserable existence. In other words, the density of population there cannot exceed 1–2 persons per square kilometre.

Each "development reserve" should strive to make a rational use of the resource potential of its ecosystem in order to establish a fairly integrated local economy, selectively linked with the outside world. Since colonial times the Amazon region has been regarded as a source of exportable raw materials and products, not as a place where many more people could live comfortably. Given the distances that separate it from the south of Brazil and from external markets, the Amazon region will always be at a disadvantage in terms of transportation costs, except for products with high value-added per unit of weight.

The variety of the Amazonian ecosystems has often been underestimated and the debate conducted as if it were a homogeneous area. The development of the Amazon will lead to multiple configurations of a biomass-based civilization in which different systems of agroforestry and of aquaculture will play a dominant role.

Agroforestry and aquaculture, eventually combined in integrated production systems, appear thus as a major priority for research and experimentation, not only in the Amazon region, but in all the countries with extensive tropical rain forests. The "blue revolution" has not as yet come of age, in so far as aquaculture is responsible for a modest share of fish and other water-grown food or animal feed; hunting and gathering still predominate.

Making cities more livable in the twenty-first century

By the beginning of the twenty-first century, the majority of the world's population will be living in cities. No visible signs suggest a significant decrease of the urbanization rates in the South within the next few decades. Even the most optimistic assessment of the prospects for biomass/biotechnology-based industrialization do not lead to the conclusion that rural-urban migrations will stop.

Most of the oil is consumed and greenhouse gases are produced in cities [68], and in terms of quality of life for the populations concerned, the disruption of the urban environment is by far the most difficult problem faced in the mega-cities of the South. The apocalyptic description of Mexico by the well-known novelist Carlos Fuentes applies to many other large, and smaller, third world cities:

The pulverized shit of three million human beings without latrines.
The dung in powder of ten million animals that defecate in the streets.
Eleven thousand tonnes of chemical waste per day.
The deadly fumes of three million engines that spew out uncontrollably gusts of pure poison, sooty miasmas; trucks, taxis, cars, each one eructating and contributing to the extinction of trees, lungs, throats and eyes. [26]

The situation in eastern Europe is also chaotic: in the former Soviet Union, 50 million people live in cities where air pollution exceeds the national standard by more than a factor of 10; half of

Poland's cities, including Warsaw, do not treat their waste at all; only 30 per cent of the sewage produced in the former Soviet Union is treated; 300 cities and towns in Hungary must rely on bottled or piped water because local water has been contaminated by fertilizer run-off; life expectancy in the polluted regions of Czechoslovakia is five years lower than in the cleaner parts of the country [25].

The situation in the cities of the North is less dramatic. Even so, action to protect urban environments suffering from pollution of all kinds is badly needed and requires imaginative new policies [72]. Such action is all the more necessary because the Northern cities are menaced by a potentially explosive combination of environmental and social problems arising out of exclusion, segregation, and lack of opportunities for young people and minorities.

In neither the South nor the North will these serious problems be resolved by investment and technologies alone. The North has the wherewithal; it is a matter of political will. In the South, lack of funding for the urban infrastructures and their maintenance makes the problems even more intractable.

The cities in the South need inexpensive and efficient technologies for sanitation, mass transport, and housing. A conference organized in São Paulo in 1978 on new technologies for the cities reached the conclusion that virtually nothing, affordable by the third world cities, was available. In sanitation, little progress has been made since ancient Rome.

Mention should be made of the imaginative, though not always practical, ideas put forward by Richard Meier in his formulation of the concept of "resource-conserving cities" for the third world, which blend the most advanced and traditional techniques [62, 63]. Meier's fundamental premise is that any pale imitation of advanced urbanization in the South would require a much greater consumption of energy, water, and human effort than is available.

Closely related to Meier's concerns are the attempts to define an ecodevelopment strategy in the urban context [87]. A city is also an ecosystem and as such is a potential resource. In every city there exist latent, idle, underutilized, and wasted resources: land that can be put under cultivation, at least temporarily; waste that can be collected and recycled; energy and water that can be saved; infrastructures, buildings, and equipment whose life cycle can be extended through proper maintenance. All these activities are fairly labour-intensive, and jobs created in this way may pay for themselves through the saving of resources.

331

People and citizens' associations have a major role to play in such ecodevelopment strategies. The same is true of "self-help building" and rehabilitation of shanty towns. The pioneering work of John Turner led to a major discussion of these matters and ultimately contributed to a welcome shift in urban policies, reducing the emphasis on the supply policies largely based on the use of industrial techniques in building and promoting instead the "enabling policies" designed to support local initiatives by making available the resources and techniques that cannot be mobilized locally (for a review, see Sachs [80], World Bank [107]).

To be implemented properly, this new approach requires a redirection of science and technology policy. The South will have to invent new cities quite unlike the models in the North, as these cannot be replicated at either the scale or the pace required by urbanization trends, nor are they in their present form a commendable blueprint for livable cities.

Pro-active and innovative strategies must address simultaneously the following aspects: institutional and managerial models; new forms of partnership between civil society, enterprises, and public authorities; shifting from supply policies to enabling policies in order to stimulate initiative and resourcefulness; continuous efforts for resource saving and elimination of wastefulness; skilful management of technological pluralism and intensified research for new technological solutions, both affordable and accessible to developing countries.

Cities are like people. They belong to the urban species but they have their unique personality. The response to the urban challenge must take into account the singular configurations of natural, cultural, and socio-political factors, as well as of the historical past and tradition of each city. Instead of proposing across-the-board, homogenizing solutions, the diversity of cities should be considered as a cultural value of paramount importance.

Concluding remarks: Disentangling Prometheus

Paraphrasing Salomon [89], one may say that Prometheus is caught in a double bind.

On the one hand, he faces an ever growing gap between the potential of science and technology and the accumulated backlog of unsolved human needs – the "social debt," as it is called in Latin America. This contradiction reaches its height when science and technology are put at the service of death, not life, in the form of sophisticated weaponry with the attendant compulsion to test their killing and destructive efficiency by putting them to actual use. Each war – a perversion of Schumpeter's "creative destruction" necessary to fuel the modernization drive – renews the demand for a new and costlier generation of weapons and for the reconstruction of what has been destroyed.

On the other hand, borrowing the metaphor from Serres [93], Prometheus must seek a "natural contract" capable of overcoming the contradiction between Man and Nature, exacerbated by the predatory use of natural resources and the overloading of the capacity of the biosphere to act as a sink. In other words, the present destructive action of human parasites on their host – Nature – should be transformed into a symbiotic relationship. The parasite will live only as long as the host continues to serve as its life support. Behind ecocide looms genocide.

To face these formidable challenges, Prometheus has two options. Either he reaffirms his blind faith in the power of science and technology to find, in time, solutions to the problems created by their progress, which means that he continues to steer the present course, in which the tool guides the hand. (As scientism is fundamentally optimistic, it tends to minimize the risk of heading towards social or ecological catastrophe.) Or he strives to get the tool under control, to harness science and technology for societal development, subordinated to the three criteria of social equity, ecological prudence, and economic efficiency.

In broad institutional terms, this means going back to Polanyi's enquiry into ways in which the economy is embedded in the society and, as far as the market-oriented economies are concerned, addressing the problématique of the social construction of markets [4]. In operational terms, it is necessary to learn how to make decisions through explicit harmonization of three distinct types of logic: the ethical, the technical, and the political [37].

The environment has been discussed mainly as a constraint and a cost. But it is possible to look at it from a positive angle as a potential asset to be used for rational purposes and through rational methods. This puts an enormous task before science and technology, while at the same time confronting the South with daunting challenges.

For obvious reasons (as we have seen in the case of biotechnology), the South cannot tolerate a situation of total dependence on imports of "black-box" technologies from industrialized countries on monopolistic conditions reinforced through an ever broader definition of intellectual property rights. But the call for Southern self-reliance, abusively interpreted in terms of autarky, is unrealistic.

All the countries in the world, including the most advanced scientifically and the richest, need a science and technology strategy with three components:
1. purchase abroad and use of black-box technologies;
2. opening of the imported technological packages and their adaptation (only then may we speak of "technology transfer");
3. domestic invention.
The proportions of these three components (as well as the balance of trade in technology) depend on the size of the country, the condition of its R&D establishment, and the financial situation.

Self-reliance should be interpreted in a narrower manner, as the ability to be selective in the choice of technologies, to strike a changing balance between the three components under discussion, and to transform the condition of latecomer into an advantage by seizing the rare opportunities for leap-frogging. Even selectivity in imports is hard to achieve, in so far as it presupposes the availability of trained manpower, access to up-to-date sources of scientific and technological information, a truly competitive international market, and institutional mechanisms to carry out effective national science and technology policies.

How many countries in the South have reached the stage of self-reliant science and technology policies? Is it realistic to expect that – except for the giants: Brazil, India, China – they will ever get there? Can South-South collective self-reliance offer a way out? [97]. The discussion around the book by Salomon and Lebeau [90] shows a wide variety of opinions.

The confidence expressed by Botkin [7] and Kandel [52] about the potential of science for the management of the biosphere will be put to a very severe test unless the present political trends are reversed. The main threats to the future of humanity and eventually to that of life on Earth pertain to the realm of the sociosphere and, more specifically, to that of the political economy of environmentally sound development that reconciles governability, democracy, social justice, ecological prudence, and economic efficiency in multiple forms of mixed economy.

References

1   Agarwal, A., and S. Narain. *Global Warming in an Unequal World: A Case of Environmental Colonialism*. New Delhi: Centre for Science and Environment, 1991.

2   Ahmad, Y. et al. *Environmental Accounting for Sustainable Development*. Washington, D.C.: World Bank, 1989.

3   Arrhenius, E., and T.W. Waltz. *The Greenhouse Effect – Implications for Economic Development*. World Bank Discussion Papers, no. 78. Washington, D.C.: World Bank, 1990.

4   Bagnasco, A. *La Costruzione del Mercato Studi sullo sviluppo di piccola*. Bologna: Il Mulino, 1988.

5   Barrère, M., ed. *Terre patrimoine commun. La science au service de l'environnement et du développement*. Paris: La Découverte, 1992.

6   Berry, A. *The Next Ten Thousand Years: A Vision of Man's Future in the Universe*. New York: Saturday Review Press, 1974.

7   Botkin, D.B. *Discordant Harmonies – A New Ecology for the Twenty-first Century*. New York: Oxford University Press, 1990.

8   Brown, L. et al. *The State of the World, A Worldwatch Institute Report on Progress Toward a Sustainable Society*. New York: W.W. Norton and Company, 1990.

9   Bunders, J., ed. *Biotechnology for Small-scale Farmers: Analysis and Assessment Procedures*. Amsterdam: VU University Press, 1990.

10   Céron, J.P., and J. Baillon. *La civilisation de l'éphémère*. Grenoble: PUG, 1979.

11   Clark, W.C., and R.E. Mann, eds. *Sustainable Development of the Biosphere*. Cambridge: Cambridge University Press, 1986.

12   Cole, H.S.D. et al. *Models of Doom: A Critique of the Limits to Growth*. New York: Universe Publishing, 1973.

13   Commoner, B. *The Closing Circle – Man, Nature and Technology*. New York: Knopf, 1971.

14   Costanza, R., ed. *Ecological Economics: The Science and Management of Sustainability*. New York: Columbia University Press, 1991.

15   Dag Hammarskjöld Foundation. "What Now: Another Development?" *Development Dialogue* Nos. 1, 2 (1975).

16   Dahlberg, Kenneth A. "Towards a Theory of Regenerative Food Systems: An Overview." Paper prepared for the conference, Redefining Agricultural Sustainability, University of California, Santa Cruz, June 1990.

17   Daly, H.E. "From Empty-world to Full-world Economics: Recognizing an Historical Turning Point in Economic Development." In: Goodland et al., eds., pp. 29–38. *See* ref. 34.

18   Dobereiner, J. "Avanços recentes na pesquisa em fixação biológica de nitrogénio no Brasil." *Estudos Avançados* 4 (1990), no. 8: 144–152.

19   Dourojeanni, M.J. *Amazonia, Que hacer?* Iquitos, Peru: Centro de estudios teológicos de la Amazonia, 1990.

20   Ehrlich, P., and A. Ehrlich. *Population, Resources and Environment: Issues in Human Ecology*. San Francisco: W.H. Freeman, 1972.

21   ———. *The Population Explosion*. New York: Simon and Schuster, 1990.

22   Fei Hsiao Tung et al. *Small Towns in China*. Beijing: New World Press, 1986.

23   Flavin, C., and N. Lenssen. *Beyond the Petroleum Age: Designing a Solar Economy*. Worldwatch Paper, no. 100. Washington, D.C., December 1990.

24   Forrester, J.W. *World Dynamics*. Cambridge, Mass.: Wright Allen Press, 1971.

25   French, H.F. *Green Revolutions: Environmental Reconstruction in Eastern Europe and the Soviet Union*. Worldwatch Paper, no. 99. Washington, D.C., 1990.

26   Fuentes, C. *Christopher Unborn*. London: André Deutsch, 1989.

27   Gallopin, G. et al. *Environment and Development – A Latin American Vision*. Report to UNCED, Ecological Systems Analysis Group. Bariloche: Fundación Bariloche, June 1991.

28   Georgescu-Roegen, N. *The Entropy Law and the Economic Process*. Cambridge, Mass.: Harvard University Press, 1971.

29   Giarini, O. *Dialogue on Wealth and Welfare – An Alternative View of World Capital Formation*. Oxford: Pergamon Press, 1980.

30   Glaeser, B. *Ecodevelopment: Concepts, Projects, Strategies*. Oxford: Pergamon Press, 1979.

31   Gleick, J. *Chaos: Making a New Science*. New York: Viking, 1988.

32   Goldemberg, J. et al. *Energy for a Sustainable World*. New Delhi: Wiley Eastern Ltd., 1988.

33   Goldsmith, E. "The Future of an Affluent Society: The Case of Canada." *The Ecologist* 7 (1974), no. 5.

34   Goodland, R. et al., eds. *Environmental Sustainable Economic Development: Building on Brundtland*. Paris: Unesco, 1991.

35   Gorz, A. *Métamorphoses du travail: quête du sens*. Paris: Galilée, 1988.

36   ———. *Capitalisme, socialisme, écologie*. Paris: Galilée, 1992.

37   Goulet, D. "Tasks and Methods in Development Ethics." *Cross Currents* 38 (1988), no. 2.

38   Gowariker, V., ed. *Science, Population and Development: An Explanation of Interconnectivity and Action Possibilities in India*. New Delhi: Unmesh Publications, 1992.

39   Haavelmo, T., and S. Hansen. "On the Strategy of Trying to Reduce Economic Inequality by Expanding the Scale of Human Activity." In: Goodland et al., eds. *See* ref. 34.

40   Hall, D.O., H.E. Mynick, and R.H. Williams. *Carbon Sequestration versus Fossil Fuel Substitution – Alternative Roles for Biomass in Coping with Greenhouse Warming*. PU/CEES Report no. 255. Princeton, N.J., November, 1990.

41   Hanumantha Rao, C.H. "Technological Changes and Capitalist Relations." *Mainstream* 29 (1991), no. 16.

42   Hardoy, J. et al. *The Poor Die Young: Housing and Health in Third World Cities*. London: Earthscan Publications, 1990.

43   Helleiner, G. "Conventional Foolishness and Overall Ignorance: Current Approaches to Global Transformation and Development." *Canadian Journal of Development Studies* 10 (1989), no. 1: 107–120.

44   Herrera, A. "The Generation of Technologies in Rural Areas." *World Development* 9 (1981): 21–35.

45   Hirsch, F. *Social Limits to Growth*. Cambridge, Mass.: Harvard University Press, 1977.

46   Illich, I. *The Right to Useful Unemployment and Its Enemies*. London: M. Boyars, 1978.

47   ILO. *The Re-emergence of Small Enterprises – Industrial Restructuring in In-*

dustrialised Countries. Geneva: International Institute for Labour Studies, 1990.

48   Ingelstam, L., and G. Backstrand. "How Much is *Lagom*? Sweden as a Case in the Quest for Appropriate Development." In: Dag Hammarskjöld Foundation. *See* ref. 15.

49   Janicaud, Dominique. *La puissance du rationnel*. Paris: Gallimard, 1985.

50   Kahn, H., W. Brown, and L. Martel. *The Next Two Hundred Years*. New York: Morrow, 1976.

51   Kalecki, M. *Selected Essays on the Economic Growth of the Socialist and the Mixed Economy*. Cambridge: Cambridge University Press, 1972.

52   Kandel, R. *Le devenir des climats*. Paris: Hachette, 1990.

53   Kothari, Rajni. "Environment, Technology and Ethics." In: J.R. Engel and J.G. Engel, eds. *Ethics of Environment and Development – Global Challenge, International Response*. Tucson: University of Arizona Press, 1990, pp. 27–49.

54   Kuznetsov, B. *Nauka v 2000 godu*. Moscow: Nauka, 1969.

55   Lester, B. "When Government Fails, Will the Market Do Better? The Privatization/Market Liberalization Movement in Developing Countries." *Canadian Journal of Development Studies* 12 (1991), no. 1: 159–172.

56   Lovelock, J. *Gaia*. Milton Keynes: Open University Press, 1987.

57   Lovins, A. *Soft Energy Paths: Towards a Durable Peace*. Cambridge, Mass.: Ballinger, 1977.

58   Lundholm, B. *Renewable Natural Resources – A Swedish Research Programme for the Future*. Stockholm: Swedish Council for Planning and Coordination of Research, 1982.

59   McNeely, J.A. et al. *Conserving the World's Biological Diversity*. Prepared and published by the International Union for Conservation of Nature and Natural Resources, World Resources Institute, Conservation International, World Wildlife Fund-US, and the World Bank, 1990.

60   Mathews, W., ed. *The Outer Limits and Human Needs: Resources and Environmental Issues of Development Strategies*. Uppsala: The Dag Hammarskjöld Foundation, 1976.

61   Meadows, D.M. et al. *The Limits to Growth: A Report for the Club of Rome's Project on the Predicament of Mankind*. New York: Universe Books, 1972.

62   Meier, R.L. *Planning for an Urban World: The Design of Resource-conserving Cities*. Cambridge, Mass.: MIT Press, 1974.

63   Meier, R.L., and A.S.M. Abdul Quim. "A Sustainable State for Urban Life in Poor Societies: Bangladesh." *Futures* 23 (1991), no. 2: 128–145.

64   Monosowski, E. "L'évaluation et la gestion des impacts sur l'environnement de grands projets de développement: le barrage de Tucurui en Amazonie, Brésil." Ph.D. diss., Paris: EHESS, 1991.

65   Morin, E. *Introduction à la pensée complexe*. Paris: ESF, 1990.

66   National Academy of Sciences. *Underexploited Tropical Plants with Promising Economic Value*. Report of Commission on International Relations, Washington, D.C., 1975.

67   National Research Council. *Board on Agriculture, Alternative Agriculture*. Washington, D.C.: National Academy Press, 1989.

337

68   Newman, P. "Greenhouse, Oil and Cities." *Futures* 23 (May 1991), no. 4: 335–348.

69   Nicholson. *The Environmental Revolution: A Guide for the New Masters of the World*. London: Hodder & Stoughton, 1970.

70   Nordhaus, W.D. "Greenhouse Economics, Count before You Leap." *The Economist* (7–13 July 1990): 19–22.

71   OECD. *Economic Instruments for Environmental Protection*. Paris: OECD, 1989.

72   ———. *Environmental Policies for Cities in the 1990s*. Paris: OECD, 1990.

73   Panikhar, P.G.K., and C.R. Soman. *Health Status of Kerala*. Trivandrum, India: Centre for Development Studies, 1984.

74   Parikh, J. *Farm Gate to Food Plate*. Paris: The Food Energy Nexus Programme, UNU, 1985.

75   Piore, M.J., and C. Sabel. *The Second Industrial Divide*. New York: Rinehart & Winston, 1984.

76   Raj, K.N. et al. *Poverty, Unemployment and Development Policy – A Case Study of Selected Issues with Reference to Kerala*. New York: UN, 1975.

77   Richta, R. *La civilisation au carrefour*. Paris: Seuil, 1974.

78   Ruddle, K., and G. Zhong. *Integrated Agriculture – Aquaculture in South Korea: The Dike Pond System of the Zhujiang Delta*. Cambridge: Cambridge University Press, 1988.

79   Ruffolo, G. *Potenza e Potere*. Bari: Laterza, 1988.

80   Sachs, C. *São Paulo – Politiques publiques et habitat populaire*. Paris: Editions de la Maison des Sciences de l'Homme, coll. Brasilia, 1990.

81   Sachs, I. "A Welfare State in Poor Countries." *Economic and Political Weekly* (Bombay) 6 (January 1971), nos. 3, 4.

82   ———. "Environnement et style de développement." *Annales, Economies, Sociétés, Civilisations* (Paris) 3 (1974): 553–570.

83   ———. *Stratégies de l'écodéveloppement*. Paris: Economie et Humanisme et les Editions Ouvrières (Développement et civilisations), 1980.

84   ———. *Development and Planning*. Cambridge: Cambridge University Press, 1987.

85   ———. "Ressources, emploi et financement du développement: produire sans détruire." *Cahiers du Brésil Contemporain* (Paris) 6 (1989).

86   ———. "Transition Strategies for the 21st Century." *Nature and Resources* (Paris) 28 (1992), no. 1.

87   Sachs, I., and D. Silk. *Final Report of the Food Energy Nexus Programme of the United Nations University, 1983–1987*. Paris: UNU-FEN, 1991.

88   Sachs, I., and K. Vinaver. "Integration of Technology in Development Planning: A Normative View." In: F.R. Sagasti and A. Araoz, eds. *Science and Technology for Development: Planning in the STPI Countries*. Ottawa: International Development Research Centre, 1979.

89   Salomon, J.-J. *Prométhée empêtré – La résistance au changement technique*. Paris: Anthropos, 1984.

90   Salomon, J.-J., and A. Lebeau. "Science, Technology and Development." *Social Science Information* (London), no. 29 (April 1990): 841–858. Reprinted as the Appendix to Salomon and Lebeau, *Mirages of Development*. See ref. 91.

91   ——. *Mirages of Development*. Boulder, Colo.: Lynne Rienner, 1993. Originally published in French as *L'écrivain public et l'ordinateur*. Paris: Hachette, 1988.

92   Science Advisory Council to the Prime Minister. *Perspectives in Science and Technology*, vol. 2. Delhi: Har-Anand Publications, 1990, pp. 395–419.

93   Serres, M. *Le contrat naturel*. Paris: François Bourin, 1990.

94   Sioli, H. "A ecologia paisagista da Amazônia e as perspectivas de uma utilização racional dos recursos." In: *1° Simpósio do Trópico Úmido: Anais Proceedings Anales* (Brasilia) 6 (1986): 31–41.

95   ——. "Introdução ao simpósio "Amazônia": desflorestamento e possiveis efeitos." *Interciência* 14 (1989), no. 6: 286–290.

96   ——. "Introdução ao simpósio internacional sobre Grandes Rios Latino-Américanos." *Interciência* 15 (1990), no. 6: 331–333.

97   South Commission. *Challenge to the South*. London: Oxford University Press, 1990.

98   Streeten, P. "From Growth to Basic Needs." *Finance and Development* 25 (1988), no. 3: 28–31.

99   Swaminathan, M.S. In: McNeely et al., eds., pp. 9–10. *See* ref. 59.

100   Tinbergen, J., and R. Hueting. "GNP and Market Prices." In: Goodland et al., eds., pp. 51–57. *See* ref. 34.

101   Toye, J. *Dilemmas of Development: Reflections on the Counter-revolution in Development Theory and Policy*. Oxford: Basil Blackwell, 1987.

102   Tricart, J., and J. Killian. *L'éco-géographie*. Paris: Maspero, 1979.

103   UNDP. *Report on Human Development*. New York: UN, 1990.

104   WCED (World Commission on Environment and Development). *Our Common Future*. The Brundtland Report. London: Oxford University Press, 1987.

105   Wisner, B. *Power and Need in Africa – Basic Human Needs and Development Policies*. London: Earthscan Publications, 1988.

106   World Bank. *Agricultural Biotechnology – The Next "Green Revolution"?* World Bank Technical Paper, no. 133. Washington, D.C.: World Bank, 1990.

107   ——. *World Development Report 1990*. Washington, D.C.: World Bank, 1990.

108   World Resources Institute. *World Resources 1990–1991: A Guide to the Global Environment*. New York: Oxford University Press, 1990.

109   Young, John E. *Discarding the Throwaway Society*. Worldwatch Paper, no. 101. Washington, D.C., January 1991.

Part 3: The policy dimension

Part 3 looks at the range of policies being used and advocated with regard to science and technology. In chapter 10, Atul Wad offers an assessment of the field of science and technology policy – a policy comprised of collective measures taken by a government in order, on the one hand, to encourage the development of scientific and technical research and, on the other, to exploit the results of this research to achieve desired general social, economic, and political objectives – and its application in the context of a dramatically changing world order characterized by a host of pressing challenges, rapid technological change, and globalization. Its importance to developing countries has, if anything, increased. He explains the rationale for science and technology policy, its historical evolution conceptually (stressing the distinctions between science policy and technology policy) and in practical terms – reviewing the range of specific policy instruments – and concludes with a description of the shortcomings of science and technology policy to date: it has produced elaborate, often overly bureaucratic, systems of science and technology in many developing countries but has had little impact on the "bottom line" of real technical change and technology decision-making at the level of the enterprise. To illustrate his analysis and the wide variety of approaches to science and technology policy, he reviews the experiences of various countries and regions, traces the role of the United Nations system in this field, and concludes that, for science and technology policy, the key contemporary issues centre around concerns over the use of technology to achieve competitive advantage, access to technology, new forms of government intervention to promote technological development at the firm level and greater participation in world markets – and all of this within the new principles of an emerging techno-economic paradigm.

Amitav Rath examines one important policy dimension: technology transfer and diffusion. He starts by describing the main elements and mechanisms of technology transfer, vertical and horizontal, and concludes that all of the channels are valuable and developing country strategies must ensure that the full mix of channels and mechanisms are used optimally; he points out that the dominant mechanism for technology flows is in the form of capital goods. In tracing the historical background, the author distinguishes two phases where twin economic and political objectives have influenced the concerns of research and policy related to technology transfer, sometimes reinforcing each other and at other times being antagonistic: post-war to the mid-1960s, and mid-1960s to the early 1980s. The main concern in the 1970s was the excessive costs of technology transactions and the many restrictive clauses that were imposed on the recipient by the supplier, thereby limiting the benefits to the recipient firm and country. Besides the implication of market imperfections, other negative impacts from technology transfer were stressed: dependency, inappropriateness, etc.

By the beginning of the 1980s, most developing countries had enacted regulatory mechanisms and rules governing investments and technology. Revisions to the framework, under the pressure of the changing international economic, technological, and policy environments for technology transfer, highlighted several aspects: transaction costs and terms, variations in technological elements and price, and new perceptions of the actors. The greater the involvement of the supplier and the recipient, the more successful is the technology transfer. Production efficiency is highly correlated with the macroeconomic policies and market structures of the recipient country. To conclude, Amitav Rath argues that excessive politicization of the issues has definitely been harmful to the interests of developing countries.

This recent technology debate has brought out into the open a dilemma facing developing countries: what mix of new, conventional, and traditional technologies should they use, and what is the appropriate balance between importing new technologies and using conventional and indigenous ones? Ajit Bhalla addresses technology choice as a crucial dimension of the development process that evolved in relation to shifts in development thinking. In the 1950s and 1960s, the issue of technology choice was secondary to that of maximizing growth. The recommended option invariably favoured the most capital-intensive and advanced technology because it contributed to maximizing savings rates and investment. In the 1970s, the criterion for choosing technology was no longer solely the reinvestible surplus of growth; the employment and income generated, the reduction of inequalities, and output generation were also important factors. The 1970s also witnessed the emergence of the concept of appropriate technology. Its protagonists highlighted the need to widen the set of technological options by developing alternative technologies in a labour-intensive direction to suit the factor endowments of developing countries. The decade of the 1980s is associated with the macroeconomics and political economy of technology decisions, intersectoral linkages to promote technology improvements and reduce technology gaps between modern and informal sectors, and the emergence of new technologies and blending. The author discusses the sparse employment and distributional implications of new technologies, the potentials for developing countries for leap-frogging and technology blending, and whether the use of new technologies and greater scope for "flexible specialization" can improve the efficiency of craft production and thus expand output as well as employment. By highlighting the differences between the debates on the concepts of "appropriate technology" and of "technology blending," Bhalla points to the emergence of a new debate, technological capability-building as a major long-term goal of development of the third world, and speculates on the issues for the 1990s.

Paulo Rodrigues Pereira discusses the breakthroughs in new technologies and assesses the opportunities and threats they represent for developing countries under the new techno-economic paradigm, as defined by Freeman and Perez, where technological development is increasingly becoming the dominant factor in determining a country's capacity to compete in world markets. Information technology allows a new technological system, in which far-reaching changes in the trajectories of electronic, computer, and telecommunication technologies converge and offer a range of new technological options to virtually all branches of the economy; moreover, this new system forms the basis for a reorganization of industrial society and the core of the emerging techno-economic paradigm. The reason for the pre-eminence of the new technological system clustered around information technology over the equally new technological systems clustered around biotechnology and new materials is the fact that information

343

activities of one kind or another make up a part of every activity within an industrial or commercial sector, as well as in our working and domestic lives. Almost all productive activities have a high information intensity, so information technology is capable of offering strategic improvements in productivity and competitiveness, by integration of functions, of virtually any economic and social activity. In assessing the implications for developing countries, the author concludes that the general tendency points to a widening of the information technology gap, both between industrialized and developing countries and within developing countries. Biotechnology, a science-led technology, induces important structural changes in the economy and has widespread applications in different industrial sectors: food and agricultural production, livestock husbandry and animal health, pharmaceuticals and chemical processing, medical treatment. One of the main advantages of these innovations in biotechnology has been the possibility of their economic use on a small scale, without large infrastructure requirements, and their application at different levels of complexity, investment, and effort. However, these opportunities should be weighed against the environmental risks and the interrelated social and economic costs. As regards new and advanced materials, the main impact of the present trends is likely to be felt by developing countries in the medium term, through the loss of competitive power of many of their manufactured products, which will increasingly have to compete with innovative products presenting higher functional integration or offering novel functions and services, manufactured by "multi-material" firms in industrialized countries. But the potential does exist for developing countries to produce materials with the higher purity necessary for high technology industries and it should be exploited.

Under the new techno-economic paradigm where technological innovation is the driving force, assessing its impacts is crucial. For Harvey Brooks, whereas in the industrialized countries, technology assessment is viewed predominantly in the context of anticipating and avoiding unintended social costs of economic growth and of technologies as their scale of application increases and spreads, in developing countries it is seen more as a means of building up an indigenous capability for wise technology choice. The costs and benefits they generate can potentially be large, and their incidence may differ significantly for different groups in society. Hence there is need for analysis so that the mismatches, the wrong investments, and the possible social conflicts can be minimized, while at the same time the beneficial effects and opportunities can be fully exploited. In this context, the issue of technology assessment is viewed as a continuing process of informing the people concerned, generating constructive public debate, and encouraging public understanding and involvement. In his chapter, Brooks first reviews the historical background of the concept and then draws lessons from over 20 years of institutionalization of technology assessment in the United States to derive a typology: project assessment, generic technology assessment, problem assessment, policy assessment, global problématique. Brooks concludes that if technology assessment is seen as a cumulative process of "social learning," it calls for very wide participation of virtually all the stakeholders. Drawing on the experience in industrialized countries with stakeholder participation in technology assessment, he derives principles that can contribute to the success of stakeholder dialogues in developing countries.

10

Science and technology policy

Atul Wad

After the Second World War, country after country in Asia and Africa was granted or achieved independence from its former colonial rulers. In many cases, the new leaders of these countries – Nehru in India, Kenyatta in Kenya, Nasser in Egypt – subscribed to an emancipated and modern view of science and saw science and technology as essential to the development of their nations. At the same time, they were strong nationalists and believed in the paramount role of the state in building up their societies. As a result, many of these countries, very shortly after independence, gave top priority to scientific and technological activities in the form of education, the establishment of government bodies dedicated to science and technology (e.g. the Council of Scientific and Industrial Research [CSIR] in India), and the promotion of science and technology at all levels in society. As they embarked upon the challenging road of economic development, small circles of academics concerned with science policy issues began to form in these countries. Studies by third world scientists began to appear, often couching the problem of science and technology in developing countries within the larger problématique of development and the structural inequalities of the post-colonial (or neocolonial) period. These scientists included Antoine Zahlan, Ziauddin Sardar, Abdul Rahman, Homi Bhabha, Amilcar Herrera, and M.A. Qurashi, who spoke directly to the problem of science and technology, as well as a number of development economists and political economists who integrated science and technology into their broader analyses of the development process; they include Samir Amin, Fernando Enrique Cardoso, Thestonio dos Santos, Prebisch,

Gunnar Myrdal, Dudley Seers, Immanuel Wallerstein, and Andre Gunder Frank. Ideological imperatives were often rampant in these latter analyses, nevertheless they represented an important period in the post-war era for the third world.

Also during this period, interest began to grow in the structure and dynamics of the scientific communities in these countries. All studies stressed the frustration and alienation of scientists operating in developing countries and generally suggested the need for international mechanisms of cooperation to support these communities. One such mechanism, the International Council of Scientific Unions, has been particularly active in this way. Another significant event was the establishment of the Third World Academy of Sciences.

Science and technology policy: Rationale and issues

The justifications

Science and technology policy (STP) represents the articulation of how the modern state and society at large view the relationships and instrumentalities between scientific and technological change and social and economic development. The effectiveness of STP is essentially a function of how realistic and comprehensive the understanding of decision makers is of these interactions and relationships. On another level, STP reflects the tremendous optimism that is still present today regarding the potential of science and technology, properly developed and applied, to solve the pressing problems of humanity.

The concept and practice of STP is based on the presumption that direct and indirect intervention by the state in scientific and technological activities and processes is necessary in order to achieve desired social, economic, and political goals. The justification for STP and state intervention derives from certain principles:

1. Technological progress may not proceed in the desired direction without influence by government, leading to poor technology choices, inappropriate allocation of resources, and distorted patterns of industrialization.

2. The returns from scientific research are too long term to expect market forces to encourage private investments in R&D in areas beneficial to society. The difficulty in "appropriating" the returns from R&D also reduces the incentive to invest by the private firm. Government must therefore intervene to rectify this "market failure," by investing itself, or by enacting policies to encourage private investment [30, 10].

3. Forsyth [11] adds that the pressures of competition in international trade can push developing countries towards labour-intensive techniques in a narrow range of products, even though a diversified industrial base is preferable. Government intervention is needed to protect and nurture those components of the industrial base that are unlikely to evolve spontaneously. This is the "infant industry" argument [3].

4. Certain areas of technology are unlikely to develop by themselves, for example the service sectors (health, education, etc.) and therefore require a direct role by the state. This is particularly important in developing countries, with their large populations and severe income inequalities. The state, as the guardian of the social well-being of the population, becomes obliged to try to channel scientific and technological activities so as to improve the living conditions of the people. For developing countries, with a pressing need to address social problems such as food, housing, employment, and health, science was seen as a panacea at independence, and in many countries STP was initially undertaken with considerable hope and enthusiasm.

5. There is also a strong political rationale to science and technology policy, particularly in the industrialized countries. The Manhattan Project and the development of the atomic bomb, which many see as a watershed in the evolution of science policy, established science as a "national asset." Nations undertook scientific projects to achieve political, and often military, goals. In the post–Second World War environment of growing international competition, science policy emerged as a strategic weapon for countries. There was an obvious correlation between the emergence of international crises and the increases in expenditures on R&D. In this regard, STP clearly derived to a major extent from military needs and priorities.

These arguments form the basis for the evolution of STP in both the industrialized and developing countries. There are debates within the field around these and related issues, and as we shall see later, there are new arguments being put forward both in favour of and against the need for STP. It is within the broad context of STP that debates continue over such issues as brain drain, technology transfer, intellectual property, and the relative importance of basic versus applied science.

The distinction between science policy and technology policy
The distinction between science policy and technology policy, once the subject of heated debate, has become less important recently as

technology has become more science-based. In the past, however, science policy tended to be given more emphasis than technology policy, the latter being seen as something that should be left to the play of the market and the private sector.

In the context of developing countries, the argument is increasingly put forward that what is needed is more technological dynamism, and that the development of scientific capabilities is less important, at least in terms of addressing the severe near-term problems of the developing countries. Given the high costs involved in developing a reasonable capability in modern science, the practical imperative to become increasingly competitive in global markets, and the successes of the technology adapting and modifying strategies of the four "tigers" of the Pacific rim and before them of Japan, there may be some merit in this argument. On the other hand, there has always been a strong case for a society to have a "culture" of science and a scientific temperament in order to achieve economic growth. In this last regard, the deep cultural importance of knowledge, science, and education and a tradition of reading and writing in the countries of the Pacific rim may well be an important factor in explaining their success.

The debate over the book by Salomon and Lebeau, *L'écrivain public et l'ordinateur – Mirages du développement* [32], addresses aspects of this issue. Salomon and Lebeau argue that science, as it is understood in the modern world, is established and defined largely by the scientific institutions of the industrialized countries and that it is naturally élitist and hence cannot realistically be expected to address the problems of developing countries. As such, science in developing countries tends to be more preoccupied with the problems defined as "important" by the international scientific community than with the problems of poverty and underdevelopment. Cooper [4] has referred to this as the "marginalization" of science in developing countries. Technology, on the other hand, being of practical relevance in an immediate sense, is where developing countries need to focus their resources and efforts.

The counter-argument of course is that sustained technological development is impossible without a reasonably strong scientific base; technology is dynamic and needs the intellectual foundation that a strong scientific tradition offers. Much of the early thinking in STP was influenced by this view – that science must precede technology. Still some think that there are more urgent needs to develop a whole country than to build up a scientific community that is isolated from the rest of the population. The debate over this issue continues.

For developing countries, the distinctions between science policy and technology policy are best described in one of the reports arising from the project on Science and Technology Project Instruments (STPI) sponsored by the International Development Research Centre (IDRC) in Ottawa (see table).

Policy for *science and policy* through *science*
A distinction is also made between policy *for* science, i.e. the encouragement of certain forms of scientific activity, and policy *through* science, which relates to the exploitation of research in areas of concern to government. These two aspects have become more complementary with the passage of time, and in particular after the Second World War. Science is both influenced by society and in turn influences social, political, and economic systems. Recently it has been argued that there is really no such thing as a separate "science and technology for development," but that science and technology are really inputs into the development of other activities, such as population control, food production, industrial development – this is reflected in the "mission" approach to development, for example in India. There may be some merit in this view, but it does not alter the basic importance of science and technology as deserving of special recognition.

The policy sciences
An entire field of "policy sciences" emerged in the post-war period. The main focus of this field is the analysis of policy-making broadly defined in all areas of government intervention. A major journal, *Policy Sciences*, was launched in 1970. Included in the scope of this field are the "philosophies, procedures, techniques and tools of the management and decision sciences – operations research, systems analysis, simulation, 'war' gaming, game theory, policy analysis, programme budgeting, and linear programming."
Such noted social and policy scientists as Amitai Etzioni, Yehezkel Dror, Harold Lasswell, Albert Hirschman, Erich Jantsch, and Marvin Cetron have been involved in this field. The main purpose of this field is to bring more systematic analytical and practical tools to bear on problems that in the past were dealt with more or less "intuitively" – to bring rationality to governmental behaviour and decision-making.
To the extent that this field covers all aspects of policy-making, not solely science and technology, it is not explicitly covered here. (For

351

Differences between national science and technology policies

Aspect	Science policy	Technology policy
Objectives	A. To generate scientific (basic and potentially useful) knowledge that may eventually have social and economic uses, and will allow understanding and keeping up with the evolution of science. B. To produce a base of scientific activities and human resources linked to the growth of knowledge throughout the world.	A. To acquire the technology and the technical capabilities for the production of goods and the provision of services. B. To acquire a national capacity for autonomous decision-making in technological matters.
Main types of activities covered	Basic and applied research that generates both basic and potentially useful knowledge.	Development, adaptation, reverse engineering, technology transfer, and engineering design, which generate ready-to-use knowledge.
Appropriation of results of activities covered	Results (in the form of basic and potentially useful knowledge) are appropriated by wide dissemination; publishing ensures ownership.	Results (in the form of ready-to-use knowledge) remain largely in hands of those who generated them; patents, secret know-how, and human-embodied knowledge ensure appropriation.
Reference criteria for performance	Primarily internal to the scientific community. Evaluation of activities is based mainly on scientific merit and occasionally on possible applications.	Primarily external to the technical and engineering community. Evaluation based mainly on contribution to social and economic objectives.
Scope of activities	Universal: activities and results have worldwide validity.	Localized (to firm, branch, sector, or national level): activities and results have validity in a specific context.
Amenability to planning	Only broad areas and directives can be programmed. Results depend on the capacity of researchers (teams and individuals) to generate new ideas. Involves large uncertainties.	Activities and sequences can be programmed more strictly. Little new knowledge generally required, and existing knowledge is used systematically. Involves less uncertainty.
Dominant time frame	Long and medium term.	Short and medium term.

Source: Ref. 15, pp. 16–17.

352

more detailed information, see the journal *Policy Sciences* and various texts by Dror, Wildavsky, Hirschman, Etzioni, and Lasswell.)

Western science and alternative models of science

Finally, there is the debate over the concept of science itself, particularly with respect to the developing world. Modern science is eminently a Western science based on Western notions of rationality and instrumentality. Some alternative forms of scientific knowledge and their relevance to developing countries are discussed in chapter 4. There is a vast body of literature on the forms of science that prevailed in regions of the world that are now part of the developing world – Joseph Needham on China [25], Claude Alvares on India [2], Nasr and Daghestani on Islamic science [24, 5], Mudimbe and Mazrui on Africa [23, 22], and also Goonatilake and Elzinga and Jamison [13, 8].

This issue is particularly important at present, when there appears to be an exhaustion of the models and strategies that have been pursued by developing countries based on Western notions of development and modernization. Indeed, the resurgence of fundamentalist movements and grass-roots initiatives may in part be seen as a response to this sense of frustration. The implications for STP are still unclear, but it would seem prudent to view the canvas of science and technology in a broader and culturally more sensitive perspective. The deep cultural underpinnings of a society clearly influence its scientific and technological capabilities and potential. Throughout history, in China, India, the Arab world, Central and South America, scientific progress has occurred within a complex and dynamic sociocultural milieu and has declined with the economic and military decline of those societies. The challenge today for the developing world is to identify what type of science and technology makes the most sense in today's politically charged, technologically infused, and global economy.

Instruments for science and technology policy

The manner in which science and technology policy is made operational is through specific policy instruments (STPI). Interest in identifying the range of instruments needed to achieve desired science and technology objectives evolved more or less in parallel with the development of an institutional context for STP in developing countries. Once formal institutions for STP began to be established in developing countries, the need emerged for instruments through which these bodies could enact their objectives and missions.

A landmark event in this area was the multi-country STPI research project undertaken with the support of the IDRC. The project's field office was in Lima and 10 research teams from Africa, Latin America, Asia, and southern Europe participated in it. The overall purpose of the STPI project was "to gather, analyze, evaluate and generate information that may help policy makers, planners, and decision makers in underdeveloped countries to orient science and technology toward the achievement of development objectives" [15].

The approach taken in the STPI project, and basically adhered to subsequently by researchers and policy makers, identified three broad categories of instruments:

1. *demand-side* instruments designed to influence the nature of demand by firms, enterprises, and organizations and the technological behaviour and decision-making of these entities;
2. *supply-side* instruments, which relate to the activities in the science and technology system that have as end products new technology and science, and to the supply of science and technology services and human resources;
3. instruments directed towards the linkage between the supply and demand sides of the equation, i.e. the links between the R&D and productive system.

The STPI project defined an instrument as "the set of ways and means used when putting a given policy into practice. It can be considered as the vehicle through which those in charge of formulating and implementing policies actualize their capability to influence decisions taken by others" [15, p. 13]. A policy instrument could be a legal device, such as a patent law or technology licensing regulations; an organizational structure, for example an R&D laboratory or a research programme involving several institutions; or a set of operational mechanisms, for example specific R&D management procedures, incentive systems, etc.

Furthermore, policies can be either explicit or implicit, with the former being articulated expressions of desired goals and objectives by high-level government officials or institutions with respect to science and technology, whereas implicit policies are directed towards areas or sectors that in turn will influence science and technology activities.

The STPI project went into great detail about various aspects of the quality and effectiveness of existing instruments in the countries

354

surveyed, and concluded that in general explicit policy instruments had little impact on technological change, particularly in the early stages of industrialization. However, they did have a significant impact on the science and technology infrastructures in these countries.

This perhaps encapsulates the shortcomings of STP to date: that it has produced elaborate, often overly bureaucratic, systems of science and technology in many developing countries, but has had little impact on the "bottom line" of real technical change and technology decision-making at the level of the enterprise.

As I discuss later, this failure to address technical change at the firm level is a result of a conceptual shortcoming in STP research itself, a shortcoming that is only recently being recognized by the research and policy community.

Another deficiency in STPI implementation has been the lack of specific and practical guidelines for policy makers. Thus, even though the broad intentions and concepts of STP were generally understood, few policy makers had a concrete sense of the specific steps that had to be undertaken in order to implement these policies. An attempt to redress this shortcoming has been made by the International Labour Office in Geneva, in the form of a manual for technology policy assessment [11].

A particularly important mechanism for the implementation of STP is the financial institution. There is a limited literature in this field (see, for example, Jecquier and Hu [16]). In general, there is little awareness of the important role that financial institutions can play in STP, especially within the science and technology community itself. Yet, because of the economic and financial rigour they bring to the assessment of a project, and the resources they can mobilize, they can be powerful actors in STP. In recent times, the emergence of "technology incubators" linked with venture capital funds for the commercialization of new technologies is seen as an essential aspect of industrial development in some countries such as the United States.

The implications of trade policy

The distinctions between science and technology policy on the one hand and trade policy on the other have recently become blurred. Increasingly, national and international trade policies have direct and indirect impacts on technological and scientific activities, and STP has an impact on trade patterns. Thus, for example, trade-related invest-

355

ment measures (TRIMS) and trade-related intellectual property rights (TRIPS) can influence the choice and acquisition of technology and the conduct of research itself.

The growing concern over the environment has also contributed to these interactions. As industrialized countries establish tighter environmental standards and specifications on products and processes, developing countries find themselves increasingly under pressure to acquire "cleaner" technologies or face the barrier of "eco-protectionism." Concern over loss of biodiversity and genetic wealth has prompted developing countries both to take stronger stands on the export and exploitation of these natural resources and to undertake research themselves to capitalize on their potential.

The interactions between STP and trade policy are complex and the subject of attention by, for example, the United Nations Conference on Trade and Development (UNCTAD) and the United Nations Centre on Transnational Corporations. Within such fora as GATT, issues related to intellectual property rights and protectionism are often debated within the same context as access to technology by developing countries and access to markets for their products.

Experiences and approaches in the third world

Since independence, and for some, well before that, the developing countries have been experimenting with a wide variety of approaches to STP and have accumulated a diverse base of experience in this respect. In this section I attempt to review some of these experiences, though obviously not exhaustively.

Latin America

The Latin American contribution to STP for development has always been significant. Perhaps having had a longer history of independence allowed for the growth of a more broadly based and deep rooted intellectual and political appreciation for the role that science and technology can and should play in society. Also, the support of the Organization of American States (OAS) in the 1960s to encourage science and technology policy research had a positive impact on the development of these capabilities. After the OAS programme declined, support was provided by the IDRC of Canada, which was somewhat unusual in the development assistance community in having a strong emphasis on the development of local capabilities for research in developing countries. (The Swedish Agency for Research

Cooperation with Developing Countries [SAREC] has also been not-able for its innovative approach to development assistance, though it has tended to be more active in Africa.) In any event, the influence of Latin American thinking on STP developments around the world, and in particular in the initiatives undertaken by the United Nations, has been significant and positive. Significant policy-level research has also come out of Asia, particularly India and Pakistan, the Arab States, and some parts of Africa.

The evolution of STP research in Latin America reflects how thinking on this subject changed during the post-war period.

Sagasti [31] identifies four overlapping phases:

(a) the "science push" phase, lasting throughout the 1950s and early 1960s; (b) the "transfer of technology and systems analysis" phase that began in the late 1960s and flourished through the 1970s; (c) the "innovation and techno-logy policy implementation" phase that began in the mid-1970s and ex-tended through the early 1980s; and (d) a phase of "politicalization of sci-ence and technology policy", which was ushered in by the 1981–1982 econo-mic crisis and also led to concerns about industrial restructuring and the im-pact of new technologies on the region.

During the "science push" phase, where the influence of Bernal's *The Social Function of Science* was evident, the main emphasis was on the establishment of a scientific and technological infrastructure consisting of laboratories, research institutes, universities, and sci-ence and technology councils. Governments responded to appeals from the scientific communities and suggestions from international organ-izations such as Unesco and agreed to large expenditures for these purposes. Hence, today there is a heritage in many developing coun-tries, not only in Latin America, of large, often unwieldy, science and technology apparatuses that are for the most part state-supported. In some instances, as in India, this system has become bureaucratized to the point where its value to the nation is highly questionable [19]. During this period, the prevailing view was that science was primary and fundamental to development. The commer-cialization of science was not seen as a problem and the focus was on ensuring autonomy for scientists so as to encourage the production of new knowledge that in turn could be used for socially productive pur-poses. It was during this period that a number of countries in Latin America and elsewhere established National Research Councils (in Latin America referred to as ONCYTs).

The second phase, that of "transfer of technology," sought to address the impact of technology flows into Latin America on the balance of payments of these countries. Technology inflows were seen as having a negative effect on their foreign exchange holdings, and the need to control or restrict these flows began to be felt. National regulatory agencies to monitor technology transfers from outside were set up as a result. In the long run, and in retrospect, these mechanisms may have had a deleterious effect on the acquisition of technological capabilities in many countries. By making access to technology a much more cumbersome, expensive, and difficult process, the path of technological development was constrained in important ways, some of which were positive, and some negative. On the positive side, there was greater pressure to undertake indigenous innovation, but this could only be done in certain areas of technology. On the negative side, access became difficult to important new areas of technology, leading to "gaps" in the technological profiles of these countries and the emergence of uncompetitive industries.

During this period, the "systems approach" to science and technology also became popular and expanded the concept of science and technology beyond the limits of R&D alone. National councils of science and technology became popular, and there was increased debate about the distinction between science policy and technology policy. The political balance began to shift in favour of the technologists and economists of technology and away from the scientists.

The third phase emerged in the late 1970s and focused attention on innovation and the development of innovative capabilities through appropriate technology policies. A much more explicit focus on the enterprise level was to be seen during this phase, with much research being conducted on the best mechanisms to generate local technological and innovation capacities. During this period, more concern began to be expressed about the linkages between environment, energy, universities, and industry, and technical cooperation between developing countries.

This wave of research was based more soundly on empirical research and was often more micro in its focus. The problems addressed dealt with the causes and consequences of technological change in developing countries and the reasons for the continued inability of these countries to develop their technological capabilities properly.

In this period, the work coming out of Latin America was significant. The 1980s began to see a decline in government support for science and technology, in large part because of the declining economic situation of these countries. The colossal debt burden and the political and economic crises facing these nations led to drastic cuts in government expenditures for science and technology, and increased debate at the political level about the proper role of science and technology in national development. In some cases, the trend was reversed after a few years, as in Brazil, with the establishment of the Ministry of Science and Technology in 1985 [1]. At the same time, research on technical change at the enterprise level continued to grow, with a broadening empirical base. The IDRC/UNDP/ECLAC research project based in Buenos Aires under the coordination of Jorge Katz began to bring new insights into the nature and dynamics of technical change at the firm level, based on detailed case-studies and empirical research. Also, there began to appear a convergence between this research and research being undertaken in the industrialized countries on the economics of technical change [18]. Primarily this was to be found not within the neoclassical economics tradition, but in the "neo-Schumpeterian" approaches to the analysis of technical change, which present an alternative, albeit more "muddy," view of how technical change occurs and how it affects society. The work of Dosi [6], Nelson and Winter [26], Freeman [12], Katz [18], and Perez [27] is significant in this regard.

Also during this period a strong shift in terms of economic philosophy was seen at the macro level. The view of the private sector, hitherto seen largely as a passive and homogeneous entity that simply responded to policy directions, was replaced by one that emphasized its entrepreneurial potential and its importance as a driving force for industrialization. "Export orientation" became fashionable, as did structural adjustment programmes (under the tutelage and with the support of the World Bank). The goal of STP was one of developing technological capabilities aimed at improving the export potential of local industry, and questions of competitive advantage, productivity, and growth emerged to the forefront. The experiences with structural adjustment programmes were, and continue to be, mixed, since the built-up inertia of years of import substitution and protectionism has been hard to overcome. But at the same time the success of the four East Asian "tigers" (South Korea, Taiwan, Hong Kong, and Singapore) prompted people to begin to enquire into how the STPs of these countries contribute to their economic success and what lessons could be learned for other countries.

Africa

At the time of independence, most African nations demonstrated a remarkable measure of enthusiasm about science and technology. Many established national STP bodies and R&D institutions. An African scientific journal was established and important political figures such as NKrumah, Nasser, and Kenyatta espoused modern science and technology as essential to the development of their new nation-states.

It was during the 1970s that many African nations established national policy mechanisms for science and technology. Algeria, Ghana, Mali, Niger, and Egypt had all set up national research councils by the end of the 1970s. Côte d'Ivoire had a Ministry of Scientific Research in 1970; Senegal had a Délégation Générale pour la Recherche Scientifique et Technologique (DGRST) by 1974. Later these were transformed into ministries for higher education and scientific research (MESRES), for instance in Senegal, Burkina Faso, Cameroon, and Benin. Nigeria established the Federal Ministry for Science and Technology in 1979, Tanzania set up the Tanzania Commission for Science and Technology in 1986, and Zimbabwe set up a National Science Council in 1986. Ethiopia's Science and Technology Commission was established in 1975; Somalia had an Academy of Sciences and Arts in 1979; Morocco established the National Centre for Co-ordinating and Planning Scientific and Technological Research in 1976; and Sudan set up its National Research Council in 1970.

The first CASTAFRICA (Conference of African Ministers Responsible for the Application of Science and Technology to Development) conference, organized under the auspices of Unesco, was held in 1974. At that time only a few African countries had explicit policies. By the time of the second conference, held at Arusha in 1987, 18 African nations had STP bodies at the ministerial level. However, this increase in numbers did not necessarily imply efficiency.

Indeed, the story of science and technology in Africa is somewhat unfortunate. Though there have been a variety of initiatives and experiments in STP, very few have borne fruit. Many countries built up science policy bodies without any scientific tradition or even infrastructure. The result was the growth of useless and irrelevant bureaucracy. Partly, the problem may have to do with the overall weakness of African states and the multitude of economic problems they face. Also in part, at least for sub-Saharan Africa, these states began with a weak science and technology infrastructure at the outset

360

– Africa's contribution to world science is the smallest in comparison to the other developing regions.

But primarily two factors seem to have affected STP in Africa most seriously: the lack of a real commitment at the national and regional levels to the development of science and technology, and the significant dependence of the economies on raw materials and commodity exports within a global system that allowed little flexibility in terms of developing domestic industrial capabilities that would enable a larger share of the value-added pie to be captured by these nations. Whether this is a conspiracy, an accident of history, or the dispassionate logic of the global market-place, Africa has not come out very well in terms of harnessing science and technology for its development.

There have been exceptions, such as the International Centre for Insect Physiology and Epidemiology (ICIPE) in Nairobi, for a short while the Kumasi Science and Technology University, WARDA (the West Africa Rice Development Agency), and ILRAD (the Institute for Livestock Research and Development) under the umbrella of the Consultative Group for International Agricultural Research (CGIAR), and to some extent the African Regional Centre for Technology in Dakar. But these are international centres with heavy international support and should be seen as the exception.

Africa has also produced many important scientists – Edward Ayensu, Aklilu Lemma, Thomas Odhiambo, etc., who have made their mark internationally. But the general trend has been gloomy.

For example, if one examines Unesco's data for Africa for the period 1974–1978, Africa's share of global R&D personnel rose from 0.4 per cent to 0.7 per cent, but the level of expenditures stayed at 0.4 per cent. On a per capita basis, the average R&D expenditure in Africa is below US$2.00 and below 1 per cent of GNP in nearly all countries.

The Unesco data also offer some perspective on the availability of science and technology personnel. Most countries in the region possess about a third of the corresponding numbers in Asia and about 3 per cent of the level in Europe, though countries such as Nigeria, Egypt, Libya, and Zambia have substantially larger manpower resources than the average. The shortage of science and technology personnel can be attributed to a number of factors – the lack of higher educational systems and research facilities, the emphasis in the past on the liberal arts and the humanities over the more applied fields of engineering, and the "brain drain." And one cannot minimize the impact of political regimes that lead intellectuals and scientists to leave.

Most African nations are, however, taking the shortage of trained personnel seriously: the average governmental expenditure on education in Africa is 15.6 per cent of total governmental functional expenditure, with some countries spending as much as 20 per cent (Botswana, Guinea-Bissau) and even 35 per cent (Côte d'Ivoire).

In recent times, there has been a growing interest in the science and technology problems of the least developed countries (as defined by the United Nations). These countries happen to be mostly in sub-Saharan Africa. For them, the options available with respect to science and technology capability development are much narrower, given their levels of poverty, shortage of financial and science and technology resources, weak infrastructure, etc. In such countries, the question must be raised as to whether science and technology, and particularly science, is not a luxury that they can ill afford. The priority may well be to find the most effective ways to use available science and technology resources from whatever sources, rather than focus on an unrealistic goal to develop local capabilities.

The Arab world

For centuries the Arab world was one of the revered centres of science and learning. In modern times, however, it has lagged far behind the industrial nations, particularly in science and technology. Although there is considerable variety in levels of development, some countries in the region remain technologically dependent on more advanced nations, with a trade structure based on importing technology and exporting primary products. Even technology produced within the Arab world is frequently the result of a design transfer, often with foreign participation in or supervision of the process. Nevertheless, attempts are being made to improve both science and technology research and planning in the region [35, 5]. Institutions engaged in such efforts include the Supreme Council of Sciences (Syria), the Royal Scientific Society (Jordan), the Foundation for Scientific Research (Iraq), and the Kuwait Institute for Scientific Research, among others.

In order to expedite these efforts by individual countries, a co-operative approach seeking to integrate the scientific endeavours and policies of the entire region was undertaken, beginning in earnest in the early to mid-1970s. First, in 1970, the Arab League Educational,

Cultural and Scientific Organization (ALESCO) was founded. It arranged a conference for Arab ministers of science in Baghdad in February 1974, which was the first time an intergovernmental meeting at the ministerial level had taken place. Although the conference adopted no plan of action, it did produce several noteworthy recommendations. Significant among them were suggestions for closer links among scientific and socio-economic bodies, preparation of science policy by high political authorities, Arab scientific cooperation, and a request for ALESCO to study the feasibility of establishing an Arab foundation for scientific research and an Arab fund for the promotion of such research [7, pp. 149–150]. This last request was taken up by the Economic Council of the Arab League at the summit meeting in Rabat in October 1974, and the results of the study were presented at the Conference of Arab Ministers Responsible for the Application of Science and Technology to Development (CASTARAB) in Rabat in August 1976.

This first CASTARAB meeting, organized by Unesco with the aid of ALESCO, did not produce markedly different recommendations on science and technology policy from the Baghdad meeting, but it went considerably further with regard to regional cooperation, detailing specific plans in certain fields and integrating Arab science efforts in general. On hearing the results of the feasibility study, delegates decided to drop plans for an Arab science foundation and concentrate instead on creating a fund for scientific and technical research [7, pp. 152–154].

However, the plans for setting up this fund were also dropped, not because of a lack of funds, but rather owing to a lack of political will. Furthermore, although preparations were made for CASTARAB II through a series of meetings in the 1980s, the conference never took place for similar reasons. In the meantime, several other meetings with experts in the field have been held, arranged primarily by Unesco in conjunction with other agencies, yet none of these has been at the ministerial level [34, p. 10].

In 1979, the Vienna Science and Technology Conference called upon countries to formulate national science and technology policies. Ten years later, not one of the Arab countries had done so. Among nations in the ESCWA region (Economic and Social Council for West Asia), only Egypt and Iraq have formulated concrete science and technology strategies that consist of five-year research plans. Most other countries in the region do not have national research plans or government bodies for science and technology, and they therefore lack comprehensive plans and policies in this field [5, p. 17].

Asia

In Asia there has been a very wide array of experiences, ranging from those of the new industrialized countries (Korea, Taiwan, Hong Kong, and Singapore) to those of the two "giants," India and China, to the poor countries of Laos, Burma, and Cambodia. With some significant variations, the new industrialized countries mostly followed a model of STP based on a significant role of the state and a major emphasis on the development of local capabilities. Their success is also partly due to the heavy export orientation of their economies, their size (especially for Singapore and Hong Kong), and the opportunities they were able to exploit as a result of development in the North. Whether these models can be easily replicated in other developing countries is highly questionable. Moreover, particularly in the cases of Korea and Taiwan, industrialization was accomplished at a massive environmental cost, and it is hard to contemplate a similar process being pursued today.

India followed a strong state interventionist model as well, but with a heavier emphasis on import substitution and the protection of domestic markets and industries. Also, it managed to develop one of the most bureaucratic and stifling science and technology systems in the world, from which it is still trying to free itself. Though India has one of the largest pools of science and technology resources in the third world, the contribution of science and technology to economic development has been far from satisfactory. But one must recognize the contribution to the defence infrastructure and establishment.

In a general sense, the Asian experience with science and technology has been more practical than theoretical – little has come out in terms of dramatic, new conceptual developments with regard to the role of science and technology in development. The contributions have been more specific – the success of the new industrialized countries, the progress made by India and China in specific areas of technology, the tremendous human resource pool that is available, the quality of science and technology education available, and the quality of the statistics available on science and technology in the region. On the other hand, science and technology is still largely seen as an élite activity primarily concerned with the generation of wealth. What becomes of significant interest, therefore, is the emergence of an incredibly dynamic and creative movement concerned with "alternative" models of science and technology.

This historical experience with STP in developing countries is important not only for itself, but also in terms of providing a deeper understanding of the issues that confront STP in today's world, which are discussed later.

The United Nations system

A discussion of STP for development would be incomplete without some description of the important and changing role of the UN system in this field.

The UN system's involvement in STP-related matters dates back to the 1963 Geneva "Conference on the Application of Science and Technology for the Benefit of Less Developed Areas," where science and technology was conceived of as a large pool of accumulated knowledge from which the developing countries could pick and choose in order to solve their development problems. The conference participants were mainly scientists and engineers and the purpose was to draw attention, especially among policy makers, to the advances that were taking place in various fields of science and technology and their relevance to the problems faced by developing countries in such sectors as agriculture, health, and transportation. Little attention was paid to non-technical matters, such as problems of acquisition and transfer, social impacts, policy issues, etc. In a sense, the entire approach at the conference was somewhat naïve in its belief that technology was a "public good" that could simply be acquired at will, given the resources. This is in sharp contrast to the prevailing viewpoints of today, where the possession of technological knowledge and its "appropriability" for private returns are of major concern to individual firms as well as matters of contention in international discussions about intellectual property rights (IPR), for example within the GATT round of talks.

Sixteen years later, a major conference was held in Vienna, the United Nations Conference on Science and Technology for Development (UNCSTD) of 1979. Here, the thinking was quite different, using what was then referred to as the "horizontal" approach to STP, meaning one that did not subscribe to sectoral categorizations but viewed science and technology "horizontally" cutting across sectors. Priority was given to the development of "endogenous capabilities" in science and technology in developing countries, a term that even today is subject to various interpretations. The conference was preceded by nearly five years of preparations, and the influence of the Latin American perspective on STP was clearly evident. Almost all countries prepared country papers summarizing the status of science and technology for the conference according to an agreed format and this in itself was a major achievement. The conference was essentially an intergovernmental event, and much attention was given to the differing needs of different types of countries and to international cooperation in science and technology. The conference ended with the adoption of the Vienna Programme of Action (VPA) on Science and Technology for Development, which became the basis for ensuing UN activities in this area and served as a frame of reference for developing countries in their individual STP efforts. A Centre for Science and Technology for Development was established at the United Nations secretariat in New York, and several other agencies created special units or divisions to deal with science and technology matters – including Unesco, UNDP, UNIDO, ILO, and the regional commissions.

Ten years later, the UNCSTD conducted the End of Decade Review to assess what had been accomplished since the VPA was adopted [33]. The VPA was essentially a broad set of guidelines regarding policy and the structural and institutional dimensions of science and technology and did not really get into specific individual situations, a deficiency that perhaps explains why one of the conclusions of the end of decade review was the substantial lack of implementation of the VPA recommendations. The review had little, however, to say about how developing countries could proceed to be more productive and effective in their STP efforts, focusing instead on newly emergent issues of concern to development. Nevertheless, it is an important document for the emphasis it gives to endogenous capability development, the impacts of new technologies, the key role of cooperation, and the changing character of the development problématique. However, as is the case with most UN reports and initiatives, it is basically a consensus document that operates at the governmental level. As such, it fails to deal in any substantial fashion with science and technology at the level of the firm or enterprise, which is basically where real processes of technical change occur and are experienced.

Nevertheless, the United Nations plays an important worldwide role in STP. Most of the arms of the United Nations are involved in specific science and technology activities. Unesco continues to be the main source of statistics on science and technology in the developing countries, though their quality may be questioned. The ILO's Technology and Employment Branch has produced an enormous number of studies and reports on a variety of topics related to science and technology and has played an important role in the appropriate technology debate. The sectoral agencies, such as FAO and UNIDO, also have strong science and technology programmes. The Centre for Science and Technology for Development continues to be an important focusing point within the United Nations for science and technology activities, and its Advisory Committee consists of leading STP experts from around the world. The World Bank, however, has played a weak role in this area. Partly due to the disciplinary bias of economics against recognizing science and technology as an important economic factor, the Bank has done relatively little in furthering our understanding of the role of science and technology in development. In recent times, the efforts of the Industry and Energy Department, however, have been more directly focused on the issues.

The knowledge base for STP

As STP has evolved, it has involved a growing number of disciplines and is today seen largely as an interdisciplinary area. But this does not imply that there is a consensus as to the base of knowledge that is now needed for STP. In particular, there is a dichotomy between the manner in which economics approaches STP and the perspective taken by the other social sciences. Many of the UN analyses, for instance, do not directly address the economic aspects of STP. This is a serious issue – what types of skill sets are required for effective STP? Rosenberg [29] addresses this issue head on:

Although research in the realms of science and technology is obviously a highly specialized activity best left to the appropriately trained professionals, science and technology policy is an entirely different matter. Insofar as interest in these subjects is due to their economic consequences, the formulation of science and technology policy is inseparable from the formulation of economic policy.

Some clarification and elaboration are in order. Putting the point negatively, it is not possible to isolate science and technology policies from economic policy making without seriously diminishing their effectiveness. In fact, it is difficult even to identify a very specific set of science-and-technology oriented programmes and label them as "Science and Techno-

367

logy Policy". The reason is that there are a great number of factors that affect the commitment of resources to science and technology and that determine the "output" that society is likely to derive from such use of resources. Science and technology are economic activities, and they represent ways of pursuing a wide range of economic goals and objectives. They are not activities that run along some parallel track to, let us say, the Departments of Energy, Transportation, Defense or Agriculture. Nor can they be readily isolated in a Department of Science or a Department of Technology. [pp. 135–136]

The issue is in fact more complicated. STP has, as discussed earlier, become the subject of interest for a number of different disciplines. It has in the past been a domain where scientists and technologists dominated as well. Today, the range of "intellectual stakeholders" in STP is vast, and this is precisely because the subject itself is so complex and multifarious. But Rosenberg's argument is an important one, precisely because the economic dimension has tended to be excluded from STP, particularly in UN circles and in the STP efforts of many developing countries. It is only recently that one finds the World Bank conducting research on technology policy with a strong economic perspective.

On the other hand, it can also be argued that the discipline of economics has not been able to address effectively many of the problems associated with science and technology in development in any meaningful sense. In most cases, technical change has been treated as a "black box" that is beyond the elegant analysis of, say, neoclassical economics. Similarly, practical matters of implementation and management are not dealt with. The tendency has been to tackle the problem at the macro level, assuming away the heterogeneity of firms and the particularities of the entire technical process, and certainly giving no recognition to either the idiosyncrasies of technical change broadly defined nor to the specificities of technical change processes in a developing country context, where many of the assumptions that are valid in an industrialized setting simply cannot be sustained. It is only among the recent and still evolving neo-Schumpeterian school of economists studying technical change that one finds a more realistic appreciation of the peculiarities of technical change and thence its relevance, both analytically and practically, for developing countries.

Therefore, though the UN style of approach to STP and similar perspectives may admittedly be deficient in their lack of attention to economic detail and the analytical rigour that economics encourages, they are still valuable for their better sense of the complex and inter-disciplinary character of the STP issue. The neo-Schumpeterian approach, by providing the rigour of the discipline but also admitting the "muddiness" of the problem, therefore offers a hope for a new and more relevant approach to STP analysis, one that is needed in the changing times that we are experiencing.

Conclusion: Key contemporary issues for STP

In effect, the issues facing STP research and analysis today consist of three elements.

The first relates to the disciplinary content of STP, with a fairly strong argument for an emphasis on economic analysis, albeit with the proviso that economics itself lacks some of the tools needed to render a realistic perspective on the complexity of issues contained in STP. The second relates to the need for a stronger integration be-tween the imperatives of STP and those of industrialization in de-veloping countries. While STP has many goals, and the argument of "science for science's sake" is still tenable, the core rationale for STP is, for developing countries, the harnessing of the potential of science and technology for an equitable and efficient process of industrializa-tion broadly defined (and not confined solely to Western models of industrialization). Finally, there is the issue of levels of analysis of STP. Most effort has been concentrated at the broad macro level, but relatively little has been done in terms of understanding firm-level processes of technical change (but see refs. 18, 14, 17, 20, 21).

The challenge for STP is to try to balance the macro and micro perspectives. This is relevant as well for the industrialized countries, but it is clear that failures and oversights will not matter for them as much as for developing countries. Moreover, the market sector is much stronger and more dynamic in industrialized countries. For all these reasons it is all the more essential for developing countries to design analytical tools that allow for the development of realistic and rigorously formulated policies.

For STP today, the major issues centre around concerns over the use of technology to achieve competitive advantage, access to tech-nology, new forms of government intervention to promote techno-logical development at the firm level and greater participation in world markets, and all of this within the new principles of an emerging techno-economic paradigm.

369

Ernst and O'Connor [9], based on extensive research on the new industrialized countries and the "latecomer" countries, suggest several policy issues for consideration in the coming decade:

1. Access to technology and acquisition strategies, including the need to diversify technology sources, assess the value of different sourcing strategies, leverage complementary assets, and counter efforts to restrict access to new technologies.

2. Technology diffusion and generation, including creating effective demand through government procurement, providing information and generating skills for effective use of new technologies, supporting financial and technical needs of small and medium-scale firms, creating a more conducive macroeconomic environment, upgrading product design and development capabilities, and speeding up the process of automation.

3. Industrial transformation, including developing policies for "sunset" industries (in sharp contrast to the protection of noncompetitive firms practised by many developing countries – the "safeguard" measures, for example), and managing the trade-off between specialization and diversification and between vertical integration and flexible supplier networks.

4. Restructuring of trade and investment relations, including avoiding "premature liberalization," export market diversification, playing the "reciprocal market" game promoting regional integration and cooperation, and reassessing policies regarding foreign direct investment (e.g. developing "backward linkages" programmes).

5. The role of the state, where the rationale for state intervention needs to be reassessed and its proper role examined for the development of indigenous technology and absorption capabilities and human technical resources.

Most of these countries had, as a result of the battle for independence and the Marxist influence afterwards, state-oriented, planned, and subsidized economies. Now they are condemned, like the ex-communist countries in Europe, to the "transition to the market." This does not imply the end of state intervention, especially in scientific, technological, and industrial matters. It does imply at least a clearer delineation between the sectors where state intervention can be useful and those that have to be left to market forces.

From within the field of STP itself, there is a need to take stock of the situation in STP in developing countries – what resources are available, what is the knowledge base, what programmes, organizations, and institutions exist, etc. Further, more research at the theoretical and conceptual levels is required with regard, for example, to the disciplinary content of STP, the linkages between science and technology and economic development, the relationships between STP and other policies, etc. Finally, a serious effort needs to be made to improve the teaching and training of STP and related subjects in developing countries [31].

STP has changed considerably during the past four to five decades. Its importance to developing countries has, if anything, increased, simply because technology has become more important to economic activity. Yet the field is still young and needs to be further developed. This is a challenge that needs to be addressed in the 1990s if there is to be a quantum leap in the contribution of science and technology to the development process.

The concerns that will face the developing world as it enters the twenty-first century are complex. Finding the proper role that science and technology can play in resolving these problems is the challenge that faces science and technology policy analysts and decision makers in both the industrialized and developing countries.

References

1   Adler, E. *The Power of Ideology: The Quest for Technological Autonomy in Argentina and Brazil*. Los Angeles: University of California Press, 1987.

2   Alvares, C. *Homo Faber: Technology and Culture in India, China and the West 1500–1972*. Bombay: Allied Publishers, 1979.

3   Bell, M., B. Ross-Larson, and L. Westphal. *Technological Change in Infant Industries: A Review of Empirical Evidence*. World Bank Working Paper. Washington, D.C.: World Bank, 1985.

4   Cooper, C. *Science, Technology and Development: The Political Economy of Technological Advance in Underdeveloped Countries*. London: Frank Cass, 1974.

5   Daghestani, F.A. *Science and Technology in the ESCWA (Economic and Social Council for West Asia) Region: End of Decade Review*. Amman: The Higher Council for Science and Technology, 1988.

6   Dosi, G. "Technological Paradigms and Technological Trajectories." *Research Policy* 11 (1982): 147–162.

7   El-Kholy, O.A. "The 1976 CASTARAB Rabat Meeting: A Review." In: Zahlan, ed. *See* ref. 35.

8   Elzinga, A., and A. Jamison. "The Other Side of the Coin: The Cultural Critique of Technology in India and Japan." In: E. Baark and A. Jamison, eds. *Technological Development in China, India and Japan*. London: Macmillan, 1986.

9   Ernst, D., and D. O'Connor. *Technology and Global Competition: The Challenge for Newly Industrialized Economies*. Paris: OECD, 1989.

10   Evenson, R., and G. Ranis. *Science and Technology: Lessons for Development Policy*. Boulder, Colo.: Westview, 1990.

11   Forsyth, D. *Appropriate National Technology Policies*. Geneva: ILO, 1989.

12   Freeman, C.L. *The Economics of Innovation*. London: Frances Pinter, 1982.

13   Goonatilake, S. *Aborted Discovery: Science and Creativity in the Third World*. New York: Zed Books, 1984.

14   Hoffmann, K. *Technological Advance and Organizational Innovation in the Engineering Industry*. World Bank Industry Series Paper no. 4. Washington, D.C.: World Bank, 1989.

15   IDRC. *Science, Technology and Development: Planning in the STPI Countries*. Ottawa: IDRC, STPI Project, 1979.

16   Jecquier, N., and Hu Yao-Su. *Banking and the Promotion of Technological Development*. New York: St. Martin's Press, 1989.

17   Kaplinsky, R. "Technological Revolution and the International Division of Labor in Manufacturing: A Place for the Third World?" *European Journal of Development Research* 1 (June 1989), no. 1.

18   Katz, J., ed. *Technology Generation in Latin American Manufacturing Industries*. London: Macmillan, 1987.

19   Lall, S. *Learning to Industrialise*. London: Macmillan, 1987.

20   ———. "Explaining Industrial Success in the Developing World." Development Studies Working Papers, University of Oxford, 1989. Mimeo.

21   ———. *Building Industrial Competitiveness in Developing Countries*. Paris: OECD, 1990.

22   Mazrui, A. *Political Values and the Educated Class in Africa*. London: Heinemann, 1978.

23   Mudimbe, V. *The Invention of Africa: Gnosis, Philosophy and the Order of Knowledge*. Bloomington: Indiana University Press, 1988.

24   Nasr, S. *Islamic Science: An Illustrated Study*. London: World of Islam Festival, 1976.

25   Needham, J. *Science and Civilisation in China*. Cambridge: Cambridge University Press, 1954.

26   Nelson, R., and S. Winter. *An Evolutionary Theory of Economic Change*. Cambridge, Mass.: Belknap Press, Harvard University, 1982.

27   Perez, C. *"Technical Change, Competitive Restructuring and Institutional Reform in Developing Countries."* World Bank Discussion Paper no. 4, Strategic Planning Review, Washington, D.C., 1989.

28   Price, D. de Solla, and I. Spiegel-Rösing, eds. *Science, Technology and Society: A Cross-Disciplinary Perspective*. London: Sage, 1977.

29   Rosenberg, N. "Science and Technology Policy for the Asian NICs." In: Evenson and Ranis, eds. *See* ref. 10.

30   Rushing, F., and C.G. Brown. *National Policies for Developing High Technology Industries*. Boulder, Colo.: Westview, 1986.

31   Sagasti, F. "Science and Technology Policy Research for Development: An Overview and Some Policies from a Latin American Perspective." *Bulletin of Science, Technology and Society* 9 (1989), no. 1.

32   Salomon, J.-J., and A. Lebeau. *Mirages of Development*. Boulder, Colo.: Lynne Rienner, 1993. Originally published in French as *L'écrivain public et l'ordinateur*. Paris: Hachette, 1988. The debate is captured in the Appendix to

the English edition, reprinted from *Social Science Information*, vol. 29, no. 4 (1990).

33   UNCSTD. *State of Science and Technology for Development in the World: Options for the Future*. New York: United Nations, 1989.

34   Unesco. *Unesco Activities in the Field of Science and Technology in the Arab Region*. Science Policy Studies and Documents, no. 65. Paris: Unesco, 1986.

35   Zahlan, A.B., ed. *Technology Transfer and Change in the Arab World: A Seminar of the United Nations Economic Commission for Western Asia*. Oxford: Pergamon, 1978.

11

Technology transfer and diffusion

Amitav Rath

The first modern reference to technology transfer was made in 1957 [10], but it was only in the early 1960s that the subject began to receive serious attention from researchers. This was followed by an explosive growth in the literature on technology transfer from the mid-1960s through the 1970s and into the 1980s. One bibliographic survey [60] cited 1,200 books and articles on the subject, and by 1980 they had increased the citations to over 2,000 items [61]. Researchers then appear to have lost interest in technology transfer issues in more recent years for reasons not entirely clear.

In spite of the voluminous literature, a number of questions, especially regarding recent trends, remain unanswered. There is no single accepted definition of technology nor of what constitutes its transfer. Furthermore, whatever definitions we choose, we have scanty information on certain channels of technology transfer and on implications for certain sectors. Even when we examine the different mechanisms of technology transfer, and improve our data on technology transfer by mechanisms, there remains a problem of aggregation, as common measures of volume or value do not exist [51]. Finally, there is also considerable disagreement about the nature of the impact of the transfer of technology on developing countries, focusing variously on the process of transfer, the conditions of transfer, and the character or sophistication of the technology transferred.

From the wide array of issues that could be covered, I shall in general restrict myself to the core issues dealt with in the literature regarding the transfer of *manufacturing technology* across national boundaries for economic development and only touch on intersecting issues in passing.

374

There have been three other extensive reviews of technology transfer issues recently. Chudnovsky [13] provides the most thorough coverage of the issues and literature for Latin America; Enos [21] focuses on the Asian experience; and Reddy and Zhao [56] focus on the actors and transactions and provide numerous references to the literature. Among the reviews, Hoffman and Girvan [25] provide a detailed treatment of strategies and the management of technology acquisition; they include an extensive bibliography on technology transfer, with full coverage of African and Caribbean literature.

Elements and mechanisms of technology transfer

"Technology" is defined here broadly to include the body of specific knowledge, the organizations and procedures, the machinery, tools, and equipment, the necessary material inputs, and human skills that are combined to produce socially desired products. It must be emphasized that technology is a "necessary input to production" and does not include *all* kinds of knowledge, human and material inputs.

Transfer of technology is said to take place when an existing technique of production is moved from one location to another. This movement may be from a research laboratory to a production location or entity, or from one production location to another. Mansfield [37] and Brooks [8] label a movement along the research, development, and production sequence as a "vertical" transfer and the movement from one production entity to another as a "horizontal" transfer. The technology transfer process is distinguished from technology diffusion in that the transfer is a "purposive movement of established technology" [64, p. 29] and diffusion is a less planned movement of technology produced more by imitation processes. Transfers can take place within a single firm where a technology used in one plant or location is transferred to another. It may take place within a country from one firm to another, very often from machinery, process, and product suppliers to user firms. Or it may take place across national boundaries, which is the only transfer I shall consider here. The processes and issues concerning vertical and horizontal transfers, which take place largely within an economy, and international transfers, which cross political and economic boundaries, are sufficiently dissimilar to require completely different treatment, though they both contribute to increasing innovation and production capacities, and the management of the transfer processes share a number of common features.

375

Cooper and Sercovitch [17] define technology transfer to cover "the transfer of those elements of technical knowledge which are normally required in setting up and in operating new production facilities or in extending existing ones." Different elements may be required at different points in time; for instance in the pre-investment phase there is a need for feasibility studies, market surveys, determining the range of relevant technologies, choosing the appropriate technique, and so on, to elements required for plant operation, maintenance, and expansion. Some of the technical knowledge will be embodied in machinery, some will be available in written form, and some will always be uncodified and reside in skilled personnel. Other categorizations of elements include technical knowledge that is general and widespread to the industry, and system- and firm-specific technical knowledge that is often patented, protected, or secret, arises from experience, and is partially uncodified.

Elements of technical knowledge can be transferred from one country to another in many different ways. These include:

1. Books, journals, and other published information such as trade literature, standards, patent information.
2. Education and training abroad.
3. Informal personal contacts and observations through travel, meetings, and conferences; visits to production sites.
4. Exchange of information and personnel through technical co-operation programmes.
5. Employment of foreign experts and consultancy arrangements.
6. Import of machinery and equipment with literature and technical information supplied.
7. Import of intermediate products, in particular those considered technology-intensive.
8. Reverse engineering.
9. Technical specifications, standards, and training provided by importers.
10. Licensing agreements to use proprietary know-how, patents, production processes, and trademarks.
11. Foreign direct investment (FDI) that brings with it all the necessary elements of technical know-how.

Some of these mechanisms are useful for transferring technical knowledge in general, as well as more specific technical information and know-how required to use a particular process or to make a given product. While items 1–5 serve more general purposes of knowledge transfer, items 6–11 tend to be mechanisms used to transfer specific components of technology required for production. There are few useful data on the magnitudes and values of technology transferred through the more general channels. Case-studies of technology transfer show that all of the channels are valuable and developing country strategies must ensure that the full mix of channels and mechanisms are used optimally.

Some of the general indicators of technology flows include the value of machinery and capital goods imported, the quantum of FDI, payments made for patents and know-how that are normally included in licence fees and royalties, the trade in technology-intensive products, and the value of bilateral and multilateral technical assistance (this last category provides a proxy for general technical flows).

We must note a number of caveats regarding the usefulness of these indicators. First, additional technology flows are usually required to make best use of capital goods and machinery that may not be captured, and the entire value of the goods cannot be considered as payment for technology. Second, licence and royalty payments often include payments for copyrights and trade marks. Third, in many countries wholly owned subsidiaries are not allowed to make additional payments for patents and know-how. Fourth, in the case of subsidiaries, tax and commercial considerations dominate the ways in which payments are made and so the amounts need not be a good indicator of the value of the technology transferred [46, p. 8]. Fifth, these statistics do not record both general and specific transfers for which monetary transactions are not made or records are not available. Finally, a problem with the literature on technology transfer is that, because of the difficulties of aggregation and measurement of the general mechanisms, it concentrates on foreign direct investment and licences as the two most significant mechanisms. We shall see below that in fact the harder to measure categories are very important for domestic technological capability, and that domestic capability in turn is the most important variable affecting the success of technology transfer. The lack of attention to, and hence data and analysis of, the other mechanisms reflect not their lower significance but the greater salience of foreign direct investment and technology contracts in the distribution of gains between the supplier and recipient.

The indicators in the table suggest, with the above caveats, that the dominant mechanism for technology flows is in the form of capital goods. The imports of capital goods by developing countries increased rapidly until the early 1980s and have stagnated since then. The value

Technology flows by type, origin, and destination (in billion US dollars, constant prices)

	1962	1972	1980	1982	1985
Capital goods					
Between industrialized countries					
exports	26.6	85.8	341.5	337.6	375.0
imports	13.7	60.4	236.1	227.6	298.9
Exports to developing countries	6.5	20.9	107.8	116.9	96.4
Imports from developing countries	0.1	1.2	12.8	15.0	28.5
Foreign direct investment					
Between industrialized countries	2.8	9.2	41.0	30.1	33.7
Industrialized to developing					
countries	1.4	4.4	10.1	10.4	7.7
Royalties and licence fees					
Between industrialized countries	—	3.9	9.9	11.2	—
Developing to industrialized					
countries	—	0.7	2.0	2.0	2.3
Technical assistance	0.7	1.8	5.5	5.4	6.0

Source: Ref. 74, pp. 88–89.

of foreign direct investment is lower by an order of magnitude, and since the 1970s almost three-quarters of it was to other OECD countries. Royalty and licence payments by developing countries represent significant amounts in foreign exchange for their budgets. Though their absolute values are very small compared with imports of machinery, their significance lies in the fact that these represent payments for know-how only and not additional goods. Finally, the value of technical assistance is more than twice the payments for production know-how.

Historical background

A common heritage of many developing countries is their experience of being colonized by Europeans in the recent past, although most Latin American countries achieved formal political independence during the mid-1800s. Naturally, there have been important differences in the nature of the colonization experience and its impacts on the political, economic, and technological developments in the Americas, Asia, and Africa.

During the colonial period, most colonies of Asia and Africa had zero or low rates of economic growth. With the rapid increases in population, particularly in Asia, the result was increased poverty during a period in which there was continuous and rapid economic growth in Europe and North America. A major concern of developing countries therefore, both before and after independence, has been to reduce the gaps between themselves and the industrialized countries.

In general, developing countries are characterized by low levels of industrial activity and low technology inputs to production. As a concomitant of economic and social processes at work, they continue to have low levels of economic output, low productivity per worker, high levels of unemployment or underemployment of labour, considerable underutilization of available natural resources, and shortages of capital and knowledge, which continue to act as constraints on their economic growth. Furthermore, their poverty and colonial history have often resulted in continued political and economic dependence on the richer, industrialized countries.

Given this background, most developing countries have accorded high priority after independence to the twin objectives of *growth* in economic output and of political *autonomy*. Sub-objectives such as eliminating poverty are seen to follow from or require the primary objective of growth to be satisfied. Similarly, objectives of increasing national control of the economy and the share of assets and incomes accruing to nationals are seen to support the primary objectives. These twin economic and political objectives have influenced the concerns of research and policy related to technology transfer, sometimes reinforcing each other and at other times being antagonistic.

Technology, technology transfer, and economic growth
Almost all economists will agree that economic output (Q) can be seen to be a function of the available (and utilized) land, labour, and capital (where land represents both physical land and other natural resources), which are constrained in specific ways depending on knowledge and institutional factors. This can be formulated as

$$Q_t = f_t \text{ (land, labour, capital)}$$

where f_t represents one particular function at time t and is dependent on the structure of the economy and available technology.

In many developing countries, not much increase in land is considered possible, and in many other countries redistribution of highly inegalitarian and inefficient landholding patterns is deemed impossible or undesirable by their leaders. In many countries, labour is thought to be in excess supply, and when employed it is so with low productivity. So the resources that are seen as the final constraints on output growth and that are required to increase the productivity of labour, as in the historical experience of the industrialized countries, are increased inputs of capital and new production technologies. This very stylized theory suggests that "at the very least . . . the prospects of raising per capita incomes in the developing countries are, in most cases, very limited unless some significant infusions of new technology take place" [58, p. 7]. This requires an increase in domestic savings, knowledge-creating and knowledge-using capacities, and the augmentation of domestic possibilities with inflows of capital and technology from outside.

First phase: Post-war to mid-1960s
The post-war to mid-1960s period was defined by the end of the Second World War, the rapid reconstruction of European economies through American financial assistance, the (almost) undisputed hegemony of the United States in the world economy, and a time of relatively stable and rapid growth. In this period the mainstream view of economic development was relatively simple. It derived from a liberal world-view strengthened by the contribution of Keynes to economic management and the experience of growth and the contribution of American capital flows in achieving the almost miraculous transformation of Europe after the devastation of the war. The other important global event during this period was the achievement of independence of most colonies in Asia and Africa, a process that was almost completed during the 1960s. These events provided considerable optimism about the prospects for economic growth and political autonomy.

The mainstream development policies suggested that an increase in capital stocks was the main source of economic growth [38, p. 71]. This view meant an emphasis on the mobilization of domestic savings and increased external capital flows through concessional and market loans and through foreign direct investments by the newly active and rapidly growing category of multinational firms. The state was assigned the role of promoting a process of modernization by shaping the appropriate attitudes and institutional structures. It was also seen to have a considerable role in undertaking investments towards improving infrastructural facilities and providing public goods that, due to externalities, would be under-provided through private initiative.

When issues of science and technology were raised in this

framework, they were resolved in two ways. It was accepted that science and knowledge were the perfect example of public goods. Once created, everyone benefited; there were no losses to one individual because another acquired them, and the cost of diffusion and transfer of knowledge was close to zero compared with developing it in the first place. Therefore, it was appropriate for developing countries to build up their domestic infrastructure for education, technical training, and research facilities. It was thought that the vast body of world knowledge was waiting as if on supermarket shelves. The increase in domestic capacity together with direct foreign investment would allow the developing countries to tap into this resource at negligible cost, allowing them to rapidly close the gap between themselves and the industrialized countries [54, p. 1434].

Second, any external costs would be taken care of by increased flows of foreign capital. As and when new technologies were needed, the foreign direct investments made by the multinational firms would bring with them the required advanced technologies. That there would be adequate flows of capital was assumed on the basis of decreasing marginal utility. As the developing countries were short of capital and the rich countries had abundant capital, the returns on capital in developing countries would be higher. So the required capital and technological flows would take place provided the infrastructural and other bottlenecks were removed.

Even within this simplified framework two troubling issues were evident. First, if domestic capital had lower rates of return than foreign direct investment over time, the shares of investible surplus and control over the domestic economy would pass increasingly to nonnational agents, nullifying the recently acquired political autonomy. The second concerned an empirical observation that most countries, as they attempted the required structural changes, faced continuous and severe foreign exchange constraints. This resulted in reduced capital inputs and reduced access to improved capital goods and world technology.

There were always a number of criticisms of this model, but in most theories, both mainstream and alternative, it was not until the late 1960s that technology and technology transfer were introduced as increasingly critical variables in the process of development.

Second phase: Mid-1960s to early 1980s
By the late 1960s, a number of factors had come together to highlight technology-related issues as central to development policies and the pattern of economic growth. Within the broad range of technology

381

policy issues, the gains and losses from technology transfer from the industrialized countries to developing countries received the greatest attention from researchers and policy makers, both in developing and industrialized countries, and also from the agencies of the United Nations, specially the UN Conference on Trade and Development (UNCTAD).

By this time, there was considerable disenchantment with the earlier approach, and continued political and economic dependence became more and more irksome. A number of observers noted that the development process followed had resulted in an increasingly dual economy, where in one small part the process of modernization, economic growth, and prosperity seemed to occur as predicted, but a much larger part of the economy appeared either to be unconnected and unchanged or the connections were negative, leading to greater impoverishment, unemployment, and inequality. Many researchers and policy makers began to believe that important reasons for the unsatisfactory development experience lay in the poor understanding of and policies related to technology issues. Questions about technology and technology transfer began to multiply in the research and policy agenda.

Developing country unhappiness with the process of technology transfer was first raised at the UNCTAD I meeting in 1964. The initial complaints of the developing countries related to their perceptions that insufficient technology was available to them for meeting their development objectives; that technology supplies were largely controlled by multinational corporations. These firms used their extensive powers, their control over patents and know-how, to enforce unequal partnerships with developing countries and firms. The agreements included restrictive clauses on the use of technology, and the companies made exorbitant profits from their transactions with their partners. These key perceptions had emerged by the time of UNCTAD II in 1968.

UNCTAD was mandated to review the role of technology imports in the industrialization efforts of member countries and supported many of the early studies to draw attention to the problem faced by developing countries in technology transfer transactions. A summary of the developing country experience is available in UNCTAD [70]. Important early studies include Vaitsos [80, 81], and Lall [32] provides a useful survey of issues. See also Chudnovsky [13], Enos [21], UNCTAD [75], Hoffman and Girvan [25].

Over a period of 15 years, UNCTAD was responsible for a large number of studies on technology transfer (some done by staff, others by consultants), and it inspired a host of studies by other researchers, national governments, and also by technology suppliers, who had to respond to an avalanche of criticism. Some of the key documents that chart the progress of work in this area include UNCTAD [68–70].

Thanks to these studies, the earlier view in which technology was equated with knowledge – the idea that, like knowledge and information, it is widely "applicable, easy to reproduce and re-use and where firms can produce and use innovations by dipping freely into a general stock or pool of technological knowledge" [51, p. 7; 22] – was increasingly seen to be mistaken. The UNCTAD view of technology pointed out that "technology is an essential input into production," and it is bought and sold in the world market in the form of "(a) capital and intermediate goods; (b) human labour, usually highly skilled and specialized manpower, with the knowledge of machinery, processes, techniques, and problem solving; and (c) as information of a technical or commercial nature, some of which is openly available and some is subject to proprietary rights and sold under specific conditions" [69, p. 5].

Following this line of argument, Vaitsos [81] and others are uncomfortable with the term "technology transfer" as it hides the sale and purchase of these items under the neutral term "transfer"; they prefer the term "technology commercialization." Once defined as an object of trade and commerce it is natural to begin to analyse the characteristics of the object traded and of the market in which such trading takes place, the characteristics of the buyers and sellers, the costs and prices, and so on. The Office of Technology Assessment of the US Congress makes a distinction between technology trade and technology transfer: technology trade refers to the commercial transactions of buying and selling a technological element; for the transfer to occur, the further element of technology absorption, assimilation, and the ability of the receiving entity to operate and maintain the facility is required [49]. This is a useful distinction, given that developing countries often criticize technology transfer operations for failing to provide them with the capacity to operate efficiently and maintain production facilities, to expand production, and to make further changes.

The technology market

Suppliers of technology
Large transnational firms own most of the production technologies available in the world. They are therefore the main technology sup-

pliers to all countries, and in particular to the developing countries. Bizec [7, p. 36] states that 80 per cent of total receipts for technology supplied by the United States accrue to multinational firms, and for European firms, the share is 50 per cent. He also states that multinationals have been more dominant in North-South transactions than in intra-OECD transactions. Large multinationals normally have as their first preference production in their home countries, whence they export to other markets.

There are two types of circumstances in which these companies are interested in locating production facilities in other countries and therefore in transferring technology. In the first type, they may be attracted by the low cost of labour, premises, and other overheads, and perhaps less stringent regulations. The other type has a longer history and applies to firms involved in the extraction of natural resources, particularly oil and minerals. Such firms are forced to operate where the resources exist; the situation is similar for tropical agricultural products such as tea, coffee, and rubber. The main concern is to extract the local raw material as cheaply as possible for export to the industrialized country markets, which are the main users. All critical decisions regarding production, marketing, and finance are usually taken at the firm's head office. Other key management and production decisions are handled by expatriates, so that nothing more than basic operational know-how is acquired by the developing country. Beyond the initial extraction, most of the further processing is done in the industrialized countries, reducing forward and backward linkage effects in the developing country. Many of the contracts and practices are rooted in the colonial past, and this form of multinational operation has been the target of considerable hostility in developing countries.

Transfer of manufacturing operations became important in the 1960s, and the factors motivating such transfers had more to do with the circumstances facing the transnational firm than with the recipient country's objectives.

A firm can serve foreign markets in three different ways: it can export from its home base; it can set up a subsidiary to manufacture and service the foreign market; or it can sell its technology to a foreign firm and receive rent on its technological assets in lieu of profits from direct sale of its products.

When the environment dictates the location of production, as in mining and plantations, the first option has to be forgone. It is also not available if legal restrictions prevent imports or if tariffs and natural factors of distance or lower costs make imports unduly expensive. The second option, setting up a subsidiary, may be chosen so as to take advantage of low wages for regional and export markets. Sometimes the move of a competitor to produce in the foreign market forces the transnational firm to follow to protect market share, although this has often been limited to assembly and packaging operations, with most components and inputs supplied from overseas.

This option may also not be feasible due to legal restrictions, lack of sufficient resources within the firm, whether financial, managerial, or in knowledge of the foreign environment. In that case, the firm may consider taking a local partner in a joint venture or even sell its technology outright for some combination of a lump sum fee and royalty. When transferring the technology, the suppliers often enter into legal contracts with developing country firms that are highly restrictive. For example, restrictions on the right of the recipient firm to make any changes in product, process, or inputs without approval; requirements to transfer to the supplier knowledge of all improvements made in the technology with no reciprocal obligation on the supplier; tied purchases of capital goods and inputs; and imposed production or export restrictions.

The technology suppliers combine their ownership of technology with dominant market power, access to large financial resources, and skilled personnel to extract agreements from developing country firms and governments that result in high costs to the developing country. These costs include high royalty, licence, and technical fees even for outdated technologies. Firms have often transferred additional incomes to themselves through tied supplies of inputs, often by charging exorbitant prices for these inputs (prices of intermediates supplied were sometimes higher than the price of the final product in the international market).

A number of specialists have analysed technology transfer from the perspective of the suppliers [20, 31, 82, 9, 66, 14, 11].

Purchasers of technology
The reappraisal of the 1970s suggested that in comparison with the suppliers, the buyers of technology suffer from a number of disadvantages. The first stems from the nature of technology itself. As Arrow [4] said, in any commercial transaction of knowledge and information, there is an inherent asymmetry between the seller, who knows what he is selling, and the buyer, who, to some degree, must remain ignorant of what is being purchased. The developing country buyer is

385

particularly disadvantaged in that usually the developing country firm is smaller, less experienced, and technologically weaker in many areas in addition to the core proprietary technology being acquired.

The overall technological and informational weaknesses of the buyer limit his search among possible technology suppliers. The buyer's ability to undertake direct technology transfers is often limited, and the transfer then has to be undertaken through an intermediary who packages the elements of required technology. The buyer suffers another information disadvantage in that there are no easily available figures of the price paid for similar transactions.

Furthermore, often owing to the protected markets of most developing countries, the private interest of recipient firms does not coincide with the social interests of the country in minimizing fees paid for technology transfer. The goods produced with the imported technology and carrying international trademarks can command high prices in the protected domestic markets, thus allowing the private firm to make sufficient profits even with high transaction costs.

Impact of market imperfections on developing countries

A key result of the work of the 1970s was that the earlier notion of a freely available technology was replaced by that of transactions in a market. The structural features of the buyers, sellers, and the commodity give rise to a market for technology that is highly imperfect, made up of many weak and poorly endowed buyers and a few large, powerful technology suppliers and marked by information gaps and information inequality between buyers and sellers. In such a market, it is more than likely that suppliers will take advantage of their monopolistic position: the price of technology will be high and the amount of technology supplied smaller than optimal.

The analysis in the 1970s focused mainly on the issues of (excessive) costs of technology transactions and the many restrictive clauses that were imposed on the recipient by the supplier, thereby limiting the benefits to the recipient firm and country (UNCTAD [70] lists 46 practices deemed by developing countries to be restrictive). Two different sets of evidence pointed to the existence of excessive payments. The direct payments for technology, i.e. royalty and licence fees, can be compared for identical transactions, though this is difficult in practice. What has been documented in many cases is that after regulations controlling and reducing payments have gone into effect, there have been substantial reductions in the fees charged, thus supporting the view that the earlier payment levels were excessive. The studies also pointed out that the total magnitude of direct payments made by developing countries towards licences, patents, know-how, trade mark, and technical services amounted to 87 per cent of profits on foreign direct investment, 56 per cent of foreign direct investment flows, 8 per cent of imports of machinery, equipment, and chemicals, and 5 per cent of total export earnings for the decade 1960–1970 [70]. For some countries, these ratios were much higher: for Mexico, for instance, direct payments amounted to almost 16 per cent of total exports [70, p. 14]. When the indirect components are added, the amounts increase significantly.

The second concern regarding excessive payments arose from the evidence of tying input supplies to the technology contracts in many cases, where the supplier insisted that the recipient must also purchase raw materials, intermediates, and/or capital goods from the supplier. In the Philippines, such clauses affected 26 per cent of the contracts, in India 15 per cent, and in the Andean Pact countries, they were found in up to 83 per cent of the contracts in one country [70, p. 16]. Tied purchases resulted in monopoly control of the inputs by the technology supplier, even of items readily available in the market. Tied purchases were found to lead to overpricing of the inputs, which ranged, in a sample of cases studied in the pharmaceutical sector, from 19 to 155 per cent in Colombia, 30 to 500 per cent in Chile, 20 to 500 per cent in Peru, and 40 to almost 1,700 per cent in Mexico [70, p. 17]. Vaitsos [81] also cites a number of examples, with overpricing amounting to a high of 3,000 per cent in one case. Similarly, Rafi [52] cites a number of examples from Iran, in one instance the figure is 1,000 per cent.

Such possibilities that exist in transactions between two independent firms become multiplied many times over when the developing country firm is a subsidiary of a multinational. In intra-firm trade, there are many items that are intermediates and so no exact replica may exist for comparisons. The subsidiary may conduct all exports and imports through the parent company. Costs and revenues are determined by internal prices, and these can be adjusted to increase input costs and reduce export revenues for the subsidiary if it is to the global advantage of the parent company.

The widespread prevalence of this transfer pricing is documented by Vaitsos [81] for the Andean Pact countries; Murray [41] for Greece in metals, minerals, and chemicals; in Britain, by the Monopolies Commission [40], and in Iran, Rafi [52] for pharmaceuticals; by Lall [33] and Kaplinsky [29] in automobile production in Malaysia and Kenya, and recently by the US government in action against Japanese electronics and automobile companies (*Newsweek*, 15 April 1991). This practice led California to develop the unitary tax system to minimize loss of tax revenue from transfer pricing. Given the much larger scale of operations of multinationals in the industrialized countries, its significance for them must be greater.

Besides overpricing, critics found foreign controls to be associated with excessive imported inputs, often tied, which increase foreign exchange costs and limit backward linkages. Studies in Mexico, Australia, Canada, and Nigeria showed that firms with greater foreign ownership and control tended to have higher import content in their products [52, p. 207]. Rafi confirms similar results for a sample of Iranian firms and states that for the foreign-controlled firms, imported materials accounted for 30–86 per cent of costs, 73 per cent of the imports were tied to the foreign firm, margins on tied imports were often over 100 per cent, and the suppliers charged an average royalty of 21 per cent [52, p. 223]. He suggests that "given the evidence of transfer prices, sales of inputs were typically a much more significant source of income to the parent than knowhow charges" [52, p. 227].

In addition to the impact on costs from royalties, fees, and tied inputs, a number of other restrictive practices reduced the benefits from technology transfer, e.g. restrictions on exports. Analysis showed that most technology-supply contracts restricted the recipient to produce for certain markets only. Often that was limited to the home market, and exports were prohibited. In other cases, exports were allowed to selected countries. All studies show that such restrictions are widespread and are based on the suppliers' need for control over the market to minimize competition and maximize profits [70].

Other clauses that were seen to reduce benefits were those limiting production amounts, type, or quality, and fixing selling prices for the product. Similarly, clauses guaranteeing a fixed payment instead of a percentage of production and calculating royalty on final sales prices instead of value-added increased the cost of technology. Clauses that prevented the recipient from making any changes to the product or process limited the capacity to innovate and adapt it to local needs. Many contracts required the recipient to grant back to the supplier any improvements made without reciprocal obligations on the supplier. Developing countries were also unhappy with contracts that barred the recipient from sub-licensing; this they felt reduced its diffusion and required them to pay repeatedly for identical knowhow.

Other negative impacts from technology transfer
While the main focus of UNCTAD's work and similarly inspired work was on the issue of imperfect markets for technology leading to excessive financial costs and unfair and restrictive practices, a host of other arguments were raised around the same time that suggested that, without careful controls, unrestricted transfers of technology can lead to many other undesirable impacts on development and costs to the developing country's economy. Additional questions raised about transferred technologies included their characteristics and longer term impacts on the structure of the economy, on domestic technological capacity, and on the political autonomy of importing countries.

INAPPROPRIATENESS. One set of arguments pointed out that technologies available in the world market had evolved over a period of time in the currently industrialized countries. The trend had been towards capital- and skill-intensive and labour-substituting production methods that were congruent with the historical growth paths of the industrialized countries. The available technologies are often designed for mass production methods meant to satisfy the demands of large markets. The material inputs similarly tend to be those available in the countries of origin and do not necessarily make good use of material resources abundant in the developing countries, where inputs are more readily supplied from abroad than from the domestic economy, leading to weak linkage effects and increased dependency on imports. Furthermore, the products that the imported technologies produce are also inappropriate as they are designed to meet the needs of higher income consumers. So the production using these technologies displaces more appropriate products but at the same time the new products can be purchased only by a small number of relatively affluent customers. This creates welfare losses and limits the potential for scale economies, expansion, and competition, placing further limits on possibilities of growth and technical efficiency [65].

The inappropriateness of the technical processes reduces employment opportunities. In addition, the more productive imported technology sector demands higher skilled labour inputs and can pay significantly higher wages to the small pool of skilled workers it uses, who in turn demand more "inappropriate" products and form the economic and political groups in favour of continued economic development along the inappropriate path. This process leads to a dual economy with one sector using modern technologies, higher skills,

389

and providing high wages side by side with growing poverty and un-employment in a large fraction of the economy cut off from the former.

DEPENDENCY. Another strand of analysis focused on the political de-pendency of developing countries that went hand in hand with the technological dependency. It was pointed out that the overwhelming body of knowledge and technology was developed in and resided in the North. Thus, both the R&D establishment and the productive en-terprises looked towards the industrialized countries for the problems and their solutions, as well as for scientific recognition. Importing knowledge, production know-how, and capital goods could be cost minimizing for individual transactions but would in the long term have negative effects for local problem-solving capacity and techno-logy development. Restrictions on technology imports would force domestic producers to pay greater attention to developing in-house R&D capacity and increase linkages to domestic input suppliers, capital goods, and research establishments. In this view, multina-tional companies transfer little know-how, stifle local initiative and problem-solving capacity, and thereby perpetuate dependency. The distribution of R&D activities by these firms between the parent company and affiliates reinforces the existing imbalance. For in-stance, in 1982 US multinationals spent US$41 billion on R&D, of which over 91 per cent was spent in the United States, a little over 8 per cent in other OECD countries, and less than 1 per cent in developing countries, while the developing countries accounted for 20 per cent of the technology receipts for the companies [78, p. 181].

The policy response

The policy responses of developing countries were guided from the 1970s by the negative implications of technology transfer practices as they were seen to prevail. The imperfect markets for technology, the tendency of suppliers to derive payments above "reasonable" levels, their resort to unfair practices, and the likely divergence between private returns to the technology-receiving firm and the costs and benefits to the national economy all suggested the need for government action. Measures were required to increase the informa-tion available to recipients, to strengthen the bargaining capacity of recipient firms, and to outlaw practices that were deemed to violate national interests. The new regulations also sought to increase the scope of national enterprises and to reduce control by foreign enterprises. They also sought to increase the possibilities for arms-length transactions and reduce transfer pricing abuses by new controls on the scope, methods, and practices of foreign direct investment.

The policy response in most countries was to promulgate laws to regulate and control foreign direct investment to specific sectors that were highly export-oriented or involved transactions accompanied by significant inputs of technology. They also established institutions to review, regulate, and register all technological contracts, and the laws generally forbade most of the restrictive practices discussed earlier, prescribed royalty levels, and reduced rates for technology payments.

The laws enacted in the 1970s in many countries were especially severe with regard to the scope for foreign direct investment. Many countries nationalized foreign-owned mining companies, declaring earlier practices and privileges to be remnants of the colonial past. This was often followed by new contractual arrangements for specific services such as exploration, management, and sales. India passed new laws in 1973 requiring wholly foreign-owned companies to reduce the share of foreign ownership. Companies providing mainly for exports or in designated high technology sectors were allowed to retain a higher percentage of foreign ownership. The Andean Pact countries inserted provisions for a "fade out" of foreign ownership over a period of 10 years [42, 3].

Besides the policy changes in individual countries, a major effort was mounted by the Group of 77 to establish an international code of conduct for the transfer of technology and for a more favourable environment for technology. This has been in the process of negotiations since 1975, but it has not yet been possible to ratify such an international code due to continuing fundamental differences in the position of the developing and the industrialized countries.

By the beginning of the 1980s, most developing countries had enacted regulatory mechanisms and rules governing investments and technology broadly following the UNCTAD guidelines and recommended by the yet to be agreed upon Code, though there were considerable differences between countries in the degree of comprehensiveness and stringency of application. The policy response of the developing countries followed largely from the composite picture of disadvantaged trade in technology that had emerged and attempted to imitate the policies of state guidance that had earlier been followed successfully by Japan [35, 50]. Briefly, the key elements of the picture include the following. The supply of technology is dominated by large multinationals. They actively pursued a global strategy for markets within which technology was a key element, and their profit maximization strategies led them to offer packages of unbundled technologies, excessive imports, especially intermediates, at high costs. The self-interest of the suppliers limited the quantity and quality of transferred know-how. The suppliers then reinforced their dominance in oligopolistic markets by their own knowledge and experience and the lack of comparable expertise among the recipients. The suppliers supported little R&D in the developing countries and inhibited local learning in a number of ways. Finally, since the marginal cost of technology transfer was close to zero, this implied that all payments were excess profits and exploitative [6, p. 31].

Towards a revised framework

Almost as soon as the picture of technology transfer sketched above became the accepted wisdom and the defensive strategies of the developing countries were put in place to counter their effects, evidence from newer studies and the changing environment began to suggest that, while the policies were often a necessary and useful step forward, they overemphasized short-run costs, were based on a static view of technology, and did not pay sufficient attention to the more basic issue of building up domestic capacity [25, p. 52].

The revisions to the framework came from various sources. First, the concern with technology issues generated new studies, some of which attempted to corroborate the validity of the framework while others attempted to examine the impact of the regulations. Still others took as given that in all developing countries, almost all modern technologies had been imported at some time and that such imports will continue and will probably increase. Therefore, they attempted to understand the factors influencing technical change, innovation, learning, and the building up of domestic technological capacity in the developing countries, with technology transfer as only one factor within a much larger set [53, pp. 15–16; 24, p. 5].

The revised framework takes as given that the nature and effects of technology transfer will vary depending on the nature of the actors involved, their characteristics, and the nature of the technological, economic, and policy environments that confront them; it then develops a more complex and varied picture of the processes and variables involved. This picture has been strongly influenced by the increased body of empirical and theoretical work along Schumpeterian lines on the factors affecting technical change, innovation, and technological capability. The final set of influences came from the changing global political and economic context, together with a rapid increase in the rate of technical change originating in the OECD countries.

Transaction costs and terms
We have reviewed the evidence of "excessive" profits made by technology suppliers from their control over technology. The government regulations, the setting of acceptable standards of royalty payments for different types of technology, and the elimination of many practices that allowed additional payments (e.g. through the supply of tied inputs) substantially reduced the payment levels. A study in Colombia [28] reports that with the new regulations, 395 contracts were reviewed between 1967 and 1971. Of these, 61 were rejected. In the remaining 334, through renegotiations, royalties were reduced by 40 per cent, or US$8 million per year. In a later period, renegotiations led to the elimination of export restrictions and of tied purchase of inputs in 90 per cent of the contracts and a reduction of minimum payments in 80 per cent of contracts. Similarly, in Mexico after the new laws were passed, 30 per cent of the contracts submitted in the first year were rejected [2, p. 153]. Aguilar cites a case where the Mexican Registry was able to obtain significantly lower rates through negotiations with technology suppliers. The Nigerian National Office of Industrial Property stated that thanks to renegotiation of terms, the country saved 115.9 million naira (at the 1990 exchange rate, approximately 10 naira = US$1) in 205 technology contracts registered. Similarly for India, Balasubramanyam [5] and Desai [19] point out that essentially the government regulations regarding payment levels forms and those regarding restrictive clauses are largely adhered to, though there are some violations. Chudnovsky [12] found that in most countries, a maximum ceiling of royalty was fixed at 5 per cent, the maximum duration was 5 years, and most restrictive clauses had been declared illegal. He found indirect evidence in Latin America that the rate of royalty payments and intra-firm royalty had been reduced with no apparent reduction in the flow of technology. And in Korea, after the liberalization of controls in 1978, a subsequent review showed that there was a sharp rise in the payments made for technology while the flow was unchanged [78, p. 185].

But a number of studies, particularly from India, suggest that government controls had a number of negative effects on technology transfer agreements. It is argued that the regulations reduced the "quality" of technology supplied to Indian firms as the price was reduced. Before examining the Indian data, we should look again at one of the major assumptions in the technology regulations debate. It had been argued that technology is knowledge and once created all can freely make use of it. Hence the only cost of transferring technology would be its marginal cost and the marginal cost would be close to zero [65, 79]. However, Teece [66] found in a study of 26 technology transfer projects that there were considerable costs involved, and they were higher the more complex the technology to be transferred, the larger the gap between the technological level of the supplier and that of the recipient. The costs are also significantly higher when the technology is transferred initially, but with experience the costs for subsequent transfers decline.

Contractor [14] confirms these findings and details the various types of costs involved for the supplier. He provides three categories of costs: the direct costs of the transfer, i.e. travel, training, personnel, documentation, and related costs; the opportunity costs of transferring technology; and the sunk, development costs. Bell and Scott-Kemmis [6] provide additional evidence of significant direct costs. Niosi and Rivard [47] suggest that the opportunity cost is of two types. One is the potential cost of market loss and competition by the recipient, but this is more often of theoretical rather than practical concern for many transactions, since in the majority of cases the suppliers do not see most recipients as capable of encroaching on their markets. The second, the opportunity cost of the skilled personnel utilized in the transfer process who could have been used more profitably in other activities, is often more important for smaller technology suppliers.

Variations in technological elements and price
The evidence from India [34, 19] suggests that the actual direct costs of technology transfer vary according to the elements being transferred. They are lowest when existing designs, drawings, specifications are copied and shipped to the recipient. The transfer costs increase as more human-embodied knowledge and skills, which are largely uncodified, are transferred. Costs also increase the greater the amount of training provided to recipient firms, and they are even higher if the process or product technologies of the supplier have to be modified for small market size in the recipient countries and for different demand patterns and factor and resource endowments. Furthermore, the direct and opportunity costs for the supplier increase to the extent

that skilled personnel are utilized to effect the transfer and if the transfer harms potential or existing export markets. The potential loss would be considered larger by the supplier where more exclusive and more recent vintage technological assets are to be transferred.

The studies of technology transfer confirm the interplay of these factors in the decisions of suppliers and recipients regarding not only whether technology will be transferred but what elements will be transferred and in what forms. Large, powerful, technologically advanced manufacturing firms, without serious constraints of human resources and capital, prefer, as they expand, exports from the home base except when either there are significant barriers to exports or an export base provides significant economies of production for regional or world markets. If the developing country location is important for market and production reasons but government policy precludes or makes difficult wholly owned subsidiaries, or if the firm has resource constraints, it will opt for joint ventures or licences for technology as the next best option.

If the market is small or the firm believes the risks of equity investment are high, it will tend to prefer contractual arrangements over equity investment. The higher the payments for technology, the more inclined the supplier will be to incur higher costs in technology transfer. If the remuneration is not high enough, it may still proceed to incur actual losses in the technology transfer if it is relevant to its long-term strategy of market penetration, knowledge, or foothold. But if the potential losses are significant it will attempt first to reduce costs by providing cheaper technology elements, and finally it will refuse to enter into any technology transfer agreements.

Studies of European technology transfer to India suggest that the reductions in payment levels have had an adverse effect on the supply of technology to Indian firms. The suppliers were less interested in providing technology that had high future profit potential, and when transfers were agreed to, the elements of production know-how included were narrow, fragmented, and did not cover all relevant knowledge. As one supplier stated, "what was transferred were dies, drawings, specifications, not 'know how' let alone 'know why'. They acquired a tiny fraction of the technology" [6, pp. 80–81].

Recipient expectations were also inhibited by payment restrictions. In many cases the Indian firm was aware that, because of its inability to pay due to regulations and/or to its small size and the small market for the product, it could not afford more expensive technological elements. So its acquisition strategy involved seeking out sources and elements of technology that could be accommodated within the limits of its capacity to pay.

Bell (1982) has categorized the elements of technological knowledge into three types: those required to operate the machinery; those required to solve production problems and undertake expansion; and those required to initiate technical change. Mlawa [39] found in a study of the textile sector in Tanzania from 1960 to 1980 only elements of the first type and even these were not comprehensive. The Tanzanian firms acquired only the basic minimum skills of operation but little in the way of engineering and product design. Again, in a larger study of technology acquisition from the United Kingdom by Indian firms, Bell and Scott-Kemmis [6] report that in many cases, the first type of flows were the minimum contractually required, could be easily provided, and were often narrow and fragmentary. There were few flows of technology representing the other elements. In one part of the sample, this appears to be due to the unwillingness of suppliers to undertake additional effort given the low rates of payments. In other cases, they were not even sought or demanded by the recipient. Sometimes the firm was satisfied with the subset of elements because it sufficed to commence production and make profits. The market did not require it to be more efficient. In other cases, the firm was not able – that is, did not have the requisite technical staff and facilities – to utilize the additional elements of knowledge and so did not seek them. There were even instances in which firms were not aware that important additional elements were available to be acquired.

New perceptions of the actors

Theory suggests that the greater the number of supplier countries and supplier firms, the greater the competition among them. The keener the competition, the lower the price at which technology would be available. This is supported by evidence in several ways. First, it should be noted that the perception that there were a small number of technology suppliers in any given product or process was probably more valid during the period immediately after the Second World War. Because of the destruction of industry in Europe and Japan, the United States was almost the sole supplier of foreign direct investment and US firms had large technological leads and were the most active in their efforts to expand internal market shares. But the subsequent decades have seen the increasing presence of European firms and the emergence of Japan as a major new source of technology.

Between 1960 and 1985, total foreign direct investment has increased from US$99 billion to US$693 billion; the United States accounted for 47 per cent of the total in 1960 but only 35 per cent in 1985, whereas Japan has increased its share from 0.7 per cent in 1960 to 11.7 per cent in 1985. Developing countries supplied 1 per cent of the total in 1960 and 2.7 per cent in 1985 [78, p. 24]. Similar global statistics on licence agreements by number, value, and sources are not available, but the data available for various recipient countries show that all countries have diversified their sources of supply over time.

The former USSR and eastern European countries also became more important sources of technology transfer during this period. Though in numbers and value their contribution was small – they accounted for less than 0.2 per cent of world foreign direct investment and less than 1 per cent of affiliated enterprises – they played a key role in transferring technologies, sometimes unavailable or available with undesirable conditions, to certain developing countries and in certain sectors. For instance, in India such transfers were critical in reducing the monopoly powers of international steel, pharmaceutical, electrical, and machinery firms and allowed initial capacity to be built up in these sectors.

Another important and noteworthy feature has been the emergence of a number of developing countries, such as South Korea, India, Taiwan, Hong Kong, Brazil, Argentina, and Mexico, as technology suppliers in the world market. These countries have been increasingly significant as exporters of smaller, more standardized, technologies, which are often scaled down and more labour-intensive than those available from the traditional sources. So it appears that while monopolistic powers have been available to suppliers, the extent of these powers over time and across sectors may have been exaggerated. In many sectors and for standard technologies, these powers appear to be declining gradually. Bell and Scott-Kemmis [6] find that in a sample of technology suppliers to India, a tiny minority could be described as monopolists, and even in oligopolistic market structures, competition among suppliers was often intense.

Increased competition among suppliers and the greater experience of the developing countries with transfers encouraged suppliers to make considerable efforts to adapt. They became more flexible and accommodating towards host country policies, and in many cases were prepared to transfer core skills provided the compensation was adequate. The response by multinationals to the restrictions on subsidiaries and direct investments has been a rapid rise in joint ventures and forms of non-equity participation such as licensing, management contracts, franchising, turnkey projects, and subcontracting [48]. These forms are often attractive to suppliers as they can produce significant revenues with lower risk.

At the same time, in certain technology-intensive sectors such as consumer electronics or pharmaceuticals, core technologies remain available normally to subsidiaries only.

The role of small supplier firms

Similarly, policy regimes driven by perceptions of the need for hard bargaining with large, powerful multinationals in *all* cases may also be overdrawn. This is not to deny the power and importance of the multinationals in certain technology-intensive sectors. For instance, 97 per cent of US employment in chemicals, 86 per cent of employment in transport equipment, and 90 per cent of US trade originate with multinationals [78]. At the same time, however, smaller companies are increasingly active in technology transfer and their significance appears to be increasing.

In a study of technology transfers to India, it was reported that "two thirds of the suppliers [from the UK], accounting for 40 percent of the agreements had 500 employees or less and a quarter of them had 50 or less." But also, a very small group, about 10 per cent of the suppliers, *were* very large (with more than 10,000 employees) and accounted for 40 per cent of the agreements [6, p. 37]. Similarly, the sample of Indian technology agreements with Belgium and the Netherlands included a large percentage of small, and larger non-transnational, firms [16, p. 112].

Other evidence suggests that the number of smaller firms entering technology transactions has increased over time and their importance is growing [47]. With the increase in smaller suppliers and the greater flexibility offered by large firms, the range of options for developing countries has increased as regards choice, terms, and control. But the exercise of these choices is accompanied by the recipient sharing a greater proportion of the risks. At the same time, we must acknowledge that in a number of sectors and in key technological areas, there are few other suppliers besides large multinationals, and for many projects the smaller suppliers complement but do not replace large suppliers.

The earlier literature had emphasized that technology transfers were initiated by the suppliers, according to their preference for equity participation, and that they were confronted by weak, passive, small, and poorly informed recipients. This too appears to have been exaggerated. It was first pointed out by Reuber [57] that often the developing country firms initiated the process of technology transfer and sought equity participation. These firms believed that with equity participation there was greater involvement by the technology supplier, greater interest to ensure that the technology could be used successfully for production, and that equity participation ensured a degree of risk sharing between the supplier and recipient. This is further documented by Balasubramanyam [5] and Contractor [14].

In many cases, the developing country firms had made a careful study of the technological options available to them; for instance, "Indian manufacturers have, generally, been quite aware of major innovations abroad and quick to take licenses" [18, p. 70]. They often entered into careful negotiations with the suppliers for favourable terms and conditions. A significant number of recipients in fact had greater experience of technology negotiations than many suppliers when the latter were smaller and without previous international experience [6, p. 41].

Thus, the idea of disadvantaged, weak, and inexperienced developing country firms needs to be qualified. Clearly, a large number of firms from the more industrialized developing countries undertake careful searches and negotiate hard over terms, especially price. Their share undoubtedly used to be smaller and is still small in the less industrialized countries.

But a more pervasive problem is that they often do not make the most of the technological knowledge made available. Bell and Scott-Kemmis [6, pp. 97–99] found in their sample of Indo-British collaborations that in 40 per cent of the cases, the suppliers believed that the recipients did not absorb the know-how made available. In one case "[the recipients] made a bit of effort," but "they wanted to make money without becoming a machinery engineering company" and "cut costs on training at all corners." In another, "they did not realize how much technology there was to absorb." A recipient states, "we under estimated how much specialized knowledge was needed" [6, p. 98].

Technology transfer is meant to provide technical knowledge lacking in a specific production environment. Training is therefore extremely important, yet this is frequently underemphasized or ignored by recipient firms and countries. For instance, in a study of over 600 petroleum exploration contracts, only 14 per cent made any provisions regarding training, employment of nationals, and local technical services [67]. Contracts in later periods were more explicit about the need for such provisions, suggesting that recipients had learned from experience. Thus we see that recipient firms and countries have all too often been unaware of their needs or unwilling to make the necessary effort towards acquiring the know-how.

Factors affecting technology transfer: New findings
The developments of the 1970s, the increased confrontations on issues of technology, and the high level of policy attention paid to the issues catalysed a number of new studies on technology transfer. The newer studies attempted to elaborate further the characteristics of the technology supplier, the recipient, and national policies and their impact. The studies attempted to examine the similarities and differences in the process by sector, by underlying firm strategies of the recipient and supplier, according to the elements of technology transferred, and the nature and amounts of payments. Finally, greater efforts were made to understand the implications of the above variables and policy choices for the economic development objectives. The 1980s saw a large number of case-studies that attempted increasingly to determine associations between the processes, the factors underlying the processes, the macroeconomic variables, and government policies, characteristics of the main agents with the "success" or "failure" of technology transfer. But then what do we mean by success?

Many definitions, concepts, and measures are used to determine success, varying in breadth and relevance. Owing to difficulties of measurement and aggregation, success or failure of technology transfer is in general more amenable to individual case-study analysis, and most aggregate, macro indicators have serious limitations.

In any transfer of technology, the prime objective is to transfer production knowledge from one location to another in order to undertake specific production activities; *the amount of output made possible* is therefore one measure of success. In addition, it is obviously important to examine *the efficiency* achieved in the new production process in terms of materials and energy used, the productivity of labour and machinery, and to compare the efficiency levels with relevant norms of best practice elsewhere, supplier standards, or with designed levels.

It would generally be agreed that a movement towards local integration represents a successful example of technology transfer, so that there should be an assessment of how far the new facility has been able to *replace imported elements over time*. Clearly a transfer that creates an efficient production facility using 100 per cent imported inputs of raw materials, skilled personnel, and know-how is significantly different from one that is at the other extreme. Then, since there should be improvements in efficiency, local content, and technological capabilities over time, we need to analyse the *speed* with which "success" is achieved. As these factors are also evolving in other countries, one may wish to compare whether and at what rate the recipient is able to *maintain or reduce the gaps* between local performance levels and rising world standards. For any economic conclusions to be drawn we need to know the magnitude of the costs incurred and benefits derived, plus those of alternative courses of action. And finally, since at least two parties are involved, we need estimates of the *costs and benefits* to each party and their degree of "reasonableness" and "equity."

The evidence clearly supports the view that *the greater the involvement of the supplier and the recipient, the more successful the technology transfer*. At the simplest level, when the recipient is a wholly owned subsidiary, the supplier has the highest potential total return, the minimum divergence of views, procedures, and objectives between the two parties and the minimum threat to future markets, so it can more easily transfer explicit, codified, and documented technical elements as well as the more implicit, uncodified human-embodied elements of knowledge. The situation is similar if the supplier owns a significant amount of equity in a joint venture. In such situations, the evidence suggests that technology transfer to the point of establishing efficient production facilities is most likely to be successful with larger equity participation by the supplier. The recipient then also has access to the ongoing results of technical change and innovation available to the supplier [23, 19, 30]. On the other hand, although firms owned by multinationals are usually more efficient [62], there is general agreement that other aspects, such as knowledge and capability related to investment, expansion, and innovations, are stunted where there is excessive foreign control.

Production efficiency, as we shall see with other aspects of technology transfer, is highly correlated with *the macroeconomic policies and market structures of the recipient country*. If the recipient environment does not promote competition but protects inefficient production, outdated processes, and products, then all forms of transfer can be seen to perform poorly when production rates, efficiencies, and costs are compared with other more competitive environments. For

401

instance, a survey of technology transfer in the textile sector in a number of African countries illustrates "the pivotal importance to which changes in competitive structures or in state policies are translated into a need to reduce costs of production" [43]. Again, in a situation of larger macroeconomic problems, variables affecting individual performance tend to get submerged. A 1985 survey of 343 manufacturing firms, most involving technology transfer both in the initial stages and often subsequently, showed that 23 per cent of them had stopped production and 57 per cent were producing at levels below their break-even point [44]. Adei [1] provides a case-study of a tyre plant in Ghana where so little production technology was transferred that, with the departure of the expatriates, plant operating levels fell to 10–20 per cent of capacity. Similarly, Lall [34], Desai [19], and others have pointed out that the excessive protection and regulations in the Indian environment reduce both incentives and opportunities for efficient production, so that Nath [45] found little change after transfer in the production performance of a group of firms that could be attributed to characteristics of the technology transfer.

With regard to the "appropriateness" of the original technology and the extent to which the technology provided is modified by the supplier to take into account different market size, factor costs, and resource endowments, it does not appear that the effects of technology transfer arise from form and ownership. In general, when the recipient market is considerably smaller, most suppliers and recipients try to make adjustments to the production technology as currently used by the supplier. This is sometimes achieved by transferring earlier vintages of the production technology used when the market in the supplier country was smaller. In some cases scaling down poses few problems, and specific exercises are undertaken for this purpose. Frequently the peripheral processes such as packing, transport, and materials handling are adapted to suit recipient capital and labour costs. Small modifications to adapt the process and machinery to the characteristics of local inputs are made less often.

Such attempts impose additional costs, require greater involvement on the part of the supplier, and greater capacity on the part of the recipient to suggest, participate in, and contribute to any modifications. A supplier that has already gained experience in adapting technologies, or that expects increased sales in the future as a result, will have greater capacity for and interest in undertaking these activities. All these factors imply that lower payments for technology, small size of suppliers, and lack of experience in other transfers will tend to inhibit modifications to the technology to make it more appropriate for the recipient economy. The evidence in Bell and Scott-Kemmis [6], Niosi and Rivard [47], and others confirms that larger, more experienced firms tend to make more adaptations, provided the direct or expected returns to them justify their additional costs. The evidence also suggests that these factors provide only the necessary conditions for adaptations of technology; the chances of their actually taking place depend heavily on the recipient's macro-economic environment, with greater competition forcing adjustments to be made. Policy environments that require domestic inputs to be used increase the need for adaptations, and greater recipient technological capacity improves the likelihood of successful adaptation.

It is probably owing to the combined effect of these variables that Reddy and Zhao [56], in their review of evidence for capital intensity of technologies and that of adaptations (if any) by supplier characteristics, find contradictory indications. Some studies find that multinationals are more likely to use labour-intensive technologies and others find the opposite to be true. Some suggest that multinationals are more willing and able to adapt technologies while others do not. Clearly, specifying the problem by supplier characteristics alone is insufficient.

The changing economic, technological, and policy environments
There is little disagreement regarding some recent trends in the world economy and their effects on issues of technology transfer; regarding others, however, there is ambiguity. The 1980s have seen a period of slow-down in the rates of world economic growth and of growth in trade, although trade growth is still far faster than economic growth, making competitiveness an important issue for all economies.

The operations of multinationals continue to grow, and they account for an even larger percentage of world production, trade, and employment than earlier. The ratio of intra-firm trade, which imposes special difficulties for national control, has increased and amounts to almost a third of all international trade. The multinationals invest larger amounts in R&D and have increased control over the research of smaller firms through take-overs.

At the same time, there is a trend towards higher rates of technical change and so larger investments are required to keep abreast of technology. The investment costs for successful innovations appear to be rising in many sectors. All of these factors have made technology a more important strategic element for firms.

For developing countries, these trends create increased barriers in some instances and increased scope in others for technology transfer. Increased scope is available from the greater number of technology suppliers and the sharper competition between suppliers. Larger numbers and sources allow developing countries to extend their options and obtain better terms; the evidence also suggests that multinationals have become more flexible and accommodating to some demands from developing countries. Willingness to transfer core technology and skills is increasingly a condition for multinationals to win contracts and is more often accorded.

At the same time, new barriers are emerging due to fiercer competition, more rapid rates of technical change, the need to have access to wider elements of knowledge for successful product and process innovation, and the need to reduce risks from higher investments. These have forced groups of firms to increase cooperative activities in technological development and production. This is shown in the increasing rates of cross-licensing of patents, joint R&D, more joint venture partnerships, and new forms of product and process development partnerships between firms simultaneously competing in other (and even related) sectors. Furthermore, the directions of technical change have cut direct labour costs in many sectors, and to that extent have reduced the attraction of developing countries as low-wage manufacturing locations. The developing countries that are attractive for foreign companies are those with large markets, such as India; close to major industrial countries, such as Mexico; and those with high growth rates, strong technological capacity, and a skilled labour force, as in East Asia.

On the other hand, economic growth in Latin America and Africa has been poor to uneven. The large debt burden has been a major drag on growth for many countries. Low growth and high debts have reduced capital flows and made foreign exchange increasingly scarce. This has reduced the inflows of foreign direct investment, imports of capital goods and machinery, technical assistance, and foreign training – all important channels for technology transfer. We have at this time almost no information on how this reduction and the other global changes have affected other issues connected with technology transfer and its effects on technological capability at the firm and national levels. These trends suggest that if developing countries were

404

ever a homogeneous group in the past, they are certainly much less so now. The "particularities" of the countries require increased particularity of analysis and policy response.

The combined effects of the diminished resource capacity of many developing countries, the greater need to attract capital flows, a recognition that regulations in the earlier period were often carried too far, and the growing threat of increased technological gaps have forced most developing countries to a degree of liberalization of technology import policies. There are good reasons for a shift away from excessive regulation, but without thereby implying that no regulations were necessary.

The 1980s saw a complete swing of the pendulum, with the triumph in the North of the view that all regulations by the state were necessarily imperfect and could not improve upon even imperfectly functioning markets. This "triumph," of what Hirschman calls "monoeconomics" and "the conviction with which the advantages of the invisible hand, and the disadvantages of the visible hand, were sought to be imposed on the faithless infidels whenever possible" (Jagdish Bhagwati), contradicts much of the evidence on developing technological capability. In the South the view had to be accepted under straitened circumstances.

In this context, the issues of unequal exchange disappeared from the international fora until a new round of bargaining to share environmental resources surfaced in the international agenda. It is interesting to note that besides all the other changes noted here, the newly perceived global environmental threats and the need to find common ground between the North and the South to confront these dangers have again forced a reappearance of the divergent views on issues of technology and its transfer to the top of the international agenda. Unfortunately, at this time both the North and the South appear locked in a sterile replay of the unresolved issues from the Code of Conduct negotiations of the 1970s, while Northern trends towards extending private property rights in completely new areas are raising a host of new concerns [55].

Concluding remarks

In some ways, excessive politicization of the issues has definitely been harmful to the interests of developing countries. It is argued that the anti-multinational rhetoric of many countries in the 1970s scared foreign investors away, leading to lower rates of foreign direct investment in later years [36]. I would suggest that, in the 1970s, the availability of loans in the international market and, in the 1980s, the increased prospects from investments in the industrialized countries and the debt crisis have been the dominant factors affecting foreign direct investment. Nevertheless, the nationalist policies certainly scared away investments in the primary sector in many countries, leading to a reduced role for multinationals. This had the perverse effect that while their market share increased, the unit prices declined. Furthermore, this forced the metals companies to move downstream and increase the technology intensity of their own activities [54, pp. 25–26]. Inefficient regulatory processes, leading to long delays in approval of technology agreements, are cited by technology suppliers in an OECD survey as the most important impediment.

In short, developing countries face a dilemma. Strict controls on technology transfer can and have improved the terms and conditions on which technology is available. They have also reduced considerably the extent of unfair restrictions imposed by suppliers. But at the same time, the regulations have reduced access by developing countries qualitatively and quantitatively to the range of technologies available to them. Cooper and Sercovitch [17] had cautioned that it would be too much to expect technology policies to resolve all dilemmas confronting the developing countries. But certainly countries that have relied on excessively defensive approaches, have placed an overriding emphasis on costs and neglected technological elements, have been concerned more with short-term production transfers than with longer term capacity building, have directed policies more towards external constraints than internal – all combined with inefficient implementation – have only accentuated their dilemmas. I have mentioned the dramatic changes in the international economic, technological, and political environment and the fact that we are woefully short on more recent studies on technology transfer to developing countries in the context of the revised framework and the new context. However, I am confident that neither "monoeconomic" nor "autarchic" policies will provide ways out of the dilemma; instead, appropriate policy balances have to be sought in each country.

References

1   Adei, S. "Technological Capability and Aborted Industrialization in Ghana: The Case of Bonsa Tyre Company." In: A. Rath, ed. "Science and Techno-

logy: Issue from the Periphery." *World Development* 18 (November 1990), no. 11: 1501–1542.

2   Aguilar, E. "Criteria for Measuring Cost-Benefits for Foreign Technology." In: R.E. Driscoll and H.W. Wallender, eds. *Technology Transfer and Development: An Historical and Geographic Perspective*. New York: Fund for Multinational Management, 1974.

3   Aguirre, C.B. "Science and Technology Policy and Instruments: The Experience of the Andean Pact." In: F.A. Daghestani et al., eds. *Science and Technology Policy for Self-reliance in the Muslim World*. Amman: Islamic Academy of Sciences, 1989, pp. 367–397.

4   Arrow, K. "Economic Welfare and the Allocation of Resources for Information." In: NBER, ed. *The Rate and Direction of Inventive Activity*. Princeton, N.J.: Princeton University Press, 1962.

5   Balasubramanyam, V.N. *International Transfer of Technology to India*. New York: Praeger, 1973.

6   Bell, M., and D. Scott-Kemmis. "Technology Import Policy: Have the Problems Changed?" In: A. Desai, ed. *Technology Absorption in Indian Industry*. New Delhi: Wiley Eastern Ltd., 1988, pp. 30–60.

7   Bizec, R.-F. *The Transfers of Technology*. New Delhi: S. Chand and Co., 1985. Original French edition published by Presses Universitaires de France, Paris, 1980.

8   Brooks, H. Paper presented at National Planning Association Conference. Quoted in W.A. Fischer, "Empirical Approaches to Understanding Technology Transfer." *R & D Management*, no. 6 (1976), p. 151.

9   Buckley, P.J., and M. Casson. *The Future of the Multinational Enterprise*. New York: Holmes and Meier, 1976.

10   Cairncross, A.K. *Factors in Economic Development*. London: George, Allen and Unwin, 1967, chap. 11, pp. 173–189. Originally published in French in the *Bulletin de la Banque Nationale de Belgique*, 1957.

11   Caves, R. *Multinational Enterprise and Economic Analysis*. Cambridge: Cambridge University Press, 1982.

12   Chudnovsky, D. "Regulating Technology Imports in Some Developing Countries." *Trade and Development*. Review no. 3. Geneva: UNCTAD, 1981.

13   ———. "North-South Technology Issues Revisited: Research Issues for the 1990s." Report to the International Development Research Centre. Ottawa: IDRC, 1990. Mimeo.

14   Contractor, F.J. *International Technology Licensing: Compensation, Costs and Negotiations*. Lexington, Mass.: D.C. Heath, 1981.

15   Contractor, F.J., and T. Sagafi-Nejad. "International Technology Transfer: Major Issues and Policy Responses." *Journal of International Business Studies* (Fall 1981): 113–135.

16   Cooper, C. "Supply and Demand Factors in Indian Technology Imports: A Case Study." In: A. Desai, ed. *Technology Absorption in Indian Industry*. New Delhi: Wiley Eastern Ltd., 1988, pp. 105–135.

17   Cooper, C., and F. Sercovitch. *The Channels and Mechanisms for the Transfer of Technology from Developed to Developing Countries*. Geneva: UNCTAD, 1971.

18   Desai, A. "Technology and Market Structure under Government Regulation: A Case Study of the Indian Textile Industry." In: IDRC, ed. *Absorption and Diffusion of Imported Technology*. Ottawa: IDRC, 1983.

19   Desai, A. "Technological Performance in Indian Industry: The Influence of Market Structures and Policies." In: A. Desai, ed. *Technology Absorption in Indian Industry*. New Delhi: Wiley Eastern Ltd., 1988, pp. 1–29.

20   Dunning, J., ed. *The Multinational Enterprise*. London: Allen and Unwin, 1970.

21   Enos, J.L. "Transfer of Technology." *Asian-Pacific Economic Literature* 3 (March 1989), no. 1: 3–37.

22   Fei, J., and G. Ranis. "Technology Transfer, Employment and Development." New Haven: Yale University Growth Center. Mimeo.

23   Fong, C.O. "Technology Acquisition under Alternative Arrangements with Transnational Corporations: Selected Industrial Case Studies in Malaysia." In: UNCTC, ed. *Technology Acquisition under Alternative Arrangements with Transnational Corporations*. Bangkok: UNCTC/ESCAP, 1987.

24   Fransman, M. "Technological Capability in the Third World: An Overview and Introduction." In: M. Fransman and K. King, eds. *Technological Capability in the Third World*. London: Macmillan, 1984.

25   Hoffman, K., and N. Girvan. *Managing International Technology Transfer: A Strategic Approach for Developing Countries*. Ottawa: IDRC, 1990.

26   Hirschman, A. *The Strategy of Economic Development*. New Haven: Yale University Press, 1959.

27   IDRC, ed. *Absorption and Diffusion of Imported Technology*. Ottawa: IDRC, 1983.

28   Junta del Acuerdo de Cartagena. *Technology Policy and Economic Development*. Ottawa: IDRC, 1976.

29   Kaplinsky, R. "Capitalist Accumulation in the Periphery – The Kenyan Case Re-examined." Brighton: University of Sussex. Mimeo.

30   Kim, L. *Technology Policy for Industrialization: Conceptual Frameworks and Korea's Experience*. World Bank Working Paper. Washington, D.C.: World Bank, 1988.

31   Kindleberger, C., ed. *The International Corporation: A Symposium*. Cambridge, Mass.: MIT Press, 1970.

32   Lall, S. "Transfer Pricing by Multinational Corporation Manufacturing Firms." *Oxford Bulletin of Economics and Statistics* 35 (1973), no. 3.

33   ———. "Transfer Pricing and Developing Countries: Some Problems of Investigation." *World Development* 7 (1979): 59–71.

34   ———. *Multinationals, Technology and Exports*. London: Macmillan, 1985.

35   Lynn, L.H. "Technology Transfer to Japan: What We know, What We Need to Know, and What We Know May Not Be So." In: Rosenberg and Frischtak, eds., pp. 255–276. *See* ref. 59.

36   Manser, W.A.P., and S. Webley. *Technology Transfer to Developing Countries*. Chatham House Papers, no. 3. London: Chatham House, 1979.

37   Mansfield, E. "International Technology Transfer: Forms, Resource Requirements, and Policies." *American Economic Review* 65 (May 1975): 372–376.

38   Mellor, J.W. "Agriculture on the Road to Industrialization." In: J.P. Lewis

and V. Kallab, eds. *Development Strategies Reconsidered*. Washington, D.C.: Overseas Development Council, 1986, pp. 67–90.

39 Mlawa, H.M. "The Acquisition of Technology, Technological Capability and Technical Change: A Study of the Textile Industry in Tanzania." Ph.D. diss., Science Policy Research Unit, University of Sussex, 1983.

40 Monopolies Commission. *Chlordiazepoxide and Diazepam*. London: HMSO, 1973.

41 Murray, R. "Transfer Pricing and the State." Paper presented at conference, Transfer Pricing, Institute of Development Studies, Sussex, 6–10 March 1978. Mimeo.

42 Mytelka, L.K. "Licensing and Technological Dependence in the Andean Pact Group." *World Development* 6 (1978): 447–459.

43 ———. "Stimulating Effective Technology Transfer: The Case of Textiles in Africa." In: Rosenberg and Frischtak, eds., pp. 77–127. *See* ref. 59.

44 ———. "The Unfulfilled Promise of African Industrialization." *African Studies Review* 32 (1989), no. 3: 77–137.

45 Nath, N.C.B. "Technology Acquisition under Alternative Arrangements with Transnational Corporations: Selected Industrial Case Studies in India." In: UNCTC, ed. *Technology Acquisition under Alternative Arrangements with Transnational Corporations*. Bangkok: UNCTC/ESCAP, 1987.

46 National Research Council. *The International Technology Transfer Process*. Washington, D.C.: National Science Foundation, 1980.

47 Niosi, J., and J. Rivard. "Canadian Technology Transfer to Developing Countries through Small and Medium Enterprises." In: A. Rath, ed. "Science and Technology: Issues from the Periphery." *World Development* 18 (November 1990), no. 11: 1529–1542.

48 Oman, C. *New Forms of International Investments in Developing Countries*. Paris: OECD Development Centre, 1984.

49 OTA. *Technology and East West Trade*. Washington, D.C.: U.S. Congress, Office of Technology Assessment, 1979.

50 Ozawa, T. "Macroeconomic Factors Affecting Japan's Technology Inflows and Outflows." In: Rosenberg and Frischtak, eds., pp. 222–254. *See* ref. 59.

51 Pavitt, K. "Technology Transfer among the Industrially Advanced Countries." In: Rosenberg and Frischtak, eds., pp. 3–24. *See* ref. 59.

52 Rafi, F. "Joint Ventures and the Transfer of Technology: The Case of Iran." In: R. Stobaugh and L.T. Wells, eds. *Technology Crossing Borders*. Boston: Harvard Business School, 1984, pp. 203–237.

53 Rath, A. "ADIT: A Review." In: IDRC, eds., pp. 13–19. *See* ref. 27.

54 ———. "Science, Technology and Policy in the Periphery: A Perspective from the Centre." In: A. Rath, ed. "Science and Technology: Issues from the Periphery." *World Development* 18 (November 1990), no. 11: 1429–1444.

55 Rath, A., and B. Herbert-Copley. *Technology and the International Environment Agenda: Lessons for UNCED and Beyond*. Ottawa: IDRC, 1992.

56 Reddy, N.M., and L. Zhao. "International Technology Transfer: A Review." *Research Policy* 19 (1990): 285–307.

57 Reuber, G.L. *Private Foreign Investment in Development*. Oxford: Clarendon Press, 1973.

58   Rosenberg, N. *The Transfer of Technology: Opportunities and Problems*. Report SS-77-11. Seoul: Korean International Economic Institute, 1977.

59   Rosenberg, N., and C. Frischtak, eds. *International Technology Transfer: Concepts, Measures and Comparisons*. New York: Praeger, 1985.

60   Saghafi-Nejad, T., and R. Belfield. *Transnational Corporations, Technology Transfer, and Development: A Bibliography*. Philadelphia: World Wide Group, Wharton School, 1976.

61   ———. *Transnational Corporations, Technology Transfer and Development: A Bibliographic Sourcebook*. New York: Pergamon Press, 1980.

62   Santikarn, M. *Technology Transfer: A Case Study*. Singapore: Singapore University Press, 1981.

63   Scott-Kemmis, D., and M. Bell. "Technological Dynamism and Technology Content of Collaborations: Are Indian Firms Missing Opportunities?" In: A. Desai, ed. *Technology Absorption in Indian Industry*. New Delhi: Wiley Eastern Ltd., 1988, pp. 71–104.

64   Spencer, D.L. *Technology Gap in Perspective: Strategy of International Technology Transfer*. New York: Spontan Books, 1970.

65   Stewart, Frances. *Technology and Underdevelopment*. London: Macmillan, 1977.

66   Teece, D. "Technology Transfer by Multinational Firms: The Resource Cost of Transferring Technological Know-how." *Economic Journal* 87 (1977): 242–261.

67   Turner, T. *Petroleum Exploration Contracts and Agreements and the Transfer of Technology*. Geneva: UNCTAD, 1982.

68   UNCTAD (United Nations Conference on Trade and Development). *Transfer of Technology, Including Know-how and Patents: Elements of a Programme for UNCTAD*. Geneva: UNCTAD, 1970.

69   ———. *Guidelines for the Study of the Transfer of Technology*. New York: United Nations, 1972.

70   ———. *Major Issues Arising from the Transfer of Technology to Developing Countries*. New York: United Nations, 1975.

71   ———. *An International Code of Conduct for the Transfer of Technology*. New York: United Nations, 1975.

72   ———. "Draft Outline for Preparation of an International Code of Conduct on Technology Transfer." Submitted by the expert from Brazil on behalf of the Group of 77. Geneva: UNCTAD, 1975.

73   ———. "Draft Outline for an International Code of Conduct on Technology Transfer." Submitted by the expert from Japan on behalf of the experts from Group B. Geneva: UNCTAD, 1975.

74   ———. *Trade and Development Report*. Geneva: UNCTAD, 1987.

75   ———. *Joint Ventures as a Channel for the Transfer of Technology*. Proceedings of a workshop organized by UNCTAD, 21–25 November 1988. New York: United Nations, 1990.

76   UNCTC. *National Legislation and Regulation Relating to Transnational Corporations*. New York: United Nations, 1976.

77   ———. *Research on the Transnational Corporations*. New York: United Nations, 1976.

78   ———. *Transnational Corporations in World Development: Trends and Prospects*. New York: United Nations, 1988.

79   Vaitsos, C. "Bargain and Distribution of Returns in the Purchase of Technology by Developing Countries." *Bulletin of the Institute of Development Studies, University of Sussex* 3 (1970), no. 1.

80   ———. *The Process of Commercialization of Technology in the Andean Pact: A Synthesis*. Washington, D.C.: Organization of American States, 1971.

81   ———. *Intercountry Income Distribution and Transnational Enterprises*. Oxford: Clarendon Press, 1974.

82   Vernon, R., ed. *The Technology Factor in International Trade*. Washington, D.C.: National Bureau of Economic Research, 1970.

12

Technology choice and development

Ajit Bhalla

The debate on technology remains a hardy perennial, though new dimensions keep emerging. In the 1950s, the debate concentrated largely on *technological determinism*, i.e. there were few technological alternatives for producing a given well-defined product; product differentiation was considered to be the same thing as a new product. The paradigm of determinism was followed by that of *technological pluralism*, thanks to the growing empirical evidence that emerged in the late 1970s indicating the existence of a fairly wide technology choice, not only in peripheral and material handling but also in manufacturing proper. Studies done at the ILO World Employment Programme, at the Yale Economic Growth Center, and at the David Livingstone Institute of Overseas Development Studies at the University of Strathclyde – among others – all pointed towards considerable technical choice in consumer goods industries, and to a lesser extent in intermediate goods. It is true that empirical evidence for technology choice in capital goods industries requiring greater precision is less clear-cut. Yet even in these industries, examples of technological adaptations have been reported, although genuine alternative choices may not exist [64].

A new dimension to the technology choice issue has been added by the advent of newly emerging technologies like micro-electronics-based innovations, new materials, telecommunications, and new biotechnologies, etc. In many ways, these "new" technologies provide a superior alternative to existing conventional and automated technologies. In other cases, they provide a complement to the conventional and, perhaps to a lesser extent, traditional technologies.

412

When the new technologies are superior to the existing ones, they will dominate and supersede the latter – a case of Schumpeter's "creative destruction" in the capitalist growth process. But Strassmann [63] has shown that the historical experience between 1850 and 1914 was one of old and new technologies coexisting without necessarily involving a replacement of the old by the new for several decades. In fact, it is this coexistence, and to some extent, combination – e.g. retrofitting – that has partly inspired the emergence of the concept of technology blending, discussed below.

Thus the recent technology debate has brought out into the open a dilemma facing the developing countries: what mix of new, conventional, and traditional technologies should they use? Another aspect of the dilemma is to determine an appropriate balance between importing new technologies (which most of them do not yet have the capacity to produce) and using conventional and indigenous technologies.

The developing countries will no doubt continue to be influenced by the technological advances being made in the industrialized countries. There is controversy as to whether developing countries (particularly those with a labour surplus) should use "high" technology, with its negative social and economic consequences like labour displacement and possible worsening of income inequalities. Some argue that using "high" technology will involve the developing countries in further technological dependence on the industrialized ones, thus hindering the process of indigenous capacity-building. Others argue that the developing countries cannot remain indifferent to the selection and utilization of high technology. For improved competitiveness in the international markets, it is imperative that these countries examine the feasibility of using these technologies on a *selective* basis.

Few developing countries today are both producers and consumers of high technology. This means that many of them will need to possess a capacity to develop new technologies in order to reduce their dependence on the industrialized countries. The fact that in general a few industrialized countries are the major sources of supply of these technologies creates a sellers' market in which the buyers (the developing countries) have a very weak bargaining position. Yet the international economic environment and the structural adjustment programmes that are being introduced in many developing countries lead to austerity in government spending. The first programmes to receive the cuts are likely to be R&D expenditures and scientific projects whose benefits are perceived to be essentially long term. Thus, under the circumstances, resort to imports of new technology is likely to be the only option for the third world.

I review the technology literature in an evolutionary perspective and group it into three phases: (1) the 1950s and 1960s, (2) the 1970s, and (3) the 1980s. In the first period, the emphasis on technology issues related mainly to investment allocation and growth-inducing influence of capital-intensive technologies through reinvestible surpluses. The 1960s also saw the beginnings of the appropriate technology concept as a reaction against the failure of heavy industrialization strategies to remove social ills like unemployment and poverty.

In the 1970s, attention therefore shifted to technology choice as an instrument of employment policy. The main concern was with technology choice and change in the context of generating employment, alleviating poverty, and satisfying basic needs. The issue of appropriate technology, which led to a long and rather sterile debate, really belongs to the class of technology choice and change issues. The protagonists of appropriate technology [54] simply highlighted the need to widen the set of technological options by developing alternative technologies in a labour-intensive direction that are more suited to the factor endowments of developing countries. This need to widen the choice also embraced such issues as choice of appropriate products and issues of consumer demand and income distribution.

In the 1980s, it was realized that too much emphasis on micro issues alone was misplaced. The issues of technology choice and transformation of developing countries needed to be placed in a macro perspective. This brought to the fore the importance of appropriate government policies to promote employment-generating technologies. The issue of the implementation of these technologies could not be left purely to the economists. Non-economic forces also influenced decision-making. This decade can therefore be associated with the macroeconomics and political economy of technology decisions, intersectoral linkages to promote technology improvements and reduce technology gaps between modern and informal sectors, and the emergence of new technologies.

In the final analysis, rational technology selection could not be made without the existence of national or indigenous technological capacity on the part of the producers and policy makers in developing countries. The subject of capability-building is therefore a major long-term goal of development of the third world.

The 1950s and 1960s: Growth, investment allocation, and technology choice

During the 1950s, 1960s, the issue of technology choice was secondary to that of maximizing growth. Therefore, technology choice was to be geared to the achievement of that objective. The choice invariably recommended was in favour of the most capital-intensive and advanced technology because it contributed to maximizing savings rates and investment. The reinvestible surpluses, it was assumed, would be higher with a capital-intensive technology than a labour-intensive one, because all profits (accruing from capital-intensive techniques) are saved, whereas most of the wages earned by labour are consumed. If a labour-intensive technology was chosen, additional employment would lead to a higher wage bill and higher consumption, thus reducing the reinvestible surplus.

The thinking along these lines of Dobb [17], Galenson and Leibenstein [26], and Sen [55, 56] dominated the development literature throughout the 1950s and the 1960s. The issue of technology choice was directly linked to the planning objectives. If growth was the major objective and capital the major constraint to development, then the choice of the most advanced technology was clear. If, on the other hand, the objective was to maximize employment or immediate output, the choice of labour-intensive technology might be rational.

During this period, empirical testing of the above hypotheses was not very fashionable. Only a few studies of a micro nature attempted to verify the validity of the assumptions that all profits are saved and all wages are consumed, and that capital-intensive techniques necessarily maximized reinvestible surplus. One such study on the textile spinning technology in India [4] estimated the orders of magnitudes of total reinvestment and total additional output and employment that could be obtained from a given initial investment made in alternative techniques. It came to the conclusion that while the factory technique (or the capital-intensive technique) maximized reinvestment, it did not maximize either output or employment. The traditional labour-intensive technique did not maximize reinvestment but it did maximize output and employment. These trade-offs between growth, output, and employment were to preoccupy scholars even until much later, well into the 1980s.

The 1970s: Technology, employment, and basic needs

In the 1950s and 1960s, it was generally believed that rapid economic growth and industrialization in developing countries would automatically remove poverty through a "trickle down" effect on the poor and the underprivileged. Despite the tremendous influence of development thinkers, in actual practice the growth maximization strategies did not lead to any substantial trickle down to make any impact on the unemployment and poverty problem. Empirical evidence generated during the 1970s also showed that in many cases, at least in Africa, even the absolute standard of living of the poor had declined. The relatively new international programmes, like the ILO World Employment Programme, and the World Bank's anti-poverty programme during the MacNamara years, therefore advocated abandoning concern with GNP and growth *per se*. The emphasis was put instead on the broader-based development strategies that gave pride of place to employment generation, human capital formation, a more egalitarian income distribution, and the satisfaction of basic human needs. In other words, what became important during the 1970s was not simply *what* was to be produced but *how* it was to be produced and for *whom*.

This new orientation towards development also meant a reorientation of the analysis of technology choice and development. The criterion for choosing technology was no longer to be the reinvestible surplus of growth, but employment and income generation and reduction of inequalities besides output generation.

The early 1970s saw a major shift towards research on socio-economic and employment implications of technology choice in developing countries. It also witnessed the emergence of the concept of appropriate technology.

Appropriate technology

Appropriate technology (AT) was defined differently by different people. It is not my purpose here to enter into a long discussion of the controversy on the subject. Suffice it to say that the concept emerged mainly as a reaction to the failure of the growth-maximizing strategy of development to alleviate the problems of unemployment and poverty. AT has been defined in terms of criteria and *objectives*: employment, basic needs and environment, etc.; and in terms of *characteristics*, i.e. simplicity, small scale of operation, labour intens-

416

ity, low skill requirements, etc. The underlying premise of the AT concept is the limited relevance of the industrialized country technologies to the different factor endowments of developing countries. There are different ways in which existing technology can be adapted and made more appropriate to the developing country conditions. Three such ways have been discussed in the literature: downscaling of large-scale technology; upgrading of traditional technology; and adaptation of imported technology.

As conditions vary between developing countries, no single technology can be considered appropriate for all countries at a given moment and over a period of time. Endowments change over time, making some technologies less appropriate in future than others. Thus AT does not represent a particular tool kit of technologies, even though they are usually related to small-scale consumer goods production. Instead, it is more useful to consider some priority areas and sectors in which the needs for gradual technological transformation are greater than others. The development of appropriate and adapted technologies needs to be concentrated on these priority areas [7, 39].

Appropriate products

The question of technology choice is closely linked with that of product choice and consumer demand. The argument for linking technology choice with product choice runs as follows. The basic goods and services consumed by low-income groups in a society (e.g. food, footwear, and clothing) tend to be more labour-intensive than those consumed by the rich (e.g. consumer durables). As goods for the masses are generally produced with simple labour-intensive techniques, a redistribution of income in favour of the poor should raise demand for these goods and thereby employment. This view presented in the ILO Comprehensive Employment Mission Report to Colombia [30] was further elaborated by James [32] and Stewart [59] and James and Stewart [35].

Empirical tests of the validity of the assumptions that the poor necessarily consume labour-intensive goods, and production of these goods will promote appropriate technology applications generating employment, were conducted on a variety of specific products and countries (e.g. soap in Barbados and Bangladesh, bicycles in Malaysia, metal household utensils in India, footwear in Ghana, furniture making in Kenya, and passenger transport in Pakistan [68]).

These studies generally confirmed the above hypothesis of an increase in demand for labour-intensive goods resulting from redistribution. However, the evidence from these and other studies [45, 66, 32] is mixed. Although consumption of basic goods does rise, this does not necessarily raise employment substantially. This limited employment impact may be due to the fact that some basic products may use capital-intensive but cheap inputs like synthetic fibre. Secondly, the employment effects may be small because the macroeconomic studies are too aggregate. Taking the Indian sugar industry as an example, James [32] shows that combining crystal sugar (capital-intensive) and *gur* (labour-intensive) underestimates the effects of changes in income distribution. If they were taken separately, the positive employment effects would increase by 50 per cent.

Furthermore, in some cases even capital-intensive goods (e.g. Bata shoes produced with modern technology) may be more appropriate for the poor than the labour-intensive goods because the former are cheaper and more durable.

Technology and employment

Far more studies have been undertaken on the effects of alternative or appropriate technology on employment than on the effects of appropriate products. The notion of technological fixity and rigid production functions dates back to the classic article by Richard Eckaus [18], in which he showed how substitution possibilities between capital and labour were limited in the industrial sector. He argued that in general there was only one efficient technique for producing a well-defined industrial product. This technique would be mostly capital-intensive and would be imported from industrialized countries. At the time when he wrote, little empirical evidence existed to challenge this assertion. However, during the 1970s and 1980s a substantial number of empirical micro-industrial case-studies dealing mainly with consumer goods, but also to a lesser extent with intermediate and capital goods sectors, clearly pointed towards wide technological choice [6, 65, 48, 71]. The range of choice is broader for crude products and for simple consumer goods industries than for those requiring high degrees of precision and quality product specifications. Nevertheless, in these latter cases, as noted above, there is evidence to suggest that the technological determinist view is exaggerated.

One attractive approach to technology and employment during the 1970s was offered by Sen [57]. Policy implications of technology choice are linked to the production and employment modes, viz. family employment, extended family, wage employment, and co-operatives. The technological sophistication increases with the mode of production/employment. For example, technologies that can be economical for wage-based firms are unlikely to be available to small household production units. With an increasing emphasis on private sector development and growth of small enterprises, a comparative analysis of technology choice by employment modes seems, on the surface, to represent a fruitful enquiry.

One has to be careful about making hasty generalizations on the basis of a very micro and heterogeneous sample of industrial and agricultural case-studies. There is clearly a dilemma here. As noticed earlier, aggregative studies tend to blur the issue of technology choice made essentially at the micro level of firms and farms; they also tend to underestimate the employment effects. But at the same time, small-scale micro studies do not lend themselves to easy generalizations.

Notwithstanding the above caveat, some general conclusions can be drawn from the wealth of empirical case-studies listed in table 1.

First, the studies show that factor price distortions (of only two factors, capital and labour), while relevant to technology decision-making, are not as important as many other factors. Furthermore, two-factor models that consider the role of factor pricing are somewhat oversimplified. In cases such as processing industries like sugar, the prices of raw material inputs may be more important in the choice of technology.

Second, even when factor pricing policies play a role as incentives or disincentives, they may not be sufficient for appropriate technology decisions. They would be more effective if combined with such measures as the establishment of appropriate institutions of technological information collection and dissemination, appropriate infrastructure and adequate planning, organization and implementation machinery, etc.

Third, the choice of technology is significantly influenced by the existing market structures and the associated issues of risk and uncertainty. The risk and uncertainty may arise due to imperfect knowledge about alternative technologies. The monopolistic advantages of a firm or industry are more likely to encourage the choice of capital-intensive technology than the more competitive structures.

Fourth, substitution between skilled and unskilled labour, and supervisory and management costs hinder the use of labour-intensive

Table 1 **Coverage of empirical studies on technology choice in manufacturing**

Product or industry	Author	Country or region[a]	Scale	Product	Skills	Raw material	Material handling, transport	Factor efficiency	Location	Energy	Employment	Environment	Used machinery
Consumer goods													
Beer brewing	ILO and Strathclyde	—		×	×	×	×	×			×		×
Bread	ILO and Strathclyde	—						×			×		×
Bread	ILO	Kenya	×			×	×	×					
Can-making	ILO	Kenya, Tanzania, Thailand				×	×	×	×				×
Cane sugar production	ILO	—	×	×	×	×	×	×					
Clothing	Michigan	Sierra Leone	×	×	×	×		×					
Coconut oil production	ILO	—	×		×	×	×	×	×		×		×
Cotton spinning	Yale	Brazil				×	×	×			×	×	
Fish preservation	ILO	—	×	×	×	×		×					
Fruit, vegetable preservation	ILO	—	×	×	×	×	×	×		×	×	×	
Gari production from cassava	ILO	—	×	×	×	×	×	×		×	×	×	
Jute processing	ILO	Kenya	×	×	×	×		×	×				
Leather shoes	ILO	Malaysia	×	×		×	×						
Leather shoes	Yale	Brazil	×	×		×	×	×		×			
Maize milling	ILO and Strathclyde	—	×		×	×	×	×		×			
Milk processing	ILO	—	×	×	×	×		×		×		×	

420

Product	Study	Country										
Rice milling	ILO and Strathclyde	Tanzania	×	×	×	×	×	×	×		×	
Salt production[b]	Enos	South-East Asia	×	×	×	×	×	×	×	×	×	×
Sugar processing	ILO	India	×	×	×	×		×	×	×	×	
Sugar processing	Strathclyde	Ghana, Ethiopia	×	×		×		×	×	×	
Textiles	ILO	United Kingdom	×	×		×	×		×	×	
Textiles	Strathclyde	Africa	×	×	×		×	×	×		
Intermediate goods												
Bricks	ILO	Malaysia	×	×	×	×	×	×	×	×	
Cement blocks	ILO	Kenya	×	×	×	×	×	×	×	×	×
Copper and aluminium	ILO	—						×	×	
Fertilizers	Strathclyde	India	×	×	×	×	×		×	
Iron foundries	Strathclyde	—	×	×		×		×		
Nuts and bolts	Strathclyde	—	×	×	×	×		×	×	×	
Capital goods												
Agric. machinery[c]	Mitra	—	×	×	×	×		×		×	
Engineering	ILO	Colombia	×	×	×	×	×	×	×	×	
Metal working	ILO	Mexico	×	×		×	×	×	×		×

Source: Ref. 6.

a. Dash indicates that the study is based on international cross-section data.

b. Ref. 20.

c. Ref. 44.

technologies. Substitutions take place not only between capital and labour but also between semi-skilled labour and skilled supervisory plus unskilled labour. Our stock of empirical knowledge about the skill implications of alternative technologies remains quite limited.

Finally, sociocultural and political forces, vested interests of decision makers, and government intervention may facilitate or hinder the use of more appropriate technologies.

The issues of energy saving and environmental conservation have also come to the forefront in recent years. This has raised the number of criteria against which technology decisions need to be judged. As noted in table 1, few of the existing studies consider environmental effects and energy consumption as important variables in technology choice [6]. Far too much emphasis in early studies was placed on the issues of employment and income distribution, although these are a major concern of developing countries.

The analysis of a relationship between technology, environment, and employment is of recent origin [46, 8]. It is therefore not surprising that even in the industrialized countries, it is difficult to find many good studies that attempt to analyse quantitatively (or even qualitatively) possible trade-offs between energy intensity, labour intensity, and pollution intensity of alternative industrial technologies. One of the major difficulties in undertaking such analyses is not so much the vagueness of definitions of environmental considerations as the lack of adequate data about polluting and non-polluting technologies and industries.

The bulk of the literature on technology choice in the 1970s was of a *static* nature, examining issues of technology choice at a point in time rather than the *dynamic* effects – social as well as economic – of technical change over a period of time. While some studies have been done to examine how technology changes take place and what effect they have on the modification of known techniques [2, 59, 61], the stock of empirical knowledge on the subject still remains relatively limited. Yet historical studies of technical change are essential to guide the planners and policy makers in making intertemporal choices regarding growth of output and employment. When the short- and long-term effects of choices differ, the policy makers are better advised about politically and socially feasible lower-cost solutions.

There is another context in which the dynamic issues of technology development are relevant. One of the major objectives of developing countries is to develop indigenous technological capacity, not only to select from existing alternatives but to widen the choice by developing new ones. A prerequisite for this is that at least some technological development activity be located in developing countries to ensure positive effects of domestic learning and to promote self-reliant attitudes. These issues came to the fore in the 1980s and are examined in the next section.

The 1980s: Macro issues, new technologies, and capabilities

In the 1980s, the evolution of development thinking shifted to more macroeconomic and sectoral aspects of technology policy and its implementation. These issues were somewhat complicated by the emergence during this period of new technological innovations like micro-electronics, telecommunications, and biotechnologies, whose potential influences on production, income distribution, and employment are not easily foreseen. I first examine issues of macroeconomic effects of technology choice and their policy implications before examining the potential effects of new technologies and technology blending.

Macroeconomic aspects of technology choice

Two sets of analyses have been undertaken to trace the macroeconomic effects of technology choice. The first has to do with the political economy considerations as a rebuttal of the neoclassical paradigm [34, 60]. The second deals with quantitative modelling based on social accounting matrices [40, 41].

Political economy considerations
In the context of technology choice, macroeconomic considerations imply an examination of the effects of government policies on technological decisions made largely at a micro level in firms and farms of different ownership and organization. The political economy aspects refer to the influences of interest groups and their power on macro decisions and the external environment in which micro units operate. The government exercises an *indirect* influence on technology decision-making through its factor and product price policy, through control or encouragement of monopolistic structures, and through distribution, credit, fiscal, and import policies, etc.

Different socio-economic groups and technology decision makers have conflicting interests and motivations. They are also likely to be affected differently by different technology choices. For example,

423

liberalization of tractor imports is likely to benefit large farmers (who can afford them) more than the small farmers. Some groups are likely to gain and others likely to lose from a given technology decision. This is illustrated in table 2 with an example of rural linkages in the Philippines.

Some have argued [28] that governments mainly represent the views of one set of interests, viz. that of "researchers, bureaucrats and capitalists" rather than the social welfare of the whole nation. A given technology is closely associated with a particular economic, social, and political structure that it is in the interest of the government to protect.

A variant of the political economy considerations of technology choice is to consider its impact/implications under alternative development strategies. Implicit in this view [34] is the assumption that a particular development strategy is shaped by the politico-economic interests of the government that has formulated it. For example, a redistributive strategy favouring an increase in incomes of the poor and fulfilment of their basic needs is more likely to be based on support from small farmers and rural masses than a purely growth-oriented strategy. The former strategy is much more likely to use technology choice as an instrument of income redistribution than the latter.

Macro-modelling
Although no one would dispute the common assertion that technology affects the entire economy and society, few macro studies have ventured to trace these effects on an economy-wide basis. Most of the empirical studies, as noted above, are of a micro nature that trace only the *direct* effects of micro decisions. There are very few studies that aggregate these effects and the *indirect* ones at a sectoral level, much less for the economy as a whole. Yet indirect effects – through backward and forward linkages – may be far more important for output and employment generation. These indirect effects could only be traced through input-output analysis of sectoral interdependence in an economy. An improvement on this analysis is the social accounting matrix (SAM) framework, which makes it possible to trace the effects of alternative technologies on macroeconomic variables and policy objectives. Khan and Thorbecke [41] use the SAM technique to examine the technology-production interactions in the energy sector of Indonesia. The production activities are classified along dualistic technological categories: viz. products that could be produced with either traditional labour-intensive technology or modern capital-

424

Table 2 **Matrix of gains and losses from policies to promote rural linkages**

			Policy change					
			Policies within agriculture			Rural infra-structure		Forward link
Interest group	Promote agric. prices/credit	Investment	Land reform	Credit/mech.	Crop composition	Elect./trans.	Credit	Support small-scale
Large landowners/farmers	G	G	LL	L	N	G	N	
Small farmers	G	GG	G	G	G	G	N	
Landless labourers	G	G	GG	G	G	G	N	
Rural industrialists	G	G	G	G	G	GG	GG	G
Elite/cronies	l	l	l	l	N	N	L	
Urban workers	LL	l	N	N	N	N	N	
Urban, informal	LL	l	N	N	N	N	N	
Aid donors	G	G	N	U	N	U	U	U
Foreign cos.	N	N	L	L	N	N	L	

G: medium gain   N: neutral
GG: large gain   U: unknown
l: small loss
L: medium loss
LL: large loss

Source: Ref. 62.

425

intensive technology. Effects of changes in output of selected dualistic production activities on aggregate output of agriculture and mining, energy and other sectors, and the whole economy are then estimated. Further, effects on factor income and employment and household income distribution are also attempted. Khan and Thorbecke [41] conclude that "the traditional technology generates greater aggregate output effects on the whole economic system than the corresponding modern technology . . ." and "the effect of the increased production of traditional technology has a greater impact on total employment and a much greater impact on the incomes of lower skilled workers than the corresponding modern alternatives" (p. 5).

These results cannot be taken as definitive; further refinements of the methodology and many more empirical studies of this kind are needed before we can be certain of their relevance for policy and practice. At present, the analysis is arbitrarily based on only two techniques for producing each of the selected products; yet there may be in practice several technological alternatives. Further micro analyses are essential to provide an improved understanding of their macroeconomic effects.

A pioneering aspect of the Khan-Thorbecke study is the use of R&D as a separate productive activity in the SAM framework. Notwithstanding the conceptual problem of reconciling the *static* nature of SAM with the *dynamic* effects of R&D, the authors have made a bold attempt to study the contribution of R&D expenditures to the development and adaptation of technology as a tool for better technology planning and policy-making. For lack of data, it was not possible to test the methodology for the Indonesian economy.

It is ironic that despite the clear recognition of the economy-wide effects of technology choice and change, and the need for understanding better these economy-wide effects, the macro-modelling has remained hampered partly by the limitations of methodology and partly by the absence of required disaggregated data.

Intersectoral linkages
Macroeconomic studies also enable an investigation of intersectoral linkages. Much of the literature on technology choice is concerned with the supply-side issues – technology development and utilization assuming that demand for technology exists. In practice, experience has shown that very few of the small-scale but improved technologies considered appropriate for many developing countries have been commercialized by the private sector on any significant scale. The

426

problem is not simply one of engineering and development of prototypes. Instead, one of the major constraints to the commercialization of alternative technologies is the low effective demand for AT from a large number of small-scale producers. The poor engaged in small-scale activities, for whom the AT devices are intended, cannot afford to purchase them even if they perceived the merit of using them for raising their productivity and incomes.

The promotion of rural and urban and farm and non-farm linkages can help relax the demand constraint. A strong and positive relationship is known to exist between agricultural growth and changes in the rural non-farm sector [29, 50].

Three types of linkages between agriculture and non-agriculture are relevant. They are: (1) backward production linkages (e.g. equipment inputs to agriculture); (2) forward production linkages (e.g. food processing); and (3) forward consumption linkages (e.g. increase in demand for industrial products induced by increased purchasing power in the agricultural sector). The third type of linkages is noted to be the most important. It depends on a number of factors like growth of agricultural output and incomes, distribution of income, state of agricultural technology, and the crop mix, etc. [50].

Increase in agricultural incomes should also provide an impetus to the demand for agricultural technologies (both biological and mechanical), assuming that they exist and that the factor price distortions do not keep them beyond the reach of those who need them most.

The linkages need not move from agriculture to non-agriculture as noted above. They are also induced from the urban to the rural sector through such mechanisms as subcontracting between large and small enterprises. Watanabe [70] noted three types of linkages: technological linkages, i.e. transfer of technology and skills; input linkages, i.e. supplies of raw materials and equipment; and market linkages between a large-scale parent firm and its small-scale subcontractors. The flow of technology and skills from the large- to the small-scale rural sector is essential to bring about narrowing of technology gaps between the modern and informal sectors of developing countries. Invariably, the rural sector does not have any internal source of technology generation and equipment supply, most of the R&D being concentrated in the urban sector. Therefore, any strategy of gradual modernization of traditional technologies in rural areas calls for external inputs, as was clearly evidenced by the experience of the Green Revolution in the 1960s.

These issues of intersectoral linkages can be handled at a macro level mainly through the input-output and SAM techniques mentioned above. But the data requirements, as already noted, are serious constraints to undertaking empirical studies in developing countries.

New technologies and blending

The issue of intersectoral linkages becomes even more important with the advent of new technologies, which are likely to exercise an increasingly pervasive influence in different economic sectors, e.g. agriculture, manufacturing, banking and financial services. The new technologies have been heralded almost as a revolution on a par with the previous ones brought about by the steam engine, steel, and electricity [24, 47]. The new technological paradigm is associated with systemic changes in embodied technology as well as organizational structures and infrastructures. The global division of labour in manufacturing is one example of the organizational or disembodied technology. Organizational innovations – changes in production organization, firm structure, labour process, etc. – are being regarded as preconditions for successful absorption of the micro-electronics-based new technologies.

The emergence of new technologies in the early 1980s first produced doomsday scenarios predicting unprecedented negative employment effects and social evils pervading the entire economies and societies of industrialized countries. Very soon, however, it was realized that the pace at which new technological breakthroughs were expected did not really materialize, partly owing to economic recession and resulting sluggishness of demand, partly to a shortage of new types of polyvalent skills required, and perhaps also to the inertia of conventional types of managements and the ignorance of policy makers about the potential benefits of new technologies.

The new technologies are heavily dependent on scientific research and development, which explains why, with few exceptions, most of the production of these technologies is concentrated in the industrialized countries. The enormous investments required are simply beyond the capacity of most developing countries. Apart from capital investments, the human resources required – scientists and engineers, systems analysts and polyvalent technicians – are often not available in developing countries. In both the industrialized and the developing countries, the introduction of new technologies is changing the occupational composition of the labour force in favour of programming and broad-based skills, and against a high degree of specialization.

428

For the above reasons, the new technologies are concentrated in and controlled by big multinational corporations with enormous resources and organization. The generation of many new technologies, particularly biotechnology research, remains in the private domain, unlike the Green Revolution breakthroughs of high yield crop varieties made possible by publicly funded research.

Some empirical knowledge on the impact of new technologies has now been accumulated, although by no means enough to make any definitive generalizations. Three sets of studies may be noted: global or "synthetic" studies, sectoral investigations, and micro analyses. Much of the work (like that on technology choice and appropriate technology reviewed earlier) is of a micro nature.

A recent study [37] considers micro-electronics-based technologies in a socio-economic framework and the global economic context of economic recession. It notes that available studies on the quantitative impact of micro-electronics on employment are undertaken at different levels of aggregation – process, plant, firm, branch, region, sector, macroeconomic and meta level – which explains why they are non-comparable and often contradictory in their conclusions. Furthermore, conflicts also arise because their results are highly sensitive to the assumptions made about growth of output and productivity, qualitative and organizational changes, and the indirect and multiplier effects that are rarely considered.

Despite the initial fears of massive unemployment due to the use of micro-electronics-based technologies, the limited experience of both industrialized and developing countries shows that the impact of these technologies on direct and indirect employment may in fact have been marginally positive [31]. The pessimistic predictions did not come true partly because of the economic recession. Furthermore, while the new technologies may displace labour in old activities, they generate additional demand for labour by creating new goods and services. Our present state of knowledge is not adequate to justify any predictions about the precise positive and negative employment effects of micro-electronics-based technologies. But one thing is clear: they are bound to affect not only the quantity of employment, occupational composition, and the labour market, but also its *quality*, in terms of flexible types of employment and its "informalization," shorter working hours, home-based work, and quality of working life.

The new technological revolution is also likely to have a tremendous potential influence on the distribution of income within and between countries [3, 33]. Yet empirical evidence of the nature

429

of these distributional implications is even more sparse than their employment implications. For example, little is known about the effects of micro-electronics technologies on the personal and size distribution of incomes, on the distribution between producers and consumers and between producers and workers. Also, it is not clear whether the effects of micro-electronics on income distribution will be different from those of the Green Revolution. James [33] argues that micro-electronics will probably have a greater impact on products than the Green Revolution did, and that the bias is likely to be weighted in favour of large producers and multinational corporations, which wield a comparative advantage in large-scale production and exports. In the case of the consumers, the gains may accrue mainly to the rich consumers in developing countries who can more easily afford the products incorporating micro-electronics (e.g. watches and clocks and passenger cars). However, there is little empirical testing of these hypotheses.

The new technologies are likely to exert a significant influence on the developing countries, both directly and indirectly. Directly, with increasing globalization of production and international competition, the export-oriented countries would be compelled to use new technologies to compete with the industrialized countries. The increasing production and use of new technologies in the industrialized countries is likely to widen the already serious technological gaps between countries. This fear of being left behind is likely to put pressures on developing countries to make a selective start with the use of new technologies. Indirectly, the use of new materials and new biotechnologies in the industrialized countries is already hurting the exports of primary materials and commodities from the developing countries. The developing countries have little control or influence on the "dematerialization of production" and the substitution of new for old commodities taking place in the industrialized countries.

As latecomers, developing countries may be able to take advantage of new technologies to leap-frog from manual methods directly to flexible manufacturing systems without having to adopt fixed automation. However, leap-frogging presupposes the existence in these countries of organizational and innovation capabilities to produce new products through the use of high technology. It also assumes that the developing countries possess technological capability at a sufficiently high level to assimilate the high technology efficiently. Yet few advanced developing countries meet these prerequisites. The majority of developing countries are at an early stage of industrialization, with small industrial sectors. For these countries, the usefulness of new technologies would not be for material industrial development, but for the development of human capabilities through, for example, the application of microcomputers for the delivery of health services to rural areas and the use of computers in education – new technologies can, for example, raise the efficiency of the traditional education system and permit training outside the school system [1, 12, 25].

In the majority of developing countries, therefore, the potential for new technologies is through what has been termed technology blending – a combination of new and traditional technologies without destroying the latter. The concept of technology blending was introduced in 1983–1984, and the idea originated from the growing recognition that the benefits of modern science and technology in the developing countries had not trickled down to the rural and the urban poor. In most developing countries, notwithstanding the development and availability of new and advanced technologies, age-old low-productivity techniques continue to be used. Can the application of new technologies to traditional activities in these countries lead to a process of gradual modernization rather than of displacement? [9, 12, 11]

Admittedly, in the process of technical change, at any given time, new and old technologies coexist. In this sense, there is nothing new – retrofitting at a micro-firm level takes place all the time. The novelty of the concept of technology blending lies instead in focusing on the potential and limitations of applying new technologies to small-scale low-income activities for meeting basic needs of the bulk of the third world's population. Three interpretations have been given to the notion of technology blending in the recent literature:
1. physical combination of old and new techniques – retrofitting;
2. application of new technologies to market-oriented traditional small-scale activities without displacing them;
3. application of new technologies to public goods and services (e.g. rural telecommunications, public health and education).

Is TECHNOLOGY BLENDING ANOTHER NAME FOR APPROPRIATE TECHNOLOGY? Like that of "appropriate technology," the concept of technology blending does not refer to a *choice* from among an existing set of techniques but rather to the *development* of new technologies that would be more suited to the needs of the poor developing countries.

Technology blending (or blends) may be best analysed in terms of

431

the *objectives* sought and the *characteristics* of the new technological variant. The objective is simple enough: it is to bring the benefits of the new technology revolution to bear on improving the standards of living of the rural and urban poor. To some extent this aim determines the required characteristics of technology blending. As a general rule, two extremes can be envisaged. At one extreme, a technology blend (as an outcome of integration of new and traditional technology) would reflect almost entirely the characteristics of the new technology. At the other extreme, a technology blend would embody the characteristics of the traditional technology. In practice, however, acceptability of technology blends to the rural and urban poor can be ensured if their characteristics are not too far removed from those of traditional technology.

As the rural and urban poor lack purchasing power (their incomes are generally very low in relation to the average income in the developing country concerned), it is clear that the technology blends cannot be too expensive, demanding cash outlays that are beyond the reach of the target beneficiaries for whom they are intended. In the initial stages, diffusion of new technologies to rural areas may have to be subsidized in the same way as were the agricultural inputs during the Green Revolution.

Moreover, infrastructure and repair and maintenance facilities and skills are very scarce in urban and rural milieus where the small producers are concentrated. The technology blends would be required to be simple, easily comprehensible, easy to maintain and repair. Thus, in terms of characteristics, technology blends are somewhat similar to the appropriate technologies considered in the previous section. It is this similarity that seems to have led some authors to describe technology blending simply as a variant of the concept of appropriate technology.

Of course, on the surface there may appear to be a parallel between the two since blending represents an intermediate stage between conventional and new technologies in a process of continuous technical change. In addition, improvements of traditional technologies is the objective common to both technology blending and appropriate technology concepts. But the similarities end there.

There are substantive differences between the two approaches to improvements of traditional technologies. First, while the concept of appropriate technology refers mainly to *incremental* innovations, the blending of new technologies with traditional activities implies a quantum leap on the part of developing countries that do not necessarily have to go through all the stages followed by the technological leaders of today. This scope for leap-frogging has been aptly summed up as follows:

Developing countries are for the first time enabled to "leapfrog" stages in the development process which have up till now always been regarded as prerequisites for the achievement of prosperity and growth. . . . This result of the current technological revolution represents a quantum leap from the debate of appropriate or intermediate technologies so central to much development thinking in the 1960s and early 1970s. [13]

Furthermore, in contrast to "appropriate technology," new technologies are being developed and applied at a very rapid rate, particularly in the industrialized countries. They entail capital costs that far exceed those involved in the development and commercialization of "appropriate technologies." Thus, the pace of development of technology blends can be more rapid and their commercialization perhaps easier if the industrialized countries – the main producers of these technologies – take account of the potential needs of the developing countries.

Whereas the developing countries are both producers and consumers of appropriate technologies, in the case of "technology blends," most of them (with very few exceptions) will remain mere consumers for the foreseeable future. The new technologies will continue to be produced mostly in the industrialized countries. While appropriate technology was inspired by a development strategy of national and local self-reliance, technology blending may tend towards eroding such self-reliance and technological autonomy. In the absence of a national/domestic capacity to develop new technologies, most developing countries will have to depend on the industrialized countries that are the main sources of supply of these technologies. This technological dependence may be particularly serious in the case of the least developed countries, with limited capabilities to create or absorb new technologies.

A differential capacity of developing countries to create and absorb new technologies may mean widening technological gaps not only *between* industrialized countries and developing countries but also *among* the developing countries (the newly industrialized countries being better endowed than the least developed ones). In other words, a growing divergence of third world interests may be explained in the future partly in terms of the technological factor.

433

THE CASE FOR "HIGH TECH" COTTAGE INDUSTRY. The micro-electronics-based high technologies are known to be miniaturized, simple to maintain, and requiring rather simple skills to operate and maintain. Miniaturization enhances the scope for flexibility in production that large-scale mass production did not offer. This "flexible specialization" and the ability to respond to fluctuating demands through small-batch production are said to make decentralized production more feasible than what was possible with the use of conventional technologies. Through the use of microcomputers, simple reprogramming enables equipment to be put to different uses without requiring any physical adjustment.

Piore and Sabel [49] have argued that mass production in the industrialized capitalist countries has come under stress owing to the limits of the model of industrial development and the economic crisis facing them. They believe that the mass production paradigm is likely to give way to craft-type specialized and customized production thanks to the advent and use of new technologies. The latter facilitate "flexible specialization" and small-batch production. This means that a shift will occur from specialized dedicated machinery to multi-purpose machinery, from a narrowly trained workforce to one with multiple skills, and from standardized products to small-batch, customized products [53]. Piore and Sabel show how in the nineteenth century, craft production was based on the use of flexible machines and varied skills to produce several products within the economic organization of "industrial districts." This craft production is shown to have re-emerged in Japan, Germany, and Italy in the twentieth century. In Japan, for example, craft industries like weaving and wood products are being revived through the use of high technology to capture "niches" in saturated markets. In Italy, the dynamic growth of small firms in the Como, Prato, and Emilia-Romagna regions in the silk, cotton textiles and garments, and ceramics industries has been quite impressive. Although not all of this dynamism can be attributed to "high" technology, there are good examples of how small firms have successfully adopted computer-based technologies to respond to rapidly changing market requirements [15, 14].

The question arises whether use of new technologies and greater scope for "flexible specialization" can improve the efficiency of craft production in developing countries and thus expand output as well as employment. There is scattered empirical evidence to show that some small firms in developing countries are taking advantage of new technologies to improve their product quality and international competitiveness. A representative survey of 19 small and medium-size firms in the mechanical industry of the state of São Paulo in Brazil showed that the majority of them adopted computer-aided design (CAD) and numerically controlled machine tools (NCMT) to attain higher quality, precision, and productivity. The survey also suggested that by and large, any direct employment losses resulting from the use of high technology seemed to be at least partly compensated for by additional demand for labour resulting from work reorganization and training activities [12].

Of course, the conditions for "flexible specialization" outlined by Piore and Sabel for the industrialized countries are different from those in the developing countries. In the former it was a response to stagnating and competitive mass markets. Although similar conditions of demand may also apply to developing countries, the supply constraints – lack of access to raw materials and spare parts due to scarce foreign exchange – are likely to impose major bottlenecks. Secondly, surplus labour conditions and resulting low labour costs in developing countries are likely to discourage the wide use of new technologies. Thirdly, to benefit most from new technologies and flexible specialization, developing countries will need to establish agglomerations of small firms, such as prevails in Kumasi (Ghana). "Small firms individually cannot attain flexible specialisation; it is the sectoral agglomeration which gives them their strength" [53].

In principle, miniaturization of high technology should facilitate decentralized production and rural industrialization. Many developing countries have in the past promoted rural industries as a means of generating employment and preventing rural-to-urban migration. But, with few exceptions, the performance of rural industry programmes has not been good. The programmes have suffered from lack of markets resulting from low local purchasing power, poor infrastructure, inadequate credit facilities, and competition from large-scale industry. The last factor has become particularly acute in recent years with privatization and liberalization of economic policies in both developing and industrialized countries.

It is not clear whether these problems faced by rural small industry can be overcome by the use of high technology. Furthermore, even if high technology were available, its cost in relation to the resources of small producers may be prohibitive – and may therefore restrict their access to it.

Little empirical evidence exists in the third world to support or refute the argument that new technologies facilitate transfer of produc-

435

tion away from the big urban centres. While some surveys of small firms have been undertaken in the industrialized countries, the same is not true of the developing countries. This is partly because the high technology is quite new, not yet widely diffused, and is perhaps also costly.

THE SCALE FACTOR. The case for the promotion of "high tech" cottage industry hinges on the premise that the use of new technologies reduces the scale of *optimal* production. In other words, economies of scale no longer matter much.

The relationship between high technology and the scale factor is not very simple or straightforward. Silberston [58] distinguished between three different dimensions of scale: life of product, plant, and equipment; cost variations for products for different rates of output and degrees of standardization among products; and scale economies for distinct levels of aggregation, e.g. for plants, firms, or industries. Generally, scale economies are considered only in relation to the size of plants and output of individual products per unit of time. There can be divergent trends in the three dimensions of scale economies noted above.

Technological trajectories towards scale economies by different types of industries are described in table 3. Mass production has generally led to growing economies of scale in both large-batch discrete products industries and continuous process industries. However, in small-batch production, the economies of scale do not seem to have become any more important. If anything, it is claimed that the replacement of the mass production paradigm by small-batch flexible production noted above [49] will lead to a decline in the importance of the scale factor.

BEGINNING OF A NEW DEBATE. The long and controversial debate on "appropriate technology" had both political and analytical connotations, whereas an emerging debate about the concept of technology blending is, at least at present, confined only to its analytical properties and limitations. The concept is being increasingly recognized as a useful practical tool for focusing attention on the relevance and application of new technologies to meet the basic needs of developing countries.

However, the critics find fault with the analytical merits or utility of the technology blending concept. The first legitimate criticism has to do with the difficulty of drawing any clear-cut dividing line be-

Table 3 **Technological trajectories towards scale economies over the past century**

Type of industry or forms of production organization	Dimensions of scale	Product economies	Plant economies of scale	Firm economies of scale
Small-batch products	discrete	static	varies by sector, but generally static	varies by sector, but generally static
Large-batch products	discrete	growing	growing	growing
Continuous	process	growing	growing	growing

Source: Ref. 39.

tween frontier technology and blended technology, although this was recognized by the proponents of the concept. The problem of boundaries cannot be solved in any process of *continuous* change. Nevertheless, to the extent that technical progress is discontinuous (which corresponds more closely to reality), the boundaries between different technology sets or trajectories can be perceived by analysing their distinctive characteristics and properties.

Secondly, some authors see utility in defining technology blending only narrowly in terms of retrofitting and not in any more macroeconomic sense. For Rosenberg [52]:

the prospects for technology blending will be very much shaped by the ease with which new technology can be introduced without having to scrap the old. Indeed, this really takes us to the essence of what blending is all about. In the extreme case, if a new technology requires the complete scrapping of an old one . . . no blending is possible. (p. 26)

Kaplinsky [38] suggests an alternative definition of technology blending. Pursuing the line that blending is a variant of AT, he identifies and distinguishes between the following sources of AT: downscaling of large-scale production, the improvement of existing/traditional technologies, and the production of new technologies. This definitional approach seems to be equally faulty since it mixes up objectives/ends (technological improvements) and means (downscaling and production of new technologies). The developing countries need not produce new technologies at a high cost if they can obtain them from abroad on reasonable terms. The whole *raison d'être* of technology blending is to adapt the imported new technologies to suit them to the developing country's conditions. The above definitional approach does not capture this participatory process, in which the consumers of new technologies (most of the developing countries) can ensure that the producers (the industrialized countries) design them so as to make them more relevant to their special requirements.

Admittedly, to some extent consumption and production of new technologies would have to be done within developing countries to ensure that learning effects are internalized.

Technological capabilities

As we noted above, the use and assimilation of new technologies presuppose the existence of a minimum of technological capabilities in developing countries to choose, acquire, generate, and apply technologies that are suited to their development objectives. Such capabilities would determine the rates and patterns of development and industrialization. Though the concept is somewhat elusive, it is clear that capabilities cannot be acquired overnight and that they will vary over time and space.

Technological capabilities can also vary between sectors [73, 21]. In the industrial sector, the elements of technology capability – production engineering, manufacture of capital goods, and research and development, etc. – are different from those essential for the services sector, for example [16]. Technological capabilities may exist in both large and small industrial sectors.

On the basis of sample surveys undertaken in capital cities and larger towns in Asia (India, Bangladesh, and Thailand), Latin America (Ecuador and Peru), and Africa (Mali and Rwanda), an examination of technological capabilities in the small-scale informal sector of developing countries found, contrary to expectations, that even very small metal-working production units possess some capacity to adapt and modify tools and equipment. In some cases, these units demonstrate an indigenous capacity and ingenuity to manufacture simple equipment [43].

In general, technology capabilities in developing countries, whether considered in macro terms or in terms of elements, would depend on such factors as: adequate number and quality of human resources with practical experience, skills, and aptitude; useful technological information on sources and conditions of technology transfer; institutions for education and training, for research and development, and for engineering design and consultancy; favourable natural environment and factor endowments, attitudes and customs, etc. [23].

A pioneering piece of work by Enos [21] puts the concept and practice of technology capability in a macro perspective of growth modelling. It reviews existing models that try to incorporate elements of technology capability. On the basis of this review, Enos develops a model that incorporates the creation of skills over time. Three fundamental components of technological capability are identified: individuals embodying skills, training, and experience and inclination; institutions within which individuals are assembled; and a "common purpose" defined in terms of objectives and motivations. The last may or may not relate to national development objectives. Since capabilities vary at different levels of aggregation – micro, sectoral, and macro – it is possible that capabilities and common purpose of a few individuals do not fully correspond to national development objectives and may even be in conflict with them.

Prospects for the 1990s

Prospects for technology work in the 1990s will depend on the scenarios regarding the strategies and prospects of development in the third world. The Fourth Development Decade of the United Nations gives a prominent place to human resources development and employment generation. Similarly, the UNDP has adopted human development as a major goal for its development efforts [67].

If the development of human capabilities and potential is the goal of the current decade, technology policies and programmes would need to be considered in the context of achieving this goal. The focus of technology research may have to shift from the embodied technical change to disembodied technical change, and from processes and products to individuals and institutions necessary to promote their capabilities.

Although some work on building technological capabilities has already been done, our understanding of the nature and the magnitude of the task of creating it is still far from adequate. Enos [21] states "the economists' efforts to promote appropriate technology took for granted the environment in the developing countries and imagined techniques to be variable: the current efforts to stimulate technological capability take technology for granted and imagine the environment to be variable." A change in this environment would be necessary to improve the latent human potential to raise the level of capabilities in the developing countries. Also, to what extent is the technological capability of a country an intermediate input to the achievement of objectives, and to what extent is it an output? Do liberal technology imports facilitate or hinder a country's efforts to develop indigenous technology capabilities? Despite some empirical research, this question remains unanswered. Further efforts are needed to document and analyse critically the developing country experiences in this regard. It may be useful to compare and contrast open and closed economies to examine their experience of building indigenous technology capabilities over time.

In the 1990s, the new technological challenge will continue, with increasing innovations and new breakthroughs in such new technologies as biotechnologies, new materials, and information technology. An analysis of these technologies and their contributions to human as well as material development (through greater food security made possible by new biotechnologies) would merit further attention.

The structural adjustment measures introduced in the 1980s are also likely to continue. As a result of these measures, in many developing countries large numbers of people are being pushed into small and micro-enterprises to make a living. Can new technologies, through greater flexibility in production (so-called flexible specialization), enable a more successful small-scale industrialization than has been possible in the third world in the past? Are flexible specialization and small-batch production universal phenomena, or are they likely to be confined, at least in the 1990s, only to the industrialized countries? Answers to these questions require further investigation.

Under the influence of new industrial technologies, new macro-economic policies and structural adjustment programmes, and new methods of industrial organization, the labour markets in both industrialized and developing countries are going to become more flexible. Informalization of the labour market and production is already being witnessed in the form of increases in casual employment, part-time and self-employment, flexible working hours, etc. The possible impact of new technologies on the informalization of work remains to be explored. Our knowledge about their impact on skill formation and substitution is also quite limited. Do new technologies have a capacity to enhance skills and resources? Under what circumstances can new technologies and blending raise overall employment? Are there trade-offs between direct and indirect employment effects of new technologies? This area of research is highly relevant to implementing the kind of strategy of human development noted above.

The enquiry into technology blending, started by the ILO World Employment Programme in the 1980s, needs further conceptual refinements and more empirical testing. What are the prerequisites for the developing countries to adopt a policy of blending? Under conditions of labour surplus, would it be economically feasible and socially desirable to apply new technologies in the place of traditional or conventional technologies?

The cost comparisons of alternative technologies including new technologies are very rare. Yet for developing country policy makers, it is important to know whether it is cost-effective to apply new technologies, which invariably have to be imported using scarce foreign exchange.

The 1990s may also witness a decline in the R&D resources allocated to technology development in the third world. This is likely to happen under the stringent application of structural adjustment programmes, since the results of R&D are essentially long term and often uncertain. What implications is this likely to have for the new technology diffusion within the developing countries and for the technological gaps between countries?

Within the third world, great heterogeneity prevails. It is therefore quite likely that technological gaps will also widen within this group of countries. The differential impact of new technologies by subgroups of developing countries is another area for future research.

References

1   Anandakrishnan, M. et al. "Microcomputers in Schools in Developing Countries." In: Bhalla and James, eds. *See* ref. 12.

2   Atkinson, A.B., and J.B. Stiglitz. "A New View of Technological Change." *Economic Journal* 79 (September 1969).

3   Bessant, J., and S. Cole. *Stacking the Chips – Information Technology and the Distribution of Income*. London: Frances Pinter, 1985.

4   Bhalla, A.S. "Investment Allocation and Technological Choice." *Economic Journal* (September 1964).

5   ———. "The Third World's Technological Dilemma." *Labour and Society* 9 (October–December 1984), no. 4.

6   ———, ed. *Technology and Employment in Industry*. Geneva: ILO, 1975, 1981, 1985.

7   ———. *Towards Global Action for Appropriate Technology*. Oxford: Pergamon, 1979.

8   ———. *Environment, Employment and Development*. Geneva: ILO, 1992.
9   Bhalla, A.S. et al., eds. *Blending of New and Traditional Technologies*. Dublin: Tycooly, 1984.
10   Bhalla, A.S., and D. James. "Technological Blending: Frontier Technology in Traditional Economic Sectors." *Journal of Economic Issues* (June 1986).
11   ———. "Integrating New Technologies with Traditional Economic Activities in Developing Countries: An Evaluative Look at 'Technology Blending'. *Journal of Developing Areas* (July 1991).
12   ———, eds. *New Technologies and Development: Experiences in Technology Blending*. Boulder, Colo.: Lynne Rienner, 1988.
13   Colombo, U. "Technology Blending as an Instrument in the Rejuvenation of Traditional Sectors: The Italian Experience and Its Relevance to the Third World." In: U. Colombo and K. Oshima, eds. *Technology Blending: An Appropriate Response to Development*. London: Tycooly, 1989.
14   Colombo, U., and D. Mazzonis. "Integration of Old and New Technologies in the Italian (Prato) Textile Industry." In: Bhalla et al., eds. *See* ref. 9.
15   Colombo, U., D. Mazzonis, and G. Lanzavecchia. "Co-operative Organisation and Constant Modernisation of the Textile Industry at Prato, Italy." In: E.U. von Weizsäcker, M.S. Swaminathan, and A. Lemma, eds. *New Frontiers in Technology Application – Integration of Emerging and Traditional Technologies*. Dublin: Tycooly, 1983.
16   Dahlman, C., and L. Westphal. "The Meaning of Technological Mastery in Relation to Transfer of Technology." *Annals of the American Academy of Political and Social Science* 458 (November 1981).
17   Dobb, M.H. "Second Thoughts on Capital Intensity of Investment." *Review of Economic Studies* 24 (1956), no. 1.
18   Eckaus, R.S. "The Factor Proportion Problem in Underdeveloped Areas." *American Economic Review* 45 (1955).
19   ———. *Appropriate Technologies for Developing Countries*. Washington, D.C.: National Academy of Sciences, 1977.
20   Enos, J.L. "More (or Less) on the Choice of Technique, with a Contemporary Example." *Seoul National University Economic Review* (December 1977): 177–199.
21   ———. *The Creation of Technology Capability in Developing Countries*. London: Frances Pinter, 1991.
22   Forsyth, D., N. McBain, and R. Solomon. "Technical Rigidity and Appropriate Technology in Less Developed Countries." *World Development* 8 (May–June 1980).
23   Fransman, M., and K. King, eds. *Indigenous Technological Capability in the Third World*. London: Macmillan, 1984.
24   Freeman, C. "Prometheus Unbound." *Futures* (1983), no. 15.
25   Galal, E.E. "Application of Micro-Computers in Primary Health Delivery Services in Egypt." In: Bhalla and James, eds. *See* ref. 12.
26   Galenson, W., and H. Leibenstein. "Investment Criteria, Productivity and Economic Development." *Quarterly Journal of Economics* (August 1955).
27   Galtung, Johan. *Development, Environment, Technology – Towards a Technology for Self-reliance*. Geneva: UNCTAD, 1979.
28   ———. *The North/South Debate: Technology, Basic Human Needs and the*

New International Economic Order. World Order Models Project Working Paper no. 12, 1980.

29   Haggblade, S., and P. Hazell. "Agricultural Technology and Farm–Non-farm Growth Linkages." *Agricultural Economics* 3 (1989).

30   ILO. *Towards Full Employment: A Programme for Colombia*. Geneva: ILO, 1970.

31   ———. *Socioeconomic Effects of New Technologies*. Report 2, ILO Advisory Committee on Technology, First Session. Geneva: ILO, 1985.

32   James, J. "Products, Processes and Incomes: Cotton Clothing in India." *World Development* (February 1976).

33   ———. *The Employment and Income Distributional Impact of Microelectronics: A Prospective Analysis for the Third World*. ILO/WEP Research Working Paper Series WEP2-22/WP153. Geneva: ILO, 1985.

34   ———. "The Role of Appropriate Technology in a Redistributive Strategy." In: James and Watanabe, eds. *See* ref. 36.

35   James, J., and F. Stewart. "New Products: A Discussion of the Welfare Effects of the Introduction of New Products in Developing Countries." *Oxford Economic Papers* (March 1981).

36   James, J., and S. Watanabe, eds. *Technology, Institutions and Government Policies*. London: Macmillan, 1985.

37   Kaplinsky, R. *Microelectronics and Employment Revisited – A Review*. Geneva: ILO, 1987.

38   ———. "Review of 'New Technologies and Development (1988)'." *Journal of International Development* (April 1989).

39   ———. *The Economies of Small: Appropriate Technology in a Changing World*. London: Intermediate Technology Publications, 1990.

40   Khan, H.A. "Technology Choice in the Energy and Textile Sectors in the Republic of Korea." In: A.S. Bhalla, ed. *Technology and Employment in Industry*. Geneva: ILO, 1985.

41   Khan, H.A., and E. Thorbecke. *Macroeconomic Effects and Diffusion of Alternative Technologies with a Social Accounting Matrix Framework*. Aldershot: Gower, 1988.

42   Lall, S. *Learning to Industrialise: The Acquisition of Technological Capability in India*. London: Macmillan, 1987.

43   Maldonado, C., and S. Sethuraman, eds. *Technological Capability in the Informal Sector: Metal Manufacturing in Developing Countries*. Geneva: ILO, 1992.

44   Mitra, A.K. "Interlinkage in Agricultural Machinery Industry for Rural Industrialization in Developing Countries." In: UNIDO, ed. *Appropriate Industrial Technology for Agricultural Machinery and Implements*, part 2. New York: United Nations, 1979.

45   Morawetz, D. "Employment Implications of Industrialisation in Developing Countries – A Survey." *Economic Journal* 84 (1974).

46   Pereira, Armand. *Technology Policy for Environmental Sustainability and for Employment and Income Generation: Conceptual and Methodological Issues*. ILO Working Paper Series WEP 2-22, no. 215. Geneva: ILO, 1991.

47   Perez, C. "Microelectronics, Long Waves and Structural Change: New Perspectives for Developing Countries." *World Development* 13 (1985), no. 3.

443

48   Pickett, James. "The Choice of Technology in Developing Countries." *World Development* 5 (special issue) (September–October 1977), nos. 9–16.

49   Piore, M.J., and C.F. Sabel. *The Second Industrial Divide: Possibility for Prosperity*. New York: Basic Books, 1984.

50   Ranis, G. "Rural Linkages and Choice of Technology." In: Stewart, Thomas, and de Wilde, eds. *See* ref. 62.

51   Rhee, Y.W., and L. Westphal. "A Micro, Econometric Investigation of Choice of Technology." *Journal of Development Economics* 4 (September 1977), no. 3.

52   Rosenberg, N.L. "New Technologies and Old Debates." In Bhalla and James, eds. *See* ref. 12.

53   Schmitz, H. *Flexible Specialisation: A New Paradigm of Small-Scale Industrialisation?* IDS Discussion Paper, no. 261. May 1989. Institute of Development Studies, University of Sussex, 1989.

54   Schumacher, E.F. *Small is Beautiful*. London: Sphere, 1973.

55   Sen, A. "Some Notes on the Choice of Capital Intensity in Development Planning." *Quarterly Journal of Economics* (November 1957).

56   ———. *Choice of Techniques*. Oxford: Clarendon Press, 1960.

57   ———. *Employment, Technology and Development*. Oxford: Clarendon Press, 1975.

58   Silberston, A. "Economies of Scale in Theory and Practice." *Economic Journal* 82 (March 1972).

59   Stewart, F. *Technology and Underdevelopment*. London: Macmillan, 1977.

60   ———, ed. *Macro-policies for Appropriate Technology in Developing Countries*. Boulder, Colo.: Westview Press, 1987.

61   ———, eds. *The Economics of New Technologies in Developing Countries*. London: Frances Pinter, 1982.

62   Stewart, F., H. Thomas, and T. de Wilde, eds. *The Other Policy – The Influence of Policies on Technology Choice and Small Enterprise Development*. London: Intermediate Technology Publications, 1990.

63   Strassmann, W. "Creative Destruction and Partial Obsolescence in American Economic Development." *Journal of Economic History* (September 1959).

64   Teitel, S. "On the Concept of Appropriate Technology for Less Industrialised Countries." *Technological Forecasting and Social Change* 11 (1978).

65   Timmer, C.P. et al. *The Choice of Technology in Developing Countries: Some Cautionary Tales*. Harvard Studies in International Affairs, no. 32. Cambridge, Mass.: Harvard University Press, 1975.

66   Tokman, V.E. "Distribution of Income, Technology and Employment: An Analysis of the Industrial Sector of Ecuador, Peru and Venezuela." *World Development* 2 (1974), nos 10–12.

67   UNDP. *Human Development Report*. London: Oxford University Press, 1990.

68   Van Ginneken, W., and C. Baron. *Appropriate Products, Employment and Technology – Case Studies on Consumer Choice and Basic Needs in Developing Countries*. London: Macmillan, 1984.

69   Watanabe, S. "Institutional Factors, Government Policies and Appropriate Technologies." *International Labour Review* 119 (March–April 1980), no. 2.

70   ———. *Technology, Marketing and Industrialisation*. New Delhi: Macmillan, 1983.

71   White, L.J. "The Evidence on Appropriate Factor Proportions for Manufac-
      turing in Less Developed Countries: A Survey." *Economic Development and
      Cultural Change* (October 1978).

72   Willoughby, Kelvin W. *Technology Choice – A Critique of the Appropriate
      Technology Movement.* Boulder, Colo.: Westview Press, 1990.

73   Zahlan, A.B. *Acquiring Technological Capacity: A Study of Arab Contracting
      and Consulting Firms.* London: Macmillan, 1990.

13

New technologies: Opportunities and threats

Paulo Rodrigues Pereira

Long waves, technological systems, and techno-economic paradigms

The discovery of cyclic phenomena of long duration in economic activity is generally attributed to the Russian economist N.D. Kondratiev, who in the 1920s described the existence of long waves in the world economy.

Kondratiev [27] based his theory on the observation of trends in the fluctuation of nineteenth-century economic indicators (mainly prices), and he explained the occurrence of long waves in terms of the durability and production period of, and amount invested in, particular types of capital goods. The specific source of the long wave was the tendency of investment in these basic capital goods to occur in clusters, due to the availability of loanable funds. Kondratiev did not explicitly include the role of technical change in his analysis of the formation of long waves, but he suggested that when a major cycle of economic expansion was under way, inventions that had remained "dormant" could find new applications. However, Kondratiev failed to make the crucial distinction between inventions and innovations. (An invention need have no practical consequences, whereas an innovation has the sanction of the market.)

The notion of long waves (or Kondratiev cycles) has since been used to refer to movements in economic variables of 50 to 60 years, in which periods of rise and decline in economic activity alternate. Nevertheless, it remains a controversial item in economic theory [16].

A contrasting interpretation to long waves was later provided by

446

Joseph Schumpeter [50] in his classic study of business cycles. For Schumpeter, the source of fluctuations in business activity is the innovation process, which he regarded as essentially discontinuous due to the nature of entrepreneurial activity. He emphasized the intrinsic instability of the capitalist growth process, rather than the general economic equilibrium and the smooth process of substitution and adaptation beloved of neoclassical economic theory. The variations in fluctuations between the three types of cycles he identified – short, intermediate, and long cycles – is accounted for by the differing impacts of different types of innovations, the long cycle being associated with fluctuations in "basic innovations," such as railroads, electricity, and motor vehicles. These "basic innovations" would cluster in the depression and early upswing phases of the Kondratiev cycle, causing surges of investment associated with "bandwagon" effects of the diffusion of new technologies related to these basic innovations. He insisted on the explosive growth of new technologies in some sectors and relative stagnation in others: once a major innovation had demonstrated profitability, this led to an imitative "swarming" behaviour as many firms rushed to get on the "bandwagon" of the new growth area. It was these spurts of innovation-related investment, caused by the simultaneously creative and destructive effects of technological innovation ("creative gales of destruction") that Schumpeter believed led to economic imbalances and particularly to cyclical phenomena in the economy as a whole and to the changing location of technological and economic leadership, both within countries and between countries. Economic growth is not merely accompanied by the introduction of new products and processes, as in neoclassical theory, but is driven by these innovations. Technical change is thus endogenous to economic progress, rather than a marginal exogenous factor. However, Schumpeter excluded social and institutional factors from the causal mechanism of cyclic economic behaviour.

The need to understand the basic forces underlying the post-war periods of economic recession, the failure of mainstream economic theories to account for them, and the recognition of the persistence and structural character of the economic crisis have revived the debate on the existence and explanation of long waves, and particularly on the role of technological innovation as the driving force behind long economic cycles.

The varieties of current theorizing about long waves in the economy range from neo-Marxists [29] to "evolutionists" [33, 34] and neo-Schumpeterians [30, 10, 17, 40]. The majority of contemporary

447

economic theorists agree with Schumpeter's emphasis on the role of innovations in explaining Kondratiev economic cycles, but differentiate between the impact of different levels of innovation, and, contrasting with Schumpeter's emphasis on *product* innovations, include the critical role of *process and service* innovations in the transition period from the depressive to the expansionary phase of a new long wave. In particular, the neo-Schumpeterian authors stress the role of diffusion of major technological breakthroughs in the stimulation of renewed economic growth and the exhaustion of older technological systems as the main forces behind the upper turning point of the long wave.

For the neo-Schumpeterians, technical change is an evolving, interactive, cumulative and institutional process that generates imbalances. Any technology thus has a history and a trajectory. Historians like Bertrand Gilles [20], Nathan Rosenberg [46, 47], and François Caron [7] have argued that technology evolves through successive formulations of technical problems and proposed solutions; these "solutions" can be accepted or refused by the economic environment, and they will move along certain "normal or technological trajectories" [33, 10] until they have nothing more satisfactory to offer. Technical and economic factors thus exert combined and complementary forces on the direction of technical change: the former determining the stock of solutions available inside a technological trajectory, the latter determining the rapidity of the transition between one or another possible solution within a certain trajectory and the distance to be covered on the trajectory before it has to be abandoned for a new and more promising technological trajectory.

Building on these concepts, Christopher Freeman [16] and Carlota Perez [40] postulate that Kondratiev cycles are not an exclusive economic phenomenon but the expression, measurable in economic terms, of major changes in the behaviour of the entire socioeconomic and institutional system. These changes are the result of strong constraints exerted on the strategies and the routine operations of firms, leading to a reorientation of industrial organization and management; for Freeman and Perez, the origin of these constraints can be traced to a number of important novel features presented by the basic technologies that underlie the existing economic cycle. Kondratiev cycles are thus associated with major technical changes. However, technical change is the result of the evolution of a whole system of technologies and innovations that do not play the same role and do not have the same impact on the economy. In order to stimulate the economy, these innovations must have several characteristics that allow them to diffuse and to infiltrate into the economic system: in this case they constitute new "technological systems" leading to new "techno-economic paradigms."

A taxonomy of innovations
Four categories of innovations have been identified by Freeman and Perez [18]:

1. INCREMENTAL INNOVATIONS. Incremental innovations occur more or less continuously in any industry or service activity, although at differing rates in different industries and different countries, depending upon a combination of demand pressures, sociocultural factors, technological opportunities and trajectories. They may often occur, not so much as the result of any deliberate R&D activity but as the outcome of inventions and improvements suggested by engineers or as the result of initiatives and proposals by users or other persons engaged in the innovation process. Although their combined effect is extremely important in the growth of productivity and in the quality improvements to products and services, particularly in the follow-through period after a radical breakthrough innovation, no single incremental innovation has dramatic effects on the economy, and they often pass unnoticed and unrecorded. However, the cumulative impact of incremental innovations can lead to productivity increases greater than those initially possible from radical innovations. Generally, the progressive evolution of incremental innovations on a technological trajectory is more or less predictable.

2. RADICAL INNOVATIONS. Radical innovations represent the introduction of truly new products and processes, an unpredictable departure from the "normal trajectory" of a technology. They are discontinuous events that cannot be attributed to the cumulative addition of incremental modifications and improvements to existing products and processes: nuclear power, for example, could never have emerged from incremental improvements to conventional power stations. Radical innovations are typically unevenly distributed over industrial sectors and over time and are usually the result of deliberate R&D activities in enterprises and/or university research laboratories originating from breakthroughs in basic research or from the search for a technical solution to an identified market need. They are important as potential springboards for the growth of new markets (radical prod-

uct innovations) or for big improvements in the cost and quality of existing products (radical process innovations). Over long periods of time, radical innovations may have fairly dramatic effects, but their immediate economic impact is relatively small and localized, unless a whole cluster of radical innovations are linked together in the rise of entirely new industries or services, in which case they constitute a new technological system.

3. NEW TECHNOLOGICAL SYSTEMS (SYSTEMIC INNOVATIONS). Systemic innovations are far-reaching changes in technology affecting several branches of the economy and giving rise to entirely new industrial sectors. They are based on a successful combination of radical and incremental innovations, together with organizational innovations that affect a great number of firms, forming "clusters" or "constellations" of technically and economically interrelated and mutually interdependent innovations. They often lead to a proliferation of radical innovations that diffuse into the economy, generating a large number of minor or incremental innovations ("bandwagon effect").

4. TECHNOLOGICAL REVOLUTIONS OR NEW TECHNO-ECONOMIC PARADIGMS. New techno-economic paradigms represent changes in technological systems that are so far-reaching in their effects that they have a major influence on the behaviour of the entire economy. They correspond to the "creative gales of destruction" (the decline of "old" industries and occupations, accompanied by an extremely uneven process of structural adaptation with substantial time lags) that lie at the heart of Schumpeter's theory of long waves in economic development. Technological revolutions involve the introduction of new technologies with the potential of transforming a vast array of economic activities, leading to a series of interrelated technological changes, including drastic reductions in the cost of many products and services, drastic improvements in the technological characteristics of many products and processes, environmental effects, and pervasive effects throughout the entire economy.

The rise of a new techno-economic paradigm implies a process of economic selection from the range of technically feasible combinations of innovations, whose diffusion throughout the economy takes a relatively long time [15]. This diffusion involves a complex interplay between technological, economic, and political forces, and above all the social and political acceptability of the new techno-economic paradigm: it is in fact a "meta-paradigm," exerting a dominant influ-

450

ence on engineers, designers, and managers over several decades and implying a radical change in the set of "common sense" or "best practice" rules and guidelines ordinarily admitted in industrial production and management (i.e. in the existing "paradigm" for the most efficient organization of production).

Freeman and Perez's conception of techno-economic paradigm is much wider than "clusters" of innovations or even of "technology systems"; it refers to "a combination of interrelated product and process, technical, organisational and managerial innovations, embodying a quantum jump in potential productivity and opening up an unusually wide range of investment and profit opportunities. Such a paradigm change implies a unique new combination of decisive technical *and* economic advantages." However, these potentialities are at first realized in only a few leading sectors; in others, such gains cannot usually be realized without profound organizational and social changes of a far-reaching character. Periods of rapid economic expansion occur when there is a good match between an emerging techno-economic paradigm's new "best practice" set of guiding principles and the socio-institutional framework; depressions represent periods of mismatch.

Table 1 sketches some of the main characteristics of successive long waves identified by Freeman and Perez.

The "regulationist" school in France has emphasized the importance of the same set of relationships, but they differentiate between what is termed the regime of accumulation and the mode of regulation. The regime of accumulation refers to the systematic division and reallocation of the social product, which achieves a match between the transformation of the conditions of production and transformations in the conditions of final consumption. The mode of regulation refers to the ensemble of institutional forms and mechanisms, including the set of values and norms, that ensure the compatibility of behaviours in the framework of the regime of accumulation, in conformity with the existing state of social relationships. Periods of crisis involve a profound mismatch between accumulation and regulation, leading to successive regimes of accumulation that dissolve and are superseded through the effects of their own internal contradictions [1]. Thus, while the regulation school does not adopt the concept of long waves, there is a strong similarity between their concept of modes of regulation of successive regimes of accumulation and Perez's notion of the socio-political infrastructures of successive techno-economic paradigms [28].

451

Table 1 A tentative sketch of some of the main characteristics of successive long waves

Approx. periodization upswing, downswing	Description	Main "carrier branches" and induced growth sectors infrastructure	Key factor industries offering abundant supply at descending price	Other sectors growing rapidly from small base	Limitations of previous techno-economic paradigm and ways in which new paradigm offers some solutions	Organization of firms and forms of cooperation and competition
1770s & 1780s to 1830s & 1840s "Industrial revolution" "Hard times"	Early mechanization Kondratiev	Textiles Textile chemicals Textile machinery Iron-working and iron castings Water power Potteries Trunk canals Turnpike roads	Cotton Pig-iron	Steam engines Machinery	Limitations of scale, process control, and mechanization in domestic "putting out" system. Limitations of hand-operated tools and processes. Solutions offering greater prospects of productivity and profitability through mechanization and factory organization in leading industries.	Individual entrepreneurs and small firms (<100 employees) competition. Partnership structure facilitates cooperation of technical innovators and financial managers. Local capital and individual wealth.
1830s & 1840s to 1880s & 1890s Victorian prosperity	Steam power and railway Kondratiev	Steam engines Steamships Machine tools Iron	Coal Transport	Steel Electricity Gas Synthetic dyestuffs	Limitations of water power in terms of inflexibility of location, scale of production, reliability and range of applications, re-	High noon of small-firm competition, but larger firms now employing thousands, rather than hundreds.

"Great depression"	Railway equipment Railways World shipping			Heavy engineering	stricting further development of mechanization and factory production to the economy as a whole. Largely overcome by steam engine and new transport system.	As firms and markets grow, limited liability and joint stock company permit new pattern of investment, risk-taking, and ownership.
1880s & 1890s to 1930s & 1940s "Belle epoque" "Great depression"	Electrical and heavy engineering Kondratiev	Electrical engineering Electrical machinery Cable and wire Heavy engineering Heavy armaments Steel ships Heavy chemicals Synthetic dyestuffs Electricity supply and distribution	Steel	Automobiles Aircraft Telecommunications Radio Aluminium Consumer durables Oil Plastics	Limitations of iron as an engineering material in terms of strength, durability, precision, etc., partly overcome by universal availability of cheap steel and of alloys. Limitations of inflexible belts, pulleys, etc., driven by one large steam engine overcome by unit and group drive for electrical machinery, overhead cranes, power tools permitting vastly improved layout and capital saving. Standardization facilitating worldwide operations	Emergence of giant firms, cartels, trusts, and mergers. Monopoly and oligopoly became typical. "Regulation" or state ownership of "natural" monopolies and "public utilities." Concentration of banking and "finance-capital." Emergence of specialized "middle management" in large firms.

Table 1 (*cont'd*)

Approx. periodization upswing, downswing	Description	Main "carrier branches" and induced growth sectors infrastructure	Key factor industries offering abundant supply at descending price	Other sectors growing rapidly from small base	Limitations of previous techno-economic paradigm and ways in which new paradigm offers some solutions	Organization of firms and forms of cooperation and competition
1930s & 1940s to 1980s & 1990s Golden age of growth and Keynesian full employment Crisis of structural adjustment	Fordist mass production Kondratiev	Automobiles Trucks Tractors Tanks Armaments for motorized warfare Aircraft Consumer durables Process plant Synthetic materials Petrochemicals Highways Airports Airlines	Energy (especially oil)	Computers Radar NC machine tools Drugs Nuclear weapons and power Missiles Micro-electronics software	Limitations of scale of batch production overcome by flow processes and assembly-line production techniques, full standardization of components and materials and abundant cheap energy. New patterns of industrial locations and urban development through speed and flexibility of automobiles and air transport. Further cheapening of mass consumption products	Oligopolistic competition. Multinational corporations based on direct foreign investment and multi-plant locations. Competitive subcontracting on "arms length" basis or vertical integration. Increasing concentration, divisionalization, and hierarchical control. "Techno-structure" in large corporations.

| 1980s & 1990s to ?* | Information and communication Kondratiev | "Chips" (microelectronics) | Computers Electronic capital goods Software Telecommunications equipment Optical fibres Robotics FMS Ceramics Data banks Information services Digital telecommunications network Satellites | "Third generation" biotechnology products and processes Space activities Fine chemicals SDI | Diseconomies of scale and inflexibility of dedicated assembly-line and process plant partly overcome by flexible manufacturing systems, "networking" and "economies of scope." Limitations of energy intensity and materials intensity partly overcome by electronic control systems and components. Limitations of hierarchical departmentalization overcome by "systemation," "networking," and integration of design, production, and marketing. | "Networks" of large and small firms based increasingly on computer networks and close cooperation in technology, quality control, training, investment planning, and production planning ("just-in-time") etc. "Keiretsu" and similar structures offering internal capital markets. |

Source: Ref. 18.
* All columns dealing with the "fifth Kondratiev" are necessarily speculative.

455

The impact of the new techno-economic paradigm on developing countries

In the new techno-economic paradigm, technological development is increasingly becoming the dominant factor in determining a country's capacity to compete in world markets. However, the industrialized and developing countries are on very different levels of scientific and technological development, and several indicators point to a widening of the economic, scientific, and technological gap between North and South, as the adoption of the key new technologies by industry proceeds at a much faster pace in the former than in the latter.

Some analysts argue that technological transition periods between Kondratiev cycles may present a particularly favourable window of opportunity for developing countries to enter the new paradigm and to reduce or eliminate their technological gap [43, 42]. Evidence from past experience, such as when countries like Germany and the United States challenged Britain's supremacy at the turn of the century (during the transition from the second to the third Kondratiev), or Japan's rise to the front rank of the industrialized countries (during the present transition to the fifth Kondratiev), suggests that the present initial diffusion stage of the new paradigm opens up encouraging development prospects for less industrialized countries (see figure).

For Perez [42], two characteristic conditions of transition periods lie behind this window of opportunity: the discontinuity in technical progress and the duration of the adaptation period for the leaders of the previous Kondratiev cycle. In other words, while the industrialized countries are still in the process of learning how to use the new technologies, and while the required new skills are not yet widely available, all entrants, small or large, more or less developed, have a chance to participate in the development of those skills and technologies.

To what extent this has already happened in biotechnology, new materials, and information technology is therefore a crucial question for the development strategies of developing countries. Developing countries face a market in key new technologies such as biotechnology and micro-electronics that is typically oligopolistic. While some more traditional technologies are available at low costs, the latest are subject to tight control, and a high cost is exacted. The characteristics of the new technologies are often inappropriate for the production and consumption needs of poorer countries, which have to make appropriate choices and adaptations of these technologies in order to reduce substantially their inappropriateness.

456

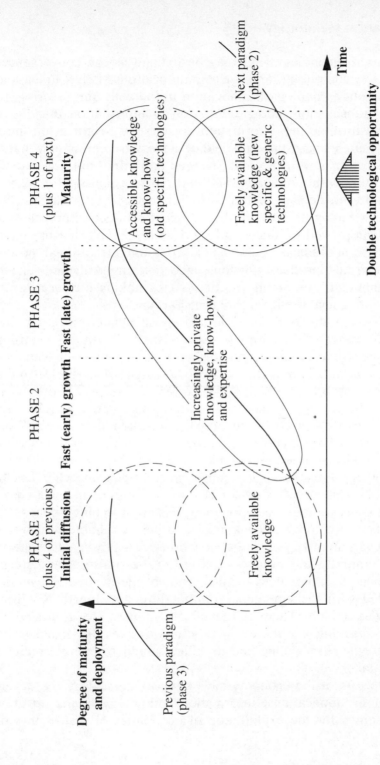

Phases (prevailing paradigm)

PHASE 1 (plus 4 of previous)	PHASE 2	PHASE 3	PHASE 4 (plus 1 of next)
Initial diffusion	Fast (early) growth	Fast (late) growth	Maturity

Degree of maturity and deployment

Time

Previous paradigm (phase 3)

Freely available knowledge

Increasingly private knowledge, know-how, and expertise

Accessible knowledge and know-how (old specific technologies)

Freely available knowledge (new specific & generic technologies)

Next paradigm (phase 2)

Double technological opportunity

Changing technological opportunities (Source: ref. 42, p. 15)

Information technology

Information technology (IT) may be defined as the convergence of electronics, computing, and telecommunications. It has unleashed a tidal wave of technological innovation in the collecting, storing, processing, transmission, and presentation of information that has not only transformed the information technology sector itself into a highly dynamic and expanding field of activity – creating new markets and generating new investment, income, and jobs – but also provided other sectors with more rapid and efficient mechanisms for responding to shifts in demand patterns and changes in international comparative advantages, through more efficient production processes and new and improved products and services (e.g. replacing mechanical and electromechanical components, upgrading traditional products by creating new product functions, incorporating skills and functions into equipment, automating routine work, making technical, professional, or financial services more transportable).

The development of IT is intimately associated with the overwhelming advances recently accomplished in micro-electronics. Based on scientific and technological breakthroughs in transistors, semiconductors, and integrated circuits ("chips"), micro-electronics is affecting every other branch of the economy, in terms of both its present and future employment and skill requirements and its future market prospects. Its introduction has resulted in a drastic fall in costs as well as dramatically improved technical performance both within the electronics industry and outside it. The continuous rise in the number of features on a single micro-electronic chip has permitted lower assembly costs for electronic equipment (each chip replacing many discrete components), faster switching speeds (thus faster and more powerful computers), and more reliable, smaller, and lighter equipment (fewer interconnections, less power and material). Similar dramatic falls in costs occurred in the transport and steel industries in the nineteenth century and in energy in the twentieth, associated with the emergence of the third and fourth Kondratiev cycles, respectively. The potential effects of micro-electronics are thus very far-reaching, for its use in production saves on virtually all inputs, ranging from skilled and unskilled labour to energy, materials, and capital.

All sectors of the economy have been influenced by the development of IT applications: information technology opens up greater opportunities for the exploitation of economies of scale and scope,

458

allows the more flexible production and use of labour and equipment, promotes the internationalization of production and markets, offers greater mobility and flexibility in capital and financial flows and services, and is frequently the precondition for the creation of innovative financial instruments. Information system developments are constantly being applied to increase the productivity, quality, and efficiency of finance, banking, business management, and public administration. In manufacturing, and to some extent in agriculture, many processes have been automated, some requiring highly flexible, self-regulating machines, or robots. The engineering industry has been transformed by computer-aided design and three-dimensional computerized screen displays.

The pace of technological change in IT will most likely accelerate the already observable growth in the interdependence of international relations – not just economic or financial, but also political and cultural. National economies have become more susceptible to the effects of policy decisions taken at the international level, and domestic economic measures are having increased impacts on economic policies of other countries. World markets for the consumption of similar goods are growing, and so are common lifestyles across national borders. The advance of telecommunications and computerization has recently enabled large companies to use information systems to transmit technical and economic information among numerous computer systems at different geographical locations, subjecting widely dispersed industrial plants to direct managerial control from a central location; this affects the international division of labour and production and international trade, changing the patterns of industrial ownership and control, altering the competitive standing of individual countries, and creating new trading partners.

It is the integration of functions that confers on information technology its real economic and social significance. More than just a gradual and incremental technological evolution leading to improved ways of carrying out traditional manufacturing processes (i.e. simply the substitution of new technologies for existing systems and the rationalization of standard activities), IT offers the opportunity for completely new ways of working through systems integration. Rather than applying one item of new technology to each of the production functions now performed at distinct stages of the production process, i.e. design, production, marketing, and distribution (in what could be called "stand-alone" improvements or "island-automation"), IT offers the possibility of linking design to production (e.g. through programmable manufacturing, measuring, and testing equipment responding to the codification of design), planning and design to marketing and distribution (e.g. through a variety of computer aids and databases that sense and collect changing market trends), production to distribution (e.g. by automatically incorporating orders and commissions by customers and suppliers into the production process), etc. The complete integration of all these production subsystems in a synergistic ensemble is still more a long-term trend than a reality, but use of automated equipment to link together individual items of equipment belonging to hitherto discrete manufacturing operations has already made IT a strategic issue for industry.

More technical advances are expected soon in the automation of telecommunications and the linkage of computers by data transmission that will enhance the possibilities of systems integration. Such "programmable automation," or computer-integrated manufacturing (CIM), has the capability of integrating information-processing with physical tasks performed by programmable machine tools or robots. CIM offers radical improvements in traditional problem areas confronting manufacturers, such as:
– reduced lead time for existing and new products;
– reduced inventories;
– more accurate control over production and better quality production management information;
– increased utilization of expensive equipment;
– reduced overhead costs;
– improved and consistent quality;
– more accurate forecasting;
– improved delivery performance [31].

These features characterize information technology as a new technological system, in which far-reaching changes in the trajectories of electronic, computer, and telecommunication technologies converge and offer a range of new technological options to virtually all branches of the economy. Moreover, IT forms the basis for a reorganization of industrial society and the core of the emerging techno-economic paradigm.

The reason for the pre-eminence of the new technological system clustered around information technology over the equally new technological systems clustered around new materials and biotechnology is the fact that information activities of one kind or another are a part of every activity within an industrial or commercial sector, as well as in our working and domestic lives. Almost all productive activities have a high information intensity (some involve little else, such as banking or education), so information technology is capable of offering "strategic" improvements in the productivity and competitiveness of virtually any economic or social activity. Information technology is universally applicable.

Probably only a fraction of the benefits derived from information technology–based innovations have so far been reaped and the rest remain to be acquired in the next decades. The shift towards systems integration to capitalize the full potential benefits of IT requires considerable adaptations, learning processes, and structural changes in existing socio-economic institutions and organizational systems. The tradition in most current organizations is still to operate in a largely "dis-integrated" fashion, reminiscent of the Ford-Taylorist management approaches that dominated the fourth Kondratiev cycle: high division of labour, increasing functional specialization/differentiation and de-skilling of many tasks, rigid manufacturing procedures and controls, long management hierarchies with bureaucratic decision-making procedures and a "mechanistic" approach to performance. Under these conditions, use of IT is restricted to piecemeal technology improvements. By contrast, information technology–based systems offer organizations the opportunity of functional integration, multi-skilled staff, rapid and flexible decision-making structures with greater delegation of responsibilities and greater autonomy of operating units, a more flexible and "organic" approach enabling a quick adjustment to changing environmental conditions. (On "flexible specialization," see Piore and Sabel [44].)

But this means that information management skills require the ability to make choices about the optimal arrangements for particular situations: unlike earlier generations of technology, IT offers not a single "best" way of organization but a set of more or less appropriate alternative organizing, staffing, and managing options that may be adopted in different organizational contexts. There is no "determinism" in the way information technology influences the socio-institutional framework: organizational innovation is a crucial part of the requirement for firms to adapt to survive [31].

Implications for developing countries
The complex interrelations between technological and institutional changes associated with information technology have significant implications for the way IT will affect the societies and economies of developing countries. The negative and positive potential impacts of IT

461

on these countries is a matter of great controversy among economists and politicians. The main short-term issues usually discussed are the potential erosion of the comparative advantages of low labour costs, particularly in relation to assembly facilities, and the effects of automation, particularly on internal markets and international competitiveness.

The first direct effect of the "micro-electronics revolution" was the location of production for export in third world countries. While production of mainframe computers continued to be located largely in industrialized countries, production of smaller computers and of micro-electronic devices, more subject to price competition, was shifted to low-wage locations, mainly in East Asia, where countries presented low wage costs as well as political stability, a docile labour force, and government incentives. Location of production for local and regional consumption followed, but the countries concerned were mainly middle income: three-quarters of US investment in third world micro-electronic industries was concentrated in 11 countries, namely the four Asian "dragons," India, Thailand, Malaysia, the Philippines, Brazil, Mexico, and Colombia [53]. Export-oriented investments in these countries were associated more with direct foreign investment from larger firms in industrialized countries than with firms producing for the local market; on the other hand, licensing was more associated with smaller firms [55].

The automation of production decreases the relative importance of labour-intensive manufacturing and cost of labour, thereby eroding the competitiveness of low labour costs. For instance, automation led to a sharp decrease in the difference between manufacturing costs of electronic devices between the United States and Hong Kong: in manual processes, manufacturing costs were three times higher in the United States, and the introduction of semi-automatic processes made the difference practically disappear [45]. Equally, the expansion of automation in Japan has contributed to a reduction of Japanese investments in the Asia/Pacific region involving firms in electronics, assembly parts, and textiles [45].

The trend to increasing systems optimization and integration is most likely to induce large producers in industrialized countries to bring back a significant share of their production located in developing countries (offshore production). This movement has been called "comparative advantage reversal." As integration increases, with functions previously obtained by assembling pieces being incorporated in the electronic components, value-added is pushed out of assembly processes into the components themselves and upwards towards servicing. In addition, the growing technological complexity of electronic devices increases the value of the parts manufactured by firms located in industrialized countries. The amount of value-added obtained in offshore assembly has thus been constantly decreasing [45]. Global factories constructed in locations of least cost, often at a considerable distance from final markets, were economically worthwhile because labour was one of the major determinants of costs. Technology and rapid responsiveness to volatile local markets are becoming more important components of competitiveness. The reduction of product cycles due to the growing resistance to obsolescence of programmable machines and equipment has led to a concentration of manufacturing investment in capital-intensive flexible manufacturing, further adding to the erosion of the comparative advantages of developing countries.

The assembly of systems will probably continue in some developing countries that have adopted protective legislation for local production targeted at particular market segments (e.g. Brazil), although this is changing very rapidly. The types of equipment produced under these circumstances are used largely in internal markets and are hardly competitive on the international level; they tend to be far more expensive than comparable equipment available abroad, and often their installation and use are also more costly because of expensive auxiliary installations, under-use, and lack of management skills. Nevertheless, they may at least provide the country with the capacity to follow the development of information technologies more closely. In other countries, assembly of equipment is taking place from components bought practically off the shelf, but as the level of hardware integration and the amount of software incorporated into the chips (firmware) grow, value-added will be taken away from the assembly process, reducing or eliminating its economic advantages.

The introduction of micro-electronics requires certain new skills of design, maintenance, and management, as well as complementary infrastructural facilities such as reliable telephone systems and power supplies. Deficiencies in these factors prevent the widespread adoption of information technology in developing countries [32]. The more advanced developing countries, with a wider basis of skills and infrastructure and a more flexible labour force, may be in a better position to adopt IT and to increase their productivity and their international competitiveness. But the less developed countries, with inadequate skills and infrastructure, low labour productivity, and lack

463

of capital resources, will find it difficult to adopt the new technologies; they are likely to suffer a deterioration in international competitiveness *vis-à-vis* both industrialized and the more advanced developing countries [53].

Quality, too, requires an adequate level of skills, infrastructure, and managerial know-how that is generally lacking in developing countries. This greatly reduces the synergies, number of options, faster responses, and more informed decisions that can be implemented in the firm by the optimization of the systems performance. In turn, the composition of the labour force existing within firms located in industrialized countries will further improve their systems performance and further reinforce the advantages derived from automation. The proportion of the labour force employed in production is constantly decreasing in the industrialized countries, implying that performances at the systems level and innovation, not manufacturing, are becoming the key to profit, growth, and survival [45].

Like biotechnology, information technology is a proprietary technology, vital technical information regarding design engineering specification, process know-how, testing procedures, etc., being covered by patents or copyrights or closely held as trade secrets within various electronic firms from industrialized countries. Many companies in the software area do not patent or copyright their products because it entails disclosing valuable information, and firms are generally reluctant to license the more recent and advanced technologies. Therefore, technology transfer takes place mainly among established or important producers, hindering the access to developing countries. Moreover, the main issue facing developing countries is not so much the access to a particular technology but to the process of technological change, because of the dynamism of this process. Recent trends in inter-firm relationships seem to indicate that this access takes place essentially through the participation in the equity of the company holding the technology [45]. The possibility of firms from developing countries doing this is small.

The general tendency thus points to a widening of the information technology gap, both between industrialized and developing countries and within developing countries. From a purely quantitative standpoint, there remain large gaps in the access to information in the world, showing that the diffusion of information technology in developing countries is still in an embryonic stage. In 1985, only 5.7 per cent of the total number of computers in the world were located in developing countries, which have so far mainly used computers for more standard functions, such as inventory control, accounting, and payroll. There are some significant exceptions: during 1981–1986, the Brazilian electronics market grew to about US$8 billion, becoming the tenth largest in the world, slightly larger than that of the Republic of Korea. Brazilian-controlled firms have rapidly expanded their production of computers, and India has been particularly successful in software, having achieved a niche in the international software market through its government policies of tax breaks and liberal foreign exchange regulations for this sector. In 1985, Brazil was by far the largest user in computers in the developing world, with about 10,000, followed by India with around 1,000 [45].

The value of data processing equipment (excluding micros) existing in developed countries was 4.5 times greater than in developing countries (including centrally planned economies) in 1983 [45]. On the other hand, the market for telecommunications equipment (i.e. telephone, telegraph, telex, data and satellite communications, mobile radio and radio telephone, radio paging, and cable TV) in industrialized countries in 1985 was 8 times greater than in developing countries [45, source: Arthur D. Little Inc.]. Increasingly severe financial problems facing the developing countries are most likely to lead to a quantitative increase in the gap.

The situation is not very different from the qualitative standpoint. The process of "informatization" of society is one in which greater amounts of knowledge and information are incorporated into goods and services. Knowledge and information are sources of wealth creation and value-added in their own right: as their amount increases, the amount of energy, materials, labour, and capital decreases. The concentration of knowledge and information-intensive services in industrialized countries is *per se* a further barrier to efforts to reduce the information technology gap between industrialized and developing countries [45]. Moreover, since systems integration and disembodied technologies are the essential locations of competitive innovation, the advantages of technology transfer by reverse engineering and technological "unpacking" are more difficult to capture [25]. For instance, even in Taiwan and Korea, all numerically controlled mechanisms are made under licence [23]; in Brazil and Argentina, computerized numerically controlled devices and motors are imported [23, 12]. India and China produce numerically controlled and computerized machine tools under licence, utilizing imported control units and motors.

There is, however, an alternative "reading" of the evidence presented above. Some commentators see IT as a powerful new opportunity for at least some developing countries to improve their competitive position in certain fields and to foster their development precisely because of their relative lack of established industrial infrastructure, meaning that there are fewer institutional barriers to the adoption of advanced systems based on information technology. Problems related to the reluctance to discard previous generations of equipment, for instance in telecommunications (i.e. outdated electromechanical and copper-wire based infrastructure), are less important in developing countries, and replacement costs are lower, allowing them to jump directly to "best practice."

This school of thought minimizes the fact that in practice, information technology is still a "black box" technology for most users in developing countries, requiring new skills to operate, to repair, and even to purchase, particularly in its integrated form, where its full benefits emerge. It is therefore most likely that information technology will be largely applied in industrialized countries to the disadvantage of the majority of developing countries, the latter remaining heavily dependent on advanced technology–based products designed in the former.

Biotechnology

Biotechnology is neither a scientific discipline nor an industry, but a rapidly developing and still diffusing field of activity that cannot be adequately described by a short definition. In a report prepared for the OECD, Bull et al. [5] propose the following "working definition" of biotechnology: "Biotechnology is the application of scientific and engineering principles to the processing of materials by biological agents to provide goods and services." This definition attempts to avoid both too narrow or too wide a view, seeing biotechnology neither as essentially genetic manipulation nor as all activities involving living materials. Thus, "scientific and engineering principles" are taken to cover a variety of disciplines, but in particular microbiology, biochemistry, genetics, and biochemical and chemical engineering; "biological agents" refer to a wide range of biological catalysts but particularly to micro-organisms, enzymes, and animal and plant cells; "materials" are taken in a broad sense to include both organic and inorganic compounds; and the essential link of scientific activity with industry is considered in the "application . . . to provide goods and services," covering a variety of products such as pharmaceuticals, biochemicals, and foodstuffs, as well as services such as water purification and waste management.

Essentially, biotechnology harnesses the catalytic power of biological systems, whether by direct use of enzymes or through the use of the intricate biochemistry of whole cells and micro-organisms. Defined in this way, biotechnology encompasses everything from the technology of bread-making to that involved in the production of human insulin from a bacterium induced to take up a non-bacterial gene and produce the protein coded by that gene. Its history goes back centuries in such activities as fermentation and brewing of alcohol or bread- and cheese-making. New scientific and technological advances in genetic engineering and other ways of transforming biological organisms in the 1970s revolutionized commercial possibilities, giving rise to a large number of applications with the development of new products and new techniques. The recent technological developments in genetic engineering, enzyme technology, and fermentation technology are often called "the second biotechnological revolution" (or the "new biotechnology"), the first being generally recognized as Pasteur's revolutionary treatment and prevention of human and animal infectious diseases through immunization in the late 1880s.

New biotechnology is typically a science-led technology, in the sense that most of the inventions and process and product innovations have emerged from breakthroughs in scientific and technological research undertaken in universities, research institutes, and industrial R&D departments. It denotes a broad and heterogeneous field of applied sciences and related strategic research, encompassing several distinct technologies utilized in a wide range of industries: agriculture, pharmaceuticals, chemicals, and even weaponry are all potential beneficiaries of the advances being made.

Industries are increasingly using biotechnology to produce industrial substitutes for natural agriculture products manufactured in large quantities (and mainly exported by developing countries). Many new substances are competing with each other as viable substitutes for a particular product (foodstuffs, flavours, additives, fragrances), a trend very similar to the one encountered in new materials. The demand for new foodstuffs and pharmaceutical products (e.g. vaccines) is becoming increasingly diversified, and biotechnology is providing industry with the opportunity to abandon commodity chemicals and move into more lucrative specialty and agricultural chemicals. Older

biotechnological techniques (e.g. fermentation) are themselves bene-
fiting from additional inputs from genetic engineering and new
enzymatic processes.

Bio-industry is reorganizing itself to respond to these trends: conscious of the economic stakes involved in the enormous potential
markets for the new biotechnological products, many chemical, pharmaceutical, petrochemical, and industrial food corporations are creating their own research laboratories in plant biology and physiology
and are investing in small venture-capital companies engaged in advanced research as well as in larger companies with R&D experience.
As new products depend heavily on new and more productive processes and call for rigorous quality standards and safety tests, bio-
industry is typically science- and capital-intensive and requires highly
qualified staff and skilled labour.

A number of biotechnology developments are having profound
technical impacts on processes and products. As with new materials,
these technical changes are inducing important structural changes in
the economy [35]:

– New commercial biotechnological devices and methods of diagnosis and prevention, based on monoclonal antibodies, biosensors,
  and gene probes, are revolutionizing the fields of health, agriculture, and environment, permitting the extension of hitherto limited
  physical and chemical measurements to the potential control and
  regulation of complex systems in the human body, in animals,
  plants, the environment, and in industrial processes.
– The specificity and diversification of biotechnological products are
  increasing, as commodity chemicals tend to be replaced by specialty and agricultural chemicals, closer to user demands. Monoclonal antibodies can be used as ultraspecific drug vectors against
  specific tissue antigens, opening the way to the introduction of
  medicines specific to individual patients (personalized therapy).
  Several distinct new biotechnology products tend to compete with
  each other as substitutes for the same traditional product: for instance, more than eight new sweeteners compete to replace sugar.
– Biotechnology contributes to a reduction in the intensity of the use
  of energy and materials: the production of chemicals through enhanced fermentation or enzymatic processes, industrial purification
  by monoclonal antibodies, and the replacement of sugar by new
  compounds with dramatically superior sweetening power may be
  mentioned as examples of this trend. New immunodiagnostic tests
  based on monoclonal antibodies and gene probes, besides being

rapid, specific, and easy to use, are sensitive to smaller quantities of test material and imply a dramatic reduction in the quantities of blood, urine, cells, etc., needed. Biotechnological processes and products present the ability to use renewable energy resources and to recover reusable or marketable by-products in the processing industry, thus increasing the productivity of all energy and materials inputs through "maximum recycling" and "minimum effluents."

– The methodologies employed in the development of new products and processes in biotechnology rely on rigorous scientific knowledge in numerous fields, thus increasing rationality and diminishing empiricism in research and industrial production through a goal-directed and systematic understanding of the processes involved. This is for instance apparent in the radical change in the methods of pharmacological research, which has shifted from the screening of a large number of molecules to the targeting of a suitable molecule to act upon the mechanism of a specific disease. This change in the paradigm of pharmacology, made possible by new biotechnological research instruments and products, has simplified and rationalized the process of innovation and profoundly affected the pharmaceutical industry: from being a drug supplier, it is becoming an "industry of function," i.e. a supplier of a wide range of therapeutic products, diagnostics, auxiliary materials, equipment, machines, biomedical systems, and technology. A similar evolution towards rationalization of the innovation process in industry can be expected in the agrochemical and food industries.

The bulk of biotechnology sales in terms of volume and value can be grouped in three main groups of products [21]:

1. Very high value medical products used in small quantities, like vitamins (B_{12}), antibiotics (cephalosporin), enzymes, novel biological products (interferon, tissue plasminogen activator – TPA), or monoclonal antibodies, which are extremely expensive and whose production in commercially viable quantities has only become possible with recent genetic engineering technologies.

2. Low value products that have to be sold in enormous quantities, usually produced by fermentation processes, and that generally compete against similar commodities produced by more traditional means, like ethanol, methane, isoglucose, and several effluent and waste treatment substances.

3. An intermediate group of organic chemicals, such as amino and organic acids (glutamic acid, lysine), fungal proteins used in novel foods, and bacterial cultures used as soil inoculants to pro-

tect plants from pests or to supply additional nitrogen to the roots, all of which also have to compete against other processes.

Some applications

Biotechnology inventions and innovations have already been applied in numerous industrial sectors.

FOOD AND AGRICULTURAL PRODUCTION. The potential of biotechnology for increasing agricultural productivity is high, in terms of both increasing the yields of cultivated plants and of obtaining foodstuffs with higher nutritional value. Many foodstuffs are produced by fermentation, and enzymes are now widely used as processing aids in food manufacturing. Acetone, citric acid, ethanol, and other chemicals are, or have been, produced industrially by fermentation. The digestion of wastes anaerobically is not only part of sewage treatment but also a way of generating methane gas as a source of energy. Biotechnology offers ways of improving even traditional fermentations like the production of silage, a fermented gas product used as cattle feed: microbial cultures are available that ensure that the correct sort of fermentation takes place. It is expected that by the year 2000, five-sixths of the annual increase in agricultural production in the world will be due to new biotechnology and other yield increases, while only one-sixth will result from the increase in the area of land used in production [37]. In the next century, about 75 per cent of all major seeds may be developed by genetic engineering or tissue culture.

Many developing countries have established programmes to incorporate biotechnology into agricultural and agro-industrial activities. Some have already successfully applied biotechnology to their production of palm coconut oil, eliminating major disease traits and thereby increasing productivity by about 30 per cent. A marked increase in production, using cloning techniques to enable the propagation of high-yielding varieties of oil-palms and cocos, would make it possible to improve the fat content of diets and thus cover the additional nutritional needs of growing populations. But the production of oil-palm and coco clones using tissue culture techniques, where the applications could benefit millions of small landholders in developing countries whose standard of living depends entirely on the productivity of their holdings and whose cultivation techniques would have to be adapted to the properties of the new clones, constitutes a break-

470

through that cannot be fully exploited before the end of the century [48].

Wood exports play an important role in the economy of many tropical developing countries. The *in vitro* micropropagation of forest tree species for their wood or paper pulp is therefore of great economic interest; this technique is for instance being studied for the large-scale production of clones of several eucalyptus species with better resistance to cold weather and greater wood yield. Similarly, the multiplication and exploitation of drought-resistant plant species of commercial interest could afford useful outlets for a number of developing countries located in arid or semi-arid zones. For instance the jojoba, cultivated today in all five continents, can tolerate temperatures up to 50°C and its roots can search for water at a depth of 30 metres. It offers the possibility of controlling desertification by fixing soils and of earning a good income from a valuable oil extracted from its seeds, thus bringing employment to the rural areas and the chance to export a multi-purpose product with a high potential demand on the world market. Jojoba oil can be used industrially as an excellent transmission fluid or lubricant for fast rotating machines under high pressures and high temperatures (replacing the strategic sperm whale oil and thus limiting the massacre of sperm whale and other cetacean populations), as a shampoo and a sun cream in the cosmetics industry, as a treatment for skin diseases and burns in the pharmaceutical industry, as a wax to replace other plant or animal waxes, and meal proteins could be extracted from it for use in animal feed [48].

Tissue culture techniques have been applied to rice, maize, wheat, barley, cabbage, lettuce, tomatoes, peas, onions, potatoes, rapeseed, tobacco, sugar cane, and cotton for such purposes as gene transfer for disease resistance and salinity tolerance, selection of plants resistant to pathogens, and recovery of immature embryos from defective seeds. Substantial research in biotechnology and genetic resources has led to the adoption of genetic selection and breeding techniques by several countries, as well as to the improvement and production of local varieties of crops with higher yields, greater pest resistance, and earlier maturation. Progress in fermentation technology for the production of feed components, single-cell protein and industrial chemicals, as well as recent developments in enzyme technology for the production of antibiotics are expected to have a large impact on industry and agriculture in several developing coun-

471

tries. Nitrogen-fixing biotechnology, which enables non-leguminous crop plants to fix atmospheric nitrogen, should permit a two to four-fold increase in corn yields.

LIVESTOCK HUSBANDRY AND ANIMAL HEALTH. Genetic engineering is already being applied in animal husbandry. Bovine embryo transfer techniques can have great zootechnical and economic advantages. Besides helping to speed up the improvement process or the preservation of superior breeds showing special characteristics (for instance, better resistance to tropical bovine diseases), embryo transfer can increase the production of meat and milk, each inseminated cow being able to give birth to up to 20 calves per year. The development of DNA probes can permit the sexing of the bovine embryos to be transplanted, thus selecting male embryos for meat production and female embryos for milk production. In some developing countries this technique could help overcome chronic milk shortages.

Genetic engineering also provides the possibility of developing and producing large quantities of new vaccines against many cattle, swine, and poultry infectious diseases that plague developing countries, like aphthous fever, theileriasis, hog cholera, colibacillus and viral diarrhoea, pseudo-rabies, coccidiosis, fowl pest, etc. Traditional vaccines against the aphthous fever virus, which is endemic in large areas in developing countries, are prepared by inactivation or attenuation of virus strains obtained from material collected from the lesions themselves, and imply the manipulation of very large quantities of virulent virus; in addition, these vaccines are unstable and must be stored under refrigeration, which is not always easy in tropical countries. The production by rDNA techniques of an effective, safe, and heat-stable vaccine against this disease will have a great economic impact in developing countries, which will be able to vaccinate their herds systematically and to increase the export of their livestock products to disease-free industrialized countries.

Fowl pest is the principal virus disease of poultry in the world, and it has devastating economic effects in several developing countries, where poultry meat and eggs form a major contribution to the human diet; most of the commonly used vaccines are relatively ineffective and must be administered on several occasions in high doses, a task rendered very difficult, particularly in countries where village poultry and small flocks predominate. A new, simple, and cheap vaccine is needed; research in genetic engineering may permit the production of massive quantities of antigen to be used for the preparation of an improved vaccine, in terms of potency and geographical utility [48].

PHARMACEUTICAL AND CHEMICAL PROCESSING. Biotechnology has been efficiently used to produce new pharmaceutical products, such as interferon, growth hormone, lymphokines, and tissue plasminogen activators. Biosynthesis of growth hormones of the main livestock species by genetically engineered micro-organisms can markedly improve their productivity and would have significant effects in intensive livestock husbandry [48]. Bovine growth hormone can increase milk production by 20 per cent at the same feed costs.

MEDICAL TREATMENT. The health care sector has attracted the most early interest for various reasons. Health care covers a large number of human activities, ranging from "formal" care provided by organized health services (clinics, hospitals, and other organizations for care, cure, or preventive medicine), "alternative" medical practitioners and self-medication or self-diagnosis products, to unpaid care of the sick and infirm. Biotechnology is particularly applicable to health care products in all these activities, including pharmaceuticals, vaccines, and diagnostic kits. It also provides ways of more rapidly screening potential pharmaceuticals, speeding up and lowering the high cost of pharmaceutical innovation.

Genetic engineering offers a way of producing on a larger scale biological molecules with therapeutic value that were formerly very scarce and therefore expensive, if available at all. Examples of these substances would include the first product of rDNA organisms for human therapy, human insulin, as well as human growth hormone, the interferons, interleukin, and other bioactive proteins. Many higher plants possess active compounds that form the starting material for a large range of drugs. The 1986 market for plant-derived pharmaceuticals was estimated at US$9 billion in the United States alone [13]. Tropical developing countries, whose pharmacopoeia is very rich and which constitute the main exporters of plant medicinal raw materials, could start from naturally occurring compounds and resort to biotechnology to isolate them and produce novel pharmaceuticals, thus reducing current imports. In addition, the amount of active product required for pharmaceutical uses of these substances is usually low and the pay-off potentially huge in many instances; however, the regulations concerning the commercialization of medicines apply equally to plant medicinal products, and since most therapeutic substances require painstaking testing, development may often be a lengthy and expensive process.

By contrast, a large number of new methods of testing human fluids and infections have been developed, based on monoclonal antibody technology. The fastest growing diagnostics markets are in immunology and microbiology. Monoclonal antibodies used in diagnostic kits offer products that, because they are not ingested by or applied to people, could be brought quickly to market and for which there is growing demand. Already, monoclonal antibody–based tests sold in pharmacies for confirming pregnancy are being established as a do-it-yourself market, and other over-the-counter products are being introduced for monitoring fertility. Monoclonal antibody products are also becoming a vital part of the growth of new types of imaging techniques, and accurate, rapid, and cheap tests based on DNA probes and biosensors are promising future developments [21].

The cost of the techniques involved are falling sharply [48], so that they are likely to become, with the improvement of current vaccines and the development of effective, safer, and cheaper new vaccines, the major instruments of public health policy in developing countries.

Recombinant DNA techniques can be used to produce large quantities of immunogenic proteins synthesized by genetically engineered micro-organisms, which are the basis for effective new vaccines. A genetically engineered vaccine requires no inactivation procedure as conventional vaccines do, facilitating its administration and reducing cost; additional economies may arise from the replacement of expensive embryo culture systems by relatively simple conventional bacterial media, from savings on high-security plants usually required in the production of conventional infectious disease vaccines, from reduced transport and storage costs, and from reduced testing, since the vaccines do not contain the disease-producing pathogen. Recombinant DNA techniques are being developed for the production of vaccines against viral hepatitis B (highly endemic in regions of Africa, Asia, and South America), rabies (a serious health problem in developing countries and still a cause of high mortality in domestic livestock), herpes, cholera, leprosy, malaria (the most widespread human infectious disease), schistosomiasis (chronic throughout tropical countries), onchocerciasis, sleeping sickness, and Chagas' disease [48].

Advantages and disadvantages

One of the main advantages of these innovations in biotechnology has been the possibility of their economic use on a small scale, without large infrastructure requirements, and their application at different levels of complexity, investment, and effort. It is in fact possible to adapt sophisticated biotechnical technologies to low-cost operations without eliminating the chances of success. This characteristic may facilitate the use of biotechnology in developing countries, provided that the promises brought to them are accurately identified, as well as the positive or negative impacts on their economy, their way of life, and their social structure. For instance, the expected growth in the market for gene synthesizers, protein fractionation equipment, or gene-splicing enzymes requires the provision of adequate infrastructure in terms of these enabling technologies, as well as culture collections and information systems. These requirements present an increasing concern to developing countries wishing to establish a sound base in biotechnology, which must therefore reconcile the spectacular progress of biotechnology with the lack of funding resources and qualified personnel needed by most sectors of bio-industry.

However, plant and animal biotechnologies may also have negative impacts in the developing countries.

Developing countries can be considered as "natural reservoirs" of wild and semi-domesticated species used to improve the crop potential [4], since the major centres of plant biological diversity are found in the tropical regions, where two-thirds of the world's plant genetic resources lie. Developing countries are therefore, directly or indirectly, the suppliers of plant genetic resources used in improving food production, in increasing yields of fibre or woody plants, and in finding new plant raw materials or pharmaceutical substances [48].

Stewart [53] has pointed to some of the major implications of the new biotechnologies for developing countries, which are much in line with the analysis presented above:

- Negative effects for primary products, as new biotechnology products substitute for exports, reducing demand for natural products traditionally produced by developing countries.
- Great increase in agricultural productivity, while economizing on most inputs such as herbicides and fertilizers. Developing countries will have to adopt the new seeds if they are to compete, but as levels of output rise, prices will tend to fall, reducing net revenues

for producers of many primary products with low elasticities of demand. New biotechnology is thus likely to shift the terms of trade against primary producers, while the cost of importing the new technologies will be high because of the trend towards privatization.

- Both production and consumption characteristics of the new biotechnology are likely to be mainly in line with the industrialized countries' demands and conditions. The scale of production of bioprocesses is large and growing, and they are highly capital-intensive compared with traditional methods of food processing. Small producers will thus find it difficult if not impossible to get access to the process, while demand for the competing products they produce will tend to fall. The trend towards large-scale farming and increased vertical coordination and control, already accentuated in the United States by the new biotechnology [38], is likely to be even greater in developing countries, where typical farm sizes are smaller and the degree of vertical integration less. Unless major modifications are made, the adoption of new biotechnology by developing countries is thus likely to worsen income distribution in those countries.
- Major consumer benefits in terms of more and better food and improved medicine. However, to realize these benefits fully it will be necessary to adapt research to the needs of the country.
- Developing countries will be pushed to undertake some R&D of their own, as independent R&D will help to hold down excessive costs of technology transfer and to ensure that research responds to their needs and conditions. But evidence from developments in the industrialized countries indicates the existence of very strong links between basic research at universities and applied work within companies at the very early stages in the innovation process [6], suggesting great difficulties for developing countries to do effective research in this area since such links are much less well developed in most developing countries. Moreover, the cost of research has been high compared to the R&D budgets of many developing nations. Competitive success requires a synergistic relationship between specialization on new biodevelopments, microprocessing, and marketing/distribution [14], which can be achieved in industrialized countries by joint ventures or close collaboration among firms, or even by uniting the three specializations in a single firm, whereas most developing countries present too "thin" an industrial structure to operate effectively over such a large area.

476

Table 2 **Impact of selected aspects of plant biotechnology**

Aspects	Positive impact	Negative impact
Genetic diversity	Quick means of germ plasm transfer; broader breeding base; genetic base for new production; reduction in losses	Increase in uniformity and vulnerability; "genetic erosion"
Germ plasm identification	Elimination of undesirable characteristics; acceleration in new cultivar development	Ignorance of local conditions, such as local pests
Cultivar dissemination	Production of a broad variety of new plants; replanting of crops feasible within a growing season	Reduced long-term biological potentials of crops
Production	Significant increase in yield	Overproduction; market instability; reduced export income
Pest problems	New and fast ways to combat pest epidemics	Alteration of natural composition of organisms with unknown consequences
Mechanization	Amenability to harvest, processing, and packaging	Unemployment; fewer product varieties
Germ plasm	Viable means for long-term storage	Storage concentrated in a few countries, with potential for discriminatory exploitation
Land use	Reduction in land area needed for production, giving room for other national purposes and redistribution of land to smaller farmers	Global overproduction; depressed economies unable to take advantage of potential benefits
Environment	Development of organisms likely to survive in difficult natural environments	Upsetting balance of Nature through release of genetically altered micro-organisms

Source: Ref. 56.

The opportunities offered by biotechnology to developing countries should therefore be weighed against its environmental risks and the interrelated social and economic implications. Table 2 attempts to summarize the main issues discussed above.

It seems probable, therefore, that developing countries that are mainly primary producers, that lack flexibility in terms of diversification potential, and that do not have a sufficient degree of indigenous scientific, technological, and industrial capacity to enter the field themselves, may suffer some major losses as a result of the new biotechnology, and will have to pay heavily for its use; their net gains are likely to be small. By contrast, developing countries capable of diversifying out of primary products and establishing some local R&D capacity will be able to use new biotechnology to raise productivity and employment.

There are also enduring regulatory uncertainties, and the whole area of intellectual property rights is one of permanent controversy, making investment in some areas extremely risky. For instance, the rules that apply to the award of patents do not apply, in most cases, to new plant varieties, and the initial national regulations in industrialized countries did not grant breeders the protection of their "inventions." But recently, important changes in the legislation regarding the protection of plant varieties, brought by the development of plant biotechnologies, have been adopted by industrialized countries under pressure from industrial companies willing to make research in plant genetics and breeding more attractive. Following a decision by the Supreme Court of the United States in 1980 (the "Chakrabarty decision"), ruling that "the relevant distinction was not between living and inanimate things, but between products of nature, whether living or not, and man-made inventions," the number of patents in genetic engineering increased twice as fast as those in other technologies. New attitudes regarding the extension of property rights to all biotechnological inventions, even those concerning genetically modified living things, are being adopted by many private organizations and even by public reseach institutions, including the extension, to the whole transformed plant, of an inventor's natural right over a new gene permitting the breeding of a new plant variety.

These attitudes, intended to protect the products of increasingly expensive plant genetic research and to make this research more profitable, together with the increasing role of the private sector in the collection, conservation, and utilization of germ plasm for commercial purposes, inevitably entail, for the developing countries, the payment of ever higher fees for the seeds of new and more productive crop varieties, often derived from species grown and improved in their own regions as the result of the efforts of many generations of farmers. As Sasson [48] puts it: "for the first time in agricultural history, there is a confrontation between the breeder's and the farmer's rights." Consequently, in the third world technological dependence is highly likely to increase. As with information technology, the impact of biotechnology on developing countries will depend largely on which political and industrial strategies are adopted, and on the appropriate choice of targets or objectives to be reached.

New and advanced materials

Materials have always played an essential role in every Kondratiev economic cycle, largely shaping every technological system. However, in the presently emerging techno-economic paradigm, their role tends to be very different: no single material seems to be associated with the paradigm, but rather a kind of global dynamics in the conception and diffusion of a vast variety of homogeneous and heterogeneous materials, in what has been called "hyperchoice" [9]. This dynamic applies not just to "recent" high-performance materials such as composites, but equally to more "traditional" materials such as metallic alloys or ceramics. It is based on increasing knowledge of the microscopic properties of matter and on mastering industrial reproduction processes of these microscopic properties, enabling different materials to be combined to make new alloys or composites and to customize their properties. The concept of "new materials" used here thus refers to substances possessing compositions, microstructures, properties, performances, or application potentials derived from the industrial reproduction of their microscopic properties. There are no "old materials," but only outdated industrial techniques, processes, and equipment. Every traditional material can become "new" through the adoption of advanced shaping and manufacturing techniques and processes permitting the control of its microscopic structure. What is in fact "new" is the unity material-process-product.

When examined under the light of the present paradigm change, the trend with the greatest force in new materials seems to be the one leading to a growing diversity in materials use. Three factors have recently acted in this direction: the increase in the relative cost in energy; the requirements of the micro-electronic components industry; the specific demands generated by the use of micro-electronics in new products and processes. The vast growth of innovation possibilities in programmable capital goods has been not only the main impetus for downstream innovations in products and services, but also a powerful impetus for upstream innovations in materials. This contrasts with the previous paradigm, in which the dynamics of innovation in the areas of materials, chemistry, and final goods set the requirements for innovation in capital goods [41]. Under the old paradigm, materials were typical examples of technical constraints imposed from the outside: designers and engineers chose a material for a principal property or physical characteristic that imposed itself technically upon the desired product; under the new paradigm, the modular character of the material, made possible by the industrial reproduction of its microscopic properties and by intensive use of computer-aided design and manufacturing, permits the *prior* identification of a technical need and the *ex post* development of a material specifically adapted to that need. In other words, from an *exogenous* constraint on industrial design and engineering, new materials have become an *endogenous* production variable [9].

The specific requirements of the micro-electronic components industry have already led to the development of a vast supplier network for semiconductive, conductive, and photosensitive materials; crystals of various types; high-purity chemicals; new ceramics and resins. The changes occurring in the functional characteristics of products and machines, i.e. the replacement of moving mechanical parts by electronic circuits and the subsequent reduction in the size of the products, reduce part of the demand for the more common engineering materials such as metals and plastics in favour of lighter ones, as well as those that present several characteristics simultaneously (e.g. the lightness of plastics plus the resistance of metals). New diverse means of interfacing with the user have required the development of new materials that are sensitive to light, to touch, to sound, to heat and others with countless special characteristics for particular purposes or particular tastes [41]. At the same time, the utilization of micro-electronics in the design and production of new materials has rejuvenated the technological trajectories in "old" materials like metals and polymers, and has created new trajectories in glass and ceramics. The convergence between these two different sets of trajectories has led to the development of several types of composite materials. In short, there is a growing richness in the information content of materials and a proliferation of alternative patterns of materials consumption, in line with the general characteristics of the new techno-economic paradigm.

The technical objectives at stake in the competition between different materials have become more numerous. Historically, competition between materials expressed itself through economic advantages that were obtained essentially by searching for alternative sources of higher quality minerals and ores, cheaper processes and transport costs. For any specific technical application, there was generally a single material that dominated more or less durably: it was the regime of "mono-choice," characterized by economies of scale and standardization. Variety, when it existed, occurred within the same family of materials (metals, plastics, ceramics, etc.), not by creating a new family. In economic terms, the lack of variety of these "commodity" materials reflected rigidity in the processes of production.

In the present transition period toward a new techno-economic paradigm, this situation is changing. Competition may still end up with a radical substitution of one material for another (in line with the "common sense" practices of the fourth Kondratiev), but increasingly frequently it is a new complementary association of materials that presents the best technical solution. The new forms of utilization are as varied as the objects to which they are applied. Often, several new materials compete with each other in offering alternative technical solutions for a particular technical device: the variety is both within and between families of materials. This movement toward diversification – or "hyperchoice" – expresses itself in the multiplication of groups, subgroups, classes, grades, and nuances of materials. Never before has mankind had available such an enormous number of materials: for instance, a limited number of basic polymers offer users countless different technical solutions by their innumerable combinations, mixtures, or alloys and by the incorporation of several liquid, solid, and fibrous additives.

The revolution in information technology greatly facilitates the rethinking and production of multi-material objects or of objects made from complex materials. Most often, the use of a new material (or a new combination of materials) involves the complete redesign of the object instead of simple piecemeal substitution. Programmable micro-electronics-based equipment assists the processing of materials that often acquire their final shape and composition within the object itself. The close integration of design and production functions within the firm has made it economically possible to produce objects that are conceived *at the same time* as their constituent materials. Attempts by firms to achieve greater flexibility in product and process design are often associated with the reconception of the object and modification of the materials used, revealing a close correlation between the will to adapt to a changing economic environment, the increase in technical flexibility of productive capital, the introduction of information technologies, and the exploitation of a large variety of materials [9]. This correlation shows that new materials technology is intrinsically coherent with the "best-practice" guidelines of the new techno-economic paradigm, and it expresses the tendency towards a deep transformation of the materials industry, which is increasingly becoming a service industry where producers sell "solutions" to a client's global problem, a sort of "kit" designed to respond to a desired "function," rather than a "material" in the proper sense.

This "functionalization" of the materials market is consequently accompanied by a "tertiarization" of employment in the materials industry, leading to a necessary integration and interfacing of know-how and skills, to a new division of labour both within and among enterprises, and to the emergence of new firms and industries. The multiplicity of variants of the same material that producers and users must learn to handle, as well as the refinement of production and processing methods of materials, require the emergence of the multi-material specialist with fluency in many skills. For instance, the transformation of plastics, which traditionally demanded mechanical knowledge alone, now requires better knowledge of chemistry for design and control of *in situ* reaction techniques, and of electronics and computer sciences for the control of automatic equipment [9].

The competitiveness of a firm is thus more and more determined by the efficiency with which it utilizes new materials. The exploitation of new materials technology has become vital for industry: it has acquired a major economic importance as a generic or "trans-sectorial" technology, spreading into all industrial sectors and affecting the production of innumerable products and services. New materials have become remarkable vectors of innovation, as advances accomplished in a particular industry through the use of a new material tend to "contaminate" one by one all other industrial sectors. They form a new technological system.

The growing variety of materials production techniques is associated with an increase in the complexity of production processes. In their effort to minimize this increasing complexity, firms are compelled to integrate production stages, i.e. to reduce the number of phases of a given process (lower stocks, less maintenance), to reduce the number of parts in the final product (lower assembly costs), or else the production time. The reduction of the number of parts results, in turn, in the integration of several simultaneous functions in the material: the final object is formally simpler, but more complex in its design, in its functions, and in the services it offers. A direct consequence of this tendency is the necessary development of new and efficient non-destructive testing methods to replace the former testing procedures based on sampling.

The best example of integration of different functions in the same material is given by composite materials ("composites" for short). Composites are best defined as "the voluntary association of non-miscible or partly miscible materials having different structures, which combine and complement their characteristics to form a heterogeneous material presenting global properties and performances superior to those of the original constituent materials and suited to required functions" [54].

Composites are generally formed by a matrix in which a different material (usually in the form of fibres) is embedded to reinforce the mechanical properties of the matrix. The most common composite is made of a polymeric matrix (epoxy resins, polyesthers, etc.) and of glass fibres: 95 per cent of the composites used in industry are of this type. Other matrix materials used for superior technical performances (mainly for aeronautic, space, and military purposes, although applications in professional sports equipment and racing cars are increasing) are metals, carbon, or ceramics; high-performance fibres are usually made of carbon, boron, or aramide (Kevlar). The role of the fibre is to absorb shocks and to give the material its mechanical resistance, whereas the matrix serves to distribute the mechanical constraints over the whole structure and to protect the fibres against environmental (mostly chemical) damage. The association fibre-matrix offers innumerable combinations of physical and chemical properties by modifying either the fibre/matrix constituent materials or the "architecture" of the composite. By employing different weaving and orientation techniques of the fibres, it is possible to control the microscopic characteristics of the composite and to obtain combinations of properties that are unconceivable with traditional materials.

Impact on developing countries

For developing countries, which are traditional suppliers of raw materials, the trends described above have both direct and indirect consequences. The direct impact is the decreasing amount of raw materials needed to manufacture a unit of industrial production. The indirect impact, which in the medium term might turn out to be by far the most significant, is the decrease in the technological innovation content of manufactured goods produced by the developing countries, i.e. the competitiveness of their industry.

The declining trend in the per unit consumption of raw materials in the industrialized countries is accelerating, through both substitution of synthetic for raw materials and development of materials-saving processing methods. Substitution still has a relatively minor impact, but it will become important in the long term because of the growing demand for enhanced performances in electronics, communications, information and data processing, transportation, energy, manufacturing, and chemical products. The largest changes are expected to occur in the replacement of metals by ceramics, polymers, and composites. The recent evolution of prices has already been extremely favourable to polymers in comparison with metals: between 1960 and 1987, the average price per unit volume of the main metals has more than trebled, whereas on average the main commodity polymers less than doubled in price in the same period [8].

These trends have strongly affected the international prices of raw materials, which have fallen sharply, with direct negative economic impacts on developing countries. But new materials tend to decrease the value of raw materials in two additional ways. First, the increased weight of processing techniques in the manufacturing process reduces the part of raw materials in the composition of the final value of the product; materials technology, not raw materials, is the main cost factor. In addition, intangible investments in software, marketing, and information technology R&D constitute a growing share of production costs. Secondly, new materials decrease the geopolitical strategic value of raw materials, for technology is able to develop appropriate substitutes.

Although some low-cost producers may be able to increase their shares in the slowly growing world market for traditional materials, developing countries will generally continue to face declining real prices. Some less developed countries may be competitive in the production of some new materials such as fibre-reinforced plastics and inorganic materials, because of the abundance of raw materials required and the labour-intensive character of parts of the production chain. Some Latin American and Asian developing countries have been carrying out research in the area of low-cost construction and building materials, as well as in alloys, polymers, and composites. Rare and rare-earth metals are and will remain essential for many scientific and technological developments, e.g. in superconductors; world markets for these materials are projected to rise many times by the year 2000 and even faster thereafter, and developing countries are their major producers. The development of ceramics is advantageous to Latin American and Asian countries, thanks to the avail-

484

ability of certain mineral resources like copper, iron, carbon, and aluminium. Developing countries in Asia and Latin America that have recently succeeded in micro-electronic components should enter into production of silicon and other semiconductor materials. African countries may develop materials for roads and for low- and middle-income housing in rural and urban areas.

In short, the potential for developing countries to produce materials of higher purity necessary for high technology industries exists and should be exploited. Producers of primary materials should therefore reconsider their policies concerning materials technology development with the objective of shifting the production of traditional raw materials to more knowledge-intensive materials. Some are already doing so.

The main impact of the present trends in new materials is most likely to be felt by developing countries in the medium term, through the loss of competitive power of many of their manufactured products, which will increasingly have to compete with innovative products presenting higher functional integration or offering novel functions and "services," manufactured by "multi-material" firms in the industrialized countries.

References

1   Aglietta, M. *A Theory of Capitalist Regulation: The US Experience*. London: New Left Books.
2   Allen, J., and D. Massey, eds. *The Economy in Question*. London: Sage, 1988.
3   Boyer, R. *The Regulation School: A Critical Introduction*. New York: Columbia University Press, 1989.
4   Brady, N.C. "Agricultural Research and US Trade." *Science* 230 (1985), no. 4725.
5   Bull, A., G. Holt, and M. Lilly. *Biotechnology: International Trends and Perspectives*. Paris: OECD, 1982.
6   Buttel, F.H., M. Kenney, and J. Kloppenburg. "From Green Revolution to Biorevolution: Some Observations on the Changing Technology Bases of Economic Transformation in the Third World." *Economic Development and Cultural Change* (1985), no. 34.
7   Caron, F. *Le résistible déclin des sociétés industrielles*. Paris: Perrier, 1985.
8   Clark, J.P. "Markets for Advanced Materials." *Materials and Society* 13 (1989), no. 3.
9   Cohendet, P., M.-J. Ledoux, and E. Zuscovitch. *Les matériaux nouveaux: dynamique économique et stratégie européenne*. Paris: Economica, 1987.
10 Dosi, G. "Technological Paradigms and Technological Trajectories: A Sug-

gested Interpretation of the Determinants and Directions of Technical Change." *Research Policy* 11 (1982), no. 4.

11   Dosi, G., C. Freeman, R. Nelson, G. Silverberg, and L.L.G. Soete, eds. *Technical Change and Economic Theory*. London: Frances Pinter, 1988.

12   Filho, A.G.A., R. Marx, and M. Zilbovicius. "Fordism and New Best Practice: Some Issues on the Transition in Brazil." In: Kaplinsky, ed. *See* ref. 25.

13   Flores, H.E., M.W. Hoy, and J.J. Pickard. "Secondary Metabolites from Root Cultures." *Trends in Biotechnology* (Cambridge: Elsevier Publications) 5 (1987), no. 3.

14   Fransman, M. *Biotechnology Generation, Diffusion and Policy: An Interpretative Survey*. Maastricht: The United Nations University, New Technologies Centre Feasibility Study, Rijksuniversiteit Limburg, 1986.

15   Freeman, C. *Technology Policy and Economic Performance: Lessons from Japan*. London: Frances Pinter, 1987.

16   ———, ed. *Long Waves in the World Economy*. London: Frances Pinter, 1984.

17   Freeman, C., J. Clark, and L.L.G. Soete. *Unemployment and Technical Innovation: A Study of Long Waves in Economic Development*. London: Frances Pinter, 1982.

18   Freeman, C., and C. Perez. "Structural Crises of Adjustment, Business Cycles and Investment Behaviour." In: Dosi et al., eds., pp. 38–61. *See* ref. 11.

19   Freeman, C., and L.L.G. Soete. *Information Technology and Employment: An Assessment*. Brighton: Science Policy Research Unit, 1985.

20   Gilles, B. *Histoire des techniques*. Paris: Encyclopédie de la Pléïade, Gallimard, 1975.

21   Green, K., and E. Yoxen. "The Greening of European Industry: What Role for Biotechnology?" *Futures* 22 (June 1990), no. 5.

22   Hippel, E. von. *The Sources of Innovation*. New York: Oxford University Press, 1988.

23   Jacobsson, S. "Technical Change and Industrial Policy: The Case of Computer Numerically Controlled Lathes in Argentina, Korea and Taiwan." *World Development* 13 (March 1985), no. 3.

24   James, J. *Microelectronics and the Third World: An Integrated Survey of Literature*. Maastricht: The United Nations University, New Technologies Centre Feasibility Study, Rijksuniversiteit Limburg, 1986.

25   Kaplinsky, R., ed. "Restructuring Industrial Strategies." *Bulletin of the Institute of Development Studies, University of Sussex* (1989).

26   Katz, R.L. "Explaining Information Sector Growth in Developing Countries." *Telecommunications Policy* (September 1986).

27   Kondratiev, N.D. "The Long Waves in Economic Life." *Review of Economic Statistics* (November 1935). Originally published in Russian in 1925.

28   Lipietz, A. "Trois crises: métamorphoses du capitalisme et mouvement ouvrier." *Cahiers CEPREMAP* 8528 (November 1985).

29   Mandel, E. *Long Waves Capitalist Development: The Marxist Interpretation*. Cambridge: Cambridge University Press, 1980.

30   Mensch, G. *Stalemate in Technology: Innovations Overcome the Depression*. New York: Ballinger, 1979.

31   Miles, I., H. Rush, K. Turner, and J. Bessant. *Information Horizons: The Long-Term Social Implications of New Information Technologies*. Hants, U.K.: Edward Elgar Publishing Ltd, 1988.

32   Munasinghe, M., M. Dow, and J. Fritz, eds. *Microcomputers for Development: Issues and Policy*. Cintec-Nas Publication, 1985.

33   Nelson, R.R., and S.G. Winter. "In Search of a Useful Theory of Innovation." *Research Policy* 6 (1977), no. 1.

34   ———. *An Evolutionary Theory of Economic Change*. Cambridge, Mass.: Belknap Press, 1982.

35   OECD. *Biotechnology: Economic and Wider Impacts*. Paris: OECD, 1989.

36   ———. *Advanced Materials: Policies and Technological Challenges*. Paris: OECD, 1990.

37   OTA (Office of Technology Assessment). *Commercial Biotechnology: An International Analysis*. OTA-BA-218, January. Washington, D.C.: Government Printing Office, 1984.

38   ———. *Technology, Public Policy and the Changing Structure of American Agriculture*. Washington, D.C.: Government Printing Office, 1986.

39   Perez, C. "Structural Change and Assimilation of New Technologies in the Economic and Social Systems." *Futures* 15 (October 1983), no. 4.

40   ———. "Microelectronics, Long Waves and World Structural Change: New Perspectives of Developing Countries." *World Development* 13 (March 1985).

41   ———. "Las nuevas tecnologías: una visión de conjunto." In: C. Ominami, ed. *La tercera revolución industrial: impactos internacionales del actual viraje tecnológico*. Buenos Aires: Grupo Editor Latinoamericano, 1986.

42   ———. *Technical Change, Competitive Restructuring and Institutional Reform in Developing Countries*. World Bank Strategic Planning and Review Department Paper no. 4. Washington, D.C.: World Bank, 1989.

43   Perez, C., and L.L.G. Soete. "Catching Up in Technology: Entry Barriers and Windows of Opportunity." In: Dosi et al., eds. *See* ref. 11.

44   Piore, M.J., and C.F. Sabel. *The Second Industrial Divide: Possibilities for Prosperity*. New York: Basic Books, 1984.

45   Rada, J. "Information Technology and the Third World." In: T. Forester, ed. *The Information Technology Revolution*. Oxford: Basil Blackwell, 1985.

46   Rosenberg, N. *Perspectives on Technology*. Cambridge: Cambridge University Press, 1976.

47   ———. *Inside the Black Box: Technology and Economics*. Cambridge: Cambridge University Press, 1982.

48   Sasson, A. *Biotechnologies and Development*. Paris: Unesco, 1988.

49   Sasson, A., and V. Costarini. *Biotechnologies in Perspective*. Paris: Unesco, 1991.

50   Schumpeter, J.A. *Business Cycles: A Theoretical, Historical and Statistical Analysis of the Capitalist Process*. New York/London: McGraw-Hill, 1939.

51   ———. *Capitalism, Socialism and Democracy*. London: Allen & Unwin, 1943.

52   Soete, L.L.G. "International Diffusion of Technology, Industrial Development and Technological Leap-frogging." *World Development* 13 (March 1985), no. 3.

53   Stewart, F. "Technology Transfer for Development." In: R. Evenson and G. Ranis, eds. *Science and Technology: Lessons for Development Policy*. Boulder, Colo.: Westview Press, 1990.

54   Thaller, R. "Pour une économie de la diffusion des innovations technologiques: l'exemple des matériaux composites." Ph.D. diss., Lyon: University of Lyon II, 1986.

55   Tigre, P.B. *The Mexican Professional Electronics Industry and Technology*. Vienna: UNIDO/IS, 1985.

56   UNCSTD (United Nations Committee on Science and Technology for Development). "End-of-Decade Review, Implementation of the Vienna Programme of Action, State of Science and Technology for Development: Options for the Future." Background paper for tenth session of the Intergovernmental Committee on Science and Technology for Development. August 1989. Mimeo.

14

Technology assessment

Harvey Brooks

Technology assessment was created in the United States as an aid in identifying and weighing the existing and probable impacts of technological applications on the natural and social environment. In the industrialized countries, the term covers activities variously described as technology assessment, environmental impact analysis, traditional policy analysis, systems analysis, operations research, social assessment of technology, technological forecasting, programme evaluation, risk analysis, or cost-benefit studies. In developing countries, technology assessment is viewed with considerable interest and hopeful expectations, primarily because of its potential in the context of development and modernization. Whereas in the industrialized countries, TA is viewed predominantly (though not exclusively) in the context of anticipating and avoiding unintended side-effects of technologies as their scale of application increases and spreads, in developing countries it is seen more as a means of building up an indigenous capability for wise technology choice in the context of a more autonomous and self-reliant development strategy [22]. While the need to monitor and control technology from a societal perspective is well recognized in developing countries, this is seen more from the point of view of avoiding the unwanted socio-economic effects of imported technologies controlled by foreign corporations, and of mastering new and emerging technologies with potential applications to development. The creation of the ATAS Bulletin published by the United Nations Centre for Science and Technology for Development is an example of this type of activity.

With the rising importance of global environmental problems, both the developed and the developing countries perceive a new mutual interest in a widely dispersed TA capability that will help avoid exacerbating these problems by non-sustainable technology choices and development strategies in the developing countries.

Historical background

Rosenberg and Birdzell [17] have shown how the industrial development of the West depended on social and technological experimentation and social learning through strong feedback mechanisms from the market. In the earliest period of the Industrial Revolution, the potential rewards for technical and managerial innovation were high and the penalties for failing to innovate were correspondingly severe. However, the risks and costs arising from the unforeseen consequences of innovation were generally treated as acts of God and were expected to be borne by society at large. Thus the substantial risks of technical or market failure borne by the innovator were not augmented by the additional risk of having to compensate society or third parties for unforeseen adverse effects, such as environmental pollution, industrial accidents, failure of companies, or displacement of labour, unless gross negligence could be shown, and the burden of proof was on plaintiffs to show both injury and fault in the case of health and environmental effects [19]. There was a general expectation that the totality of innovations elicited by the high potential rewards would have an aggregate net benefit to society far exceeding the aggregate net cost to society associated with particular adversely affected groups or individuals. This was just the "price of progress." In this situation technology assessment was important, but it focused on technical feasibility and market potential, not on possible adverse effects on people who were not direct parties in the production and consumption of the goods or services resulting from innovation. Technology assessment was the responsibility of the entrepreneur and was primarily directed at reducing entrepreneurial risk arising from technical or market failure through the exercise of better technical and economic foresight.

Even when there were adverse effects either on the environment, health, safety, or social welfare, they were expected to be short term, local, and reversible. It was thought that oversights could almost always be remedied by a technical fix at modest cost, so there was not much point in looking for dangers that, more likely than not, could be easily overcome by human ingenuity as they were encountered.

490

The development economist, Albert Hirschman, pointed out in a famous article [9] that development projects in developing countries often fail to recognize many difficulties and obstacles to success, but equally they underestimate the power of human ingenuity to surmount such obstacles, so that most projects that would not have been undertaken if the difficulties and side-effects had all been foreseen, nevertheless end by being successful because of human ingenuity and adaptive innovation. A similar belief underlay the attitude toward technological innovation in most developed countries until the mid-1960s. Indeed, in the nineteenth century, innovators were virtually free from the risk of being held accountable to society for the unforeseen social costs entailed by the widespread application of their innovations.

Thus the dominant attitude in most industrial societies of the North until very recently has been that technology should be deemed innocent until proved guilty beyond a reasonable doubt. Among most élites of the developing world, technology is generally seen as the only realistic avenue of escape from the growing disparity in power and living standards between the developed and developing countries. Indeed, if they have had any concern with the social costs of technical progress, it was more with the social costs of unemployment and displacement of labour than with environmental and health "externalities."

The rise of global problems
Throughout most of human history, societies that despoiled their environment or exhausted their mineral resources eventually declined and were lost to sight, while civilization and power reappeared in new locations where nature had been less exploited. Today, however, human society exists on a truly planetary scale, and we are in the midst of trying to sort out what this means for the sustainability of civilization as we have known it. There are fewer and fewer unexploited resources or undegraded environments in which civilizations can be reborn, allowing humanity to progress by simply abandoning its past mistakes. Thus the experimentation and social selection described by Rosenberg and Birdzell has depended historically on geographical isolation of human communities, which tended to prevent mistakes of individual institutions or nations from dragging the whole world down together.

Since the 1960s there has been a sea change in consciousness with respect to technical and material progress. For the first time in the history of humankind on the planet, human activities are becoming a major natural phenomenon in their own right comparable in potential to the natural phenomena that have altered the face of the planet in geological history. In absolute terms, population, economic activity, and technical change are still increasing, and there is a widespread recognition, at least among élites, that something has to give. The only point debated is about how soon and in what way. This new consciousness has spread from the developed world to many élites in the developing world in less than 20 years.

In fact the shift towards a more cautious attitude toward technical progress arises from a combination of two factors: growing evidence of real environmental deterioration, and rising expectations for environmental quality among those societies or parts of societies where basic material needs have been met and that have become increasingly aware of "externalities" associated with the process of economic growth. Paradoxically, much of the evidence of environmental deterioration has come from progress in science – from better understanding of natural processes and from advances in the ability to detect ever smaller traces of pollutants in the environment. For example, it was only in the early 1950s that the photochemical effects of sunlight on the gases emitted in automobile exhausts were first shown to be the main cause of atmospheric smog in the Los Angeles Basin. It was not until the 1970s that the upper atmospheric chemical reactions leading to the depletion of the life-protecting stratospheric ozone layer began to be unravelled, implicating a man-made chemical originally introduced widely because of its chemical inertness and complete lack of toxicity to humans and animals – namely the chloro-fluorocarbons (CFCs). The man-made chemical compound DDT was discovered during the Second World War to be a powerful insecticide, apparently harmless to humans. Introduced for the purpose of eliminating disease-bearing insects, such as the lice-carrying typhus, it was hailed as a miracle technology that was credited with saving nearly half a billion lives, mainly in the developing world. Yet, when applied on an even larger scale in agriculture, it was eventually found to cause such ecological damage that its use was largely banned in developed countries [23, 6].

Other chlorinated hydrocarbons were introduced that were believed to be less ecologically damaging because less persistent in the environment, but they also proved more hazardous to human health, especially in the developing world, and began to turn up in drinking water supplies and soils. Today, the use of chemicals in agriculture is

492

in increasingly bad repute because of both its cumulative ecological effects and its declining effectiveness due to adaptive evolution of the target pests. Most of these hazards were not fully anticipated, in part because, when first introduced in small amounts, the substances appeared to be relatively harmless and their benefits were immediate, obvious, and demonstrable. At the same time, nobody foresaw the rapid growth in their scale of use that their benefits to the producers would elicit, nor the new secondary ecological effects that large-scale use would generate.

Not only did scale of use produce unanticipated effects, but many chemicals introduced in small amounts constituted a "time bomb," with latent health effects that did not appear until many decades after the initial exposure of people to them. One of the most dramatic examples was occupational exposure to asbestos, much of which occurred during the Second World War, with fatal health effects appearing only many decades later. It became conceivable that a large population could breathe or otherwise ingest large quantities of a substance for years with no adverse effects, only to show up with delayed cancers or other diseases in near epidemic proportions many decades later. There were also many examples of systemic effects occurring only at the end of a long causal chain, in which the original triggering event was obscure. This also led to increased separation in both space and time between the people who enjoyed the benefits of a new technology and those who were exposed to the risks or bore the ultimate costs, even extending across generations. In all these instances, dramatic examples or anecdotes were far more influential on public opinion than statistical averages, and thus became an important factor in shifting the burden of proof against the introduction of new technology.

The response

The 1960s in most of the Western industrialized world were a period of dramatic technological and social change accompanied by a shift towards the more sceptical view of technical progress already mentioned [20]. According to one author writing during this period [2], the adverse second-order consequences of well-intended enterprises had long been recognized by historians and social critics but had tended to be regarded as inevitable and uncontrollable. Indeed, the beneficial technological by-products of even such undesirable activities as wars had frequently been commented on. However, it was said that "it is a mark of our times that we no longer accept a conclusion that we can do nothing about the unwanted consequences of our actions, or that a criterion such as 'progress' in technology is *per se* justification for imposing unpleasantness on ourselves." In fact, "particularly in the past decade, the American public seems less willing that either of these sorts of second-order consequences [i.e., beneficial or adverse] should be left to chance" [2, p. 1]. This, in essence, is the underlying motivation for the technology assessment movement as it grew up in the United States in the 1960s, and is still its main justification today. In the context of a developing country, the justification would be similar, but not identical, because a developing country would be following a different technological trajectory in a different social and cultural milieu. Therefore, technology assessments performed in industrialized countries are not necessarily readily applicable in developing countries, even when the purely technical features of the technology being assessed are similar.

Institutionalization of technology assessment

The decade of the 1970s saw a proliferation of laws and regulations throughout the industrialized world, but particularly in the United States, designed to control the social costs of economic growth and of the introduction and diffusion of technology. The objectives of these laws, as expressed in their preambles, were often extraordinarily ambitious and presumed a scientific knowledge base and an analytical capability both inside and outside government to monitor, anticipate, and assess the effects of technology, although they did not yet exist. Because tight deadlines were often imposed, regulation tended to favour "end-of-pipe" curative technologies rather than potentially more cost-effective but more long-term, high-risk-preventive ones. Slow economic growth in the industrialized world reinforced this preference since such technologies would not require replacement of incompletely depreciated plant and equipment that could not meet the new standards.

In developing countries, there would be a greater incentive to invest in preventive technology, since new production capacity would have to be built anyway and it probably would be cheaper overall than buying obsolete plant with end-of-pipe retrofits – perhaps even cheaper in its own right in some instances.

Two pieces of US legislation have influenced the social and environmental assessment of technology elsewhere: the National Environmental Policy Act of 1969 (NEPA), which introduced environmental impact statements (EIS), and the Technology Assessment Act of 1972, setting up the Office of Technology Assessment.

ENVIRONMENTAL IMPACT STATEMENTS. An Environmental Impact Statement is required from every agency of the Federal government to accompany recommendations or reports on proposals for legislation and other major Federal actions significantly affecting the quality of the human environment, with an analysis of the effect of any alternatives to the proposed action, including inaction. The draft EIS is then subject to comment by other Federal agencies and from the public where practicable. The question of public hearings and comments is often contentious because there can be a very fine line between representation of legitimate substantive concerns and the use of the EIS procedure and lead time for comments as a blocking tactic by groups objecting to the action on other grounds. The judgement as to whether this line has been crossed is subjective, depending on the values of the observer.

Many aspects of the EIS procedure are peculiar to the American political and legal system, in particular the uniquely important role that judicial review of administrative actions plays in the US regulatory system, severely limiting administrative discretion in the interpretation and enforcement of the laws passed by Congress. Nevertheless, some procedure of this type would be desirable for most developing countries. For one thing, donors of international aid – including both multinational agencies such as the World Bank and regional development banks – are increasingly imposing some sort of environmental review on investment projects, and it would be desirable for recipient countries to develop their own procedures that help ensure that local cultural and social priorities are properly taken into account in such reviews. To the extent that recipient countries demonstrate their own capabilities for technology assessment or environmental impact assessment, they are less likely to be second-guessed by donors and more likely to maintain control over their own development priorities.

A particular problem with the EIS procedure as implemented in the United States is that Congress has provided very little guidance to the agencies as to the relative weights to be given to various factors in arriving at a decision concerning the intended agency action; this difficulty is intensified by the fact that the responsible agency decision maker is often a proponent of the project being assessed. While the procedure has seemed to imply a kind of overall cost-benefit analysis – a balancing of aggregate costs against aggregate benefits for all affected interests – this has never been explicit, nor is there any guidance as to how distributional considerations are to be taken into account when benefits and costs or risks are experienced by different groups – a situation increasingly frequent as the scale and time span of effects have expanded [5, p. 146]. This lack of guidance on the relative weighting of factors – including short-term vs. long-term, intangible vs. economic, and winners vs. losers – often means in practice that environmental impact statements become catalogues of every conceivable effect with little evaluation, let alone quantification, of their relative importance, particularly because legal challenge was most likely to be based on omissions rather than commissions.

THE OFFICE OF TECHNOLOGY ASSESSMENT. The Office of Technology Assessment was set up in 1972 in response to the felt need in the Congress for independent, non-partisan technical advice drawing on the best available scientific, engineering, and other expert knowledge inside and outside government. The focus was on assessing alternative courses of action that Congress might take where scientific and technological considerations were heavily involved. Although the techniques were developed in the US context, with its sharp separation of legislative and executive powers, I believe they are widely generalizable and could be used effectively in many other settings, including some appropriate to developing countries.

In fact, the American pattern of an OTA attached to the legislative branch has been followed in several countries, especially in Europe, with various institutional differences related to the specific historical and constitutional context. For instance, in France the Office parlementaire d'évaluation des risques scientifiques et technologiques has also been designed to be a bridge between the Chamber of Deputies and the Senate, and is a non-partisan body made up of equal numbers of representatives from the parliamentary majority and the opposition. In this case, each report is prepared and signed by a representative of either the Senate or the Chamber of Deputies, with the assistance of the secretariat (for example, on nuclear waste, biotechnology, high definition TV, etc.), and is discussed before publication by the parliamentary committee covering the field.

In the United States, the research agenda is usually formulated through a fairly extended negotiation involving several Congressional committees, the Technology Assessment Board (TAB) of the Congress, and the Director of the OTA. Since many more studies are requested than the OTA can carry out, efforts are made to consolidate and reconcile requests from more than the one committee, thus ensuring that several committees feel a sense of "ownership" of the study and will be ready to take its conclusions and recommendations seriously. The staff of the OTA may take a good deal of initiative in recasting and sharpening the questions and in formulating problems in a way that enhances the non-partisan stance of the Office, so as to increase the credibility of the results to members of Congress with widely differing policy perspectives and constituency interests.

Once the final definition of a study has been formally approved by the TAB, a project advisory committee is appointed – usually 20–30 outside experts and members of the public representing a wide range of expertise and political views, chaired by an experienced person. The members of the committee bear no responsibility for the content of the final report, which is prepared by the OTA staff, but they have frequent opportunities to review and comment on drafts of sections relevant to their individual interests and expertise. A great effort is made to understand and explain the reasons for any differences of opinion, whether they be technical or derived from differing value-perspectives. This is done even though some points of view may be rejected, with reasons explained, in the final report.

On average, about half the substantive research going into a report is done "in-house" by staff and half is commissioned from outside consultants. In fact, the staff usually does not do original research, apart from reading the literature and interviewing experts; rather the purpose of the OTA studies is to synthesize, distil, and interpret existing knowledge and recast it in a form so as to fit a specific policy context. The commissioned reports are used as raw material for the final report, but the OTA staff is under no obligation to use them or to adopt their conclusions.

As a report nears completion, drafts of all or parts of it are sent to outside reviewers – as many as 100 or more – for comment and criticism. Although confidentiality is requested of reviewers, it is not always respected, but this has generally not caused problems. Frequently leaks have served to alert the Director and the TAB to political sensitivities, or errors of fact or interpretation, before the report is approved for public release by the TAB.

The non-partisan nature of an OTA report is stressed by casting the final recommendations in the form of a list of options for possible Congressional action, usually with options to satisfy both the cautious

497

and the activists. This studied neutrality is sometimes frustrating to members of Congress who would prefer stronger or more definitive policy recommendations, but it probably serves to enhance the credibility of the OTA in the long run. The OTA operates less by picking the best policy options than by screening out those that cannot be justified by the evidence at hand, while still leaving open for debate a number of conclusions and recommendations that do not fly in the face of the evidence or the best professional opinion that is available, although they may express different value judgements.

OTA reports are published in essentially three forms: a one-page summary of conclusions and recommendations intelligible to laymen; an executive summary of about 50 pages synthesizing the principal conclusions, recommendations, and arguments; and a full report – sometimes in many volumes – collecting essentially all the work that has been done in a form that can be used as an authoritative reference by Congressional professional staff and by outside specialists concerned with the problem or issue. In addition, the Directorate of the OTA and the staff who prepared the report are available for testimony to Congressional committees and for one-on-one meetings with Congressional staff or members of Congress or other government officials. Thus there is a rather extensive effort to disseminate the results of a study beyond the mere publication of a report.

The methodology and its critics

Technology assessment started by examining the technical characteristics of a given technology, such as the automobile or nuclear power, and then attempting to explore all the possible social, economic, environmental, health, and ecological effects of its application. This simple definition runs into difficulties, however, since it implies a notion of technological determinism, a unidirectional causality from technology to society. The social impact of a technology – indeed its environmental impact as well – depends on the social supporting systems and ancillary or supporting technologies that accompany its large-scale deployment. These ancillary systems may well be different in different societies and political systems, as in the case of television broadcasting. There is therefore the question of whether the term "technology" in TA refers just to a single artefact or whether it refers to the whole system of ancillary technologies and social supporting systems actually used in connection with the widespread deployment of the dominant artefact.

498

A great deal of effort and debate have gone into the methodology of TA. The field has been criticized as "non-paradigmatic" and, by inference, therefore not cumulative [22, p. 7]. In a survey of actual TA projects in the United States, Rossini et al. [18] found that TA practitioners seldom used any of the quantitative techniques that had been widely advocated in the theoretical literature. Wad and Radnor point out that in fact specialists "have a disdain for [quantitative techniques], preferring to rely on their own judgments and intuition in the selection of approaches and in the design of the TA" [22, p. 38]. This has implications for the use of TA in developing countries. As observed by Wad, "if the techniques of TA receive scant attention from practitioners within the very society from which TA evolved as a body of knowledge, it is very questionable whether they would have much relevance in other, quite different societies" [22, p. 39]. This criticism of pure technique could be an advantage if it makes TA more accessible to societies in which sophisticated knowledge is in short supply. On the other hand, a well-defined paradigm would offer a common language, making it easier to transfer TA across cultures than is the case with intuitive practices.

Another criticism is that the evaluation of technology in a society, and indeed all so-called "objective knowledge," is primarily a reflection of the power interests of various social groups and the resulting "imperatives for the reproduction and legitimation of existing social structures and [power] relationships" [7, 25, 26, p. 108]. Thus seemingly technical debates about the choice or regulation of technology are nothing but political power struggles in which science is just another instrument. There is some element of truth in this view in that political and cultural biases, often heavily weighted with perceived self-interest, can never be completely expunged from discussion of "science for policy." However, Laudan [10] argues that the role played by such social and political factors varies inversely with the uncertainty and immaturity of a scientific field. As evidence accumulates and a field matures, rationality becomes more and more of a constraint on social construction. The assessment depends on both the scientific uncertainty and the political stakes and power of the various players. The critics thus maintain that science is largely if not wholly irrelevant to actual policy outcomes. They may be correct in the implication that policy assessment and implementation, no matter how technical in nature, must be sensitive to the distributional implications of the conclusions reached, and that analysts must try to anticipate the effect of this on the implementability of their recommendations. However, this must not be taken to mean that either science or policy analyses are valueless.

A typology of technology assessment and policy analysis

Since in practice it is so difficult to separate a technology from the socio-technical system in which it is embedded, it is convenient to develop a typology of TA classified along a great many dimensions.

Dimensions of technology assessment

1. The degree of specificity of the category being assessed, e.g. a specific project such as a dam, dams in general, or earth dams in particular.
2. The scope of the system included in the assessment, e.g. the automobile as an artefact, the automobile with its whole set of supporting systems including highways, oil refineries, gasoline stations, auto insurance companies, repair shops, auto factories and the impact of all of these together on society and/or the environment.
3. The degree to which the assessment is restricted to hardware and technical characteristics, e.g. emissions, safety in a collision, fuel efficiency, driving performance, and excludes legal systems, insurance, policing, licensing systems, etc.
4. The scope of the types of impacts considered, e.g. environmental, health, safety, ecological, economic, social, psychological.
5. The geographical and temporal scope of impacts considered, e.g. local, regional, short-term, intergenerational impacts, long-term ecological productivity, etc.
6. The degree to which the likely political and behavioural responses to alternative policy prescriptions for the socio-technical system being assessed are explicitly considered.
7. The degree of "neutrality" aimed at in the assessment, e.g. whether the assessment is designed to gather evidence in support of an already chosen policy, to evaluate and compare the consequences of alternative policies, or to explore in as value-free a way as possible the probable or possible consequences of a continuation or likely evolution of existing trends.
8. The stage of development in the "life cycle" of the technology being assessed, e.g. whether it is in the R&D stage, whether it is already beginning to be deployed, or whether it is already deployed on a large scale and the challenge is to regulate or alter it

500

to reduce its secondary impacts [21]. This dimension, of course, greatly affects the degree of foresight required and the resultant uncertainties and the kind of evidence that can be assembled.

Types of technology assessment
Taking into account primarily the first three dimensions listed above, one can distinguish five types of technology assessment.

PROJECT ASSESSMENT. Here we are concerned with a concrete project such as a highway, a shopping centre, an oil pipeline, or the actual plan for construction and testing of a prototype of a new aircraft or power plant. The environmental impact statement process most often deals with such specific projects. Project assessments may be further subclassified according to the novelty or extent of previous experience with the technologies to be employed, or the degree of previous experience with the particular type of environment in which they are to be deployed. To the extent that a project presents special challenges either because of the novelty of the technology used or of the unprecedented problems of a new environment (such as was the case with the Alaska pipeline), project assessment may spill over into the next category, generic technology assessment, which must rely more on theoretical insights derived from science and less on cumulative practical experience with the technology in operation.

GENERIC TECHNOLOGY ASSESSMENT. Here the focus is on a general class of technologies without reference to a particular project or a particular site, environment, or social setting. An example of an especially important and common class of generic TAs is the assessment of medical therapies or prescription drugs for the treatment of particular medical conditions [11]. This sort of TA has come into prominence, particularly in the United States, as health care costs have continued to rise faster than the cost of living index in most of the industrialized countries, directing more and more attention to the proliferation of new medical technologies and their cost-effectiveness. In this case the system boundary is well defined, and the generic nature of the technology is obvious because it is applied repetitively under similar circumstances in many places to many different patients with similar characteristics.

Another example might be a new generation of "inherently safe" nuclear reactor designs being proposed for the next generation of nuclear power plants. Here the boundaries of the system to be considered are much less clear because nuclear power is a systemic "global technology" in which adverse experience anywhere in the world has repercussions throughout the entire system due to political reactions, and the safety and reliability of the system as a whole is very sensitive to non-technical human factors of management and regulation as well as to essential supporting technologies such as waste storage and disposal, and support of the nuclear fuel cycle. Generic TA is what one mostly has in mind when one uses the term TA, yet only a relatively small proportion of the reports produced by the OTA or other similar agencies could be said to conform to this description of generic TA.

PROBLEM ASSESSMENT. Here the approach is to examine a broad problem area such as commercial air transport and assess a variety of technologies as well as non-technical measures that might be used to cope with the problem. For example, instead of assessing a supersonic transport programme such as was proposed by the Nixon administration in the early 1970s, one might have posed the problem of future air transportation needs and considered a variety of aircraft types as well as air traffic control systems for meeting a defined social need. Indeed, a more sensible approach might be to extend the boundaries of the problem to define it as enhancing the mobility of goods and people, and including various ground transportation technologies as well. Even in project assessment, the EIS procedure requires that "alternatives to the proposed action" be fully assessed, implying consideration of alternative technologies or social actions that would achieve the same objectives as the particular project being proposed.

POLICY ASSESSMENT. Policy assessment is very similar to problem assessment, except that it takes greater account of non-technological alternatives to achieving social goals for whose realization new technology is only one of many options. A good example might be the use of various kinds of economic incentives to both electricity consumers and public utilities to reduce peak or total demand for electricity as an alternative to constructing additional power plants, or the development of more efficient or environmentally benign generation or transmission technology. Policy assessment blends imperceptibly into policy analysis, where the emphasis shifts more completely away from technology towards broader social and political measures that require less prescriptive design. An important advantage of policy assessment over generic TA or problem assessment is that it tends to be more even-handed as between technical and non-technical solutions to the problems being addressed. At the same time it is more likely to recognize that the overly conservative regulation or suppression of new technology is just as likely to have unforeseen and undesirable side-effects as the introduction of technology – once again leading to a more even-handed balance between technical and non-technical approaches [8, 24, pp. 91–94].

GLOBAL PROBLÉMATIQUE. When a number of closely interrelated social, political, economic, and technical problems coexist and are difficult to attack piecemeal, and the resulting cluster of problems affect the world as a whole considered as a single system, we call the assessment required a "global problématique." What makes it more challenging than other forms of technology assessment is the close interconnection among many of the component problems, and particularly in the interaction between the technical and political dimensions of the environmental risks concerned. What makes the "problématique" different from other forms of TA or problem assessment is that no single scientific report, no single decision, and no single nation will have the last word, or even a very important word, on how humanity ultimately comes to terms with those risks. The management of the problématique has to be a cumulative process of "social learning" with, ultimately, very wide participation of virtually all the stakeholders.

In the 1970s a fashion arose for developing computer models of the entire world designed to assess trends in and effects of various combinations of policies on food, energy, environment, population, natural resources, and even human relations. Models add no new raw information and are no better than the data and assumptions that go into them; they are nevertheless a valuable accounting device for keeping track of many more variables than can be embraced by the human mind. Such models have attempted to assess socio-technical systems of increasingly comprehensive scope until in some cases, the whole world is treated as a single coupled system.

This interest in modelling as an aid to policy-making was part of the motivation for the creation of the International Institute for Applied Systems Analysis (IIASA) as a joint East-West research institute in Laxenburg, Austria, in 1972. Among its major projects, IIASA created computer models of the world energy system [1] and the world system of agricultural production and trade. One aim of

503

such models was to explore the consequences of various national and world policies in these sectors on a global basis, e.g. the potential role of various energy sources, or the effects of more open world trade in agricultural products on the world hunger problem. There was a great deal of debate as to whether the many simplifying assumptions and subjective judgements that had to be made for the models to be manageable would largely vitiate their usefulness as policy tools.

Stakeholder participation in technology assessment

In current discussions of development stategies in developing countries, there is much interest in ways of ensuring adequate participation of "stakeholders" in arriving at social decisions about both technology choices and more broadly about alternative development paths. The term "stakeholders" covers suppliers and users of technology, including external donors or multinational companies as well as many different internal groups potentially affected by possible unforeseen side-effects of the policies chosen. Much can be learned about the *processes* for stakeholder involvement from recent experience in industrialized countries where the tradition and practice of public participation has come a long way in the last two decades, and is still evolving [12, 19].

For stakeholder participation to be more than symbolic, decision makers must be genuinely willing to allow others a say, to encourage both public understanding of the (sometimes complex) issues and constructive public debate. Indeed, one of the most debated issues regarding the social management and control of technology is the degree and form of public participation in decisions about its development, deployment, and regulation [15]. There is always the fear that, the greater the level of participation the greater the risk that any single group that perceives its particular interests or values to be adversely affected by the application of technology will be able to exercise a de facto veto over a technical enterprise almost regardless of the consequence for other affected interests or values. An intensely felt opposition from a small minority might outweigh the diffuse, unmobilized interests of a large majority. The opposite could also occur if the majority has an effective and well-connected political advocate; i.e. the legitimate rights of a minority could be overridden by a well-represented majority.

Encouraging stakeholder participation means sharing information at an early stage in the proceedings and not when it is too late to alter the outcome. Three broad categories of government response can be distinguished: some try to give the public better access to information; some try both to improve the information available to decision makers and to find out how the public feels about issues; others encourage more direct public involvement in decision-making. Examples of the first category are the Swedish study groups, or the German and Austrian public information campaigns. Many countries have established consultative bodies that fall in the second category (for instance the French Commission Informatique et Libertés or the Collège de la prévention des risques technologiques); the European Community has its FAST programme for forecasting and evaluation, and both the European Parliament and the Council of Europe have means of educating their members and the public on technological issues. The most radical approach is to hold judicial hearings or public referendums, as in Austria, Denmark, Italy, Norway, and Switzerland. The three methods obviously overlap and may be combined in various ways depending on the topic of .concern, the size as well as the political and cultural traditions of the country. None of them will deal with all the problems or bring an end to all controversy, but at least they remove some of the uncertainties.

Features for successful stakeholder dialogues
The following characteristics, based on the United States' experience, are important features of any technology assessment institution, as, for example, one designed to inform the development process within a country or within a multilateral development organization.

1. The system represents a compromise between the objectives of delivering its products within a reasonable time and securing a wide diversity of both expert and public inputs. The project advisory panel system does not allow for registration of all political interests, but it does help to ensure a broader set of perspectives than would be considered in a more strictly expert government "in-house" process. It serves to moderate and temper a technocratic process without undermining its intellectual integrity.

2. The relationship of the OTA to the substantive Congressional committees and its accountability to a non-partisan Technology Assessment Board ensures greater political relevance, and the framing of issues in a way more likely to meet the needs of the policy pro-

cess and secure a better match between the formulation of issues as perceived by the public and as perceived by experts.

3. An agenda that is negotiated iteratively between the OTA staff and the professional staffs of individual Congressional committees ensures a greater commitment to the resulting study agenda by both experts and politicians, and also tends to create a receptive audience for the resultant work products.

4. The OTA process is not necessarily geared to concrete action by either the legislature or the executive, so that it is difficult to prove how much it actually influences national policy. Such effects as there are tend to be gradual and cumulative and to occur as much through defining the terms of debate as through the direct impact of specific recommendations or policy options. However, this result is probably more consistent with the true nature of the policy process, which is more a cumulative "social learning" process than a single transaction between experts and society.

5. An early warning function of identifying problems and issues before they were high on the political agenda was one of the original purposes of the OTA. In practice this is somewhat diluted by leaving the initiative for requesting studies in the hands of Congressional committees whose interest generally signals that the issue has already arrived on the political agenda, although this disadvantage tends to lessen over time as mutual trust between the Congress and the OTA director and staff grows with experience and continual interaction.

6. The OTA process probably owes its comparative success partly to the good match of its organization and procedures to the separation of powers in the US Constitution, which in recent decades has meant that Congress has acquired a professional staff capable of assimilating and processing a great deal of specialized information and of reinterpreting it for its political masters. This makes it rather expensive and perhaps less adaptable to parliamentary systems that do not have large staff resources. On the other hand, the greater congruence of political priorities and values between the executive and legislature in parliamentary systems may make this problem less serious than it might appear at first.

7. The OTA staff has a high level of professional competence combined with a generalist orientation that does not instinctively look for answers restricted to its own domain of expertise, and is able to appeciate and make effective use of expertise outside its own boundaries. This enhances the match between the expert policy process and the political process.

Concluding remarks

The motivation for technology assessment would be somewhat different for a developing country than for a fully industrialized country, depending on its particular stage of development. In the first place, a developing country would face some of the cumulative environmental and resource depletion consequences of the previous development of the industrialized world, e.g. global environmental impacts such as global warming and stratospheric ozone depletion, or the exhaustion of the most accessible reserves of non-renewable resources. Also, in many cases the local environment is already overstressed by population densities and degrees of urbanization that have no precedent in the earlier history of the present industrialized world at a corresponding stage of its historical development. Offsetting this is the fact that the presently developing countries have access, at least potentially, to many modern technologies, such as biotechnology and information technology, that may allow them to bypass many of the development stages that the industrialized world had to go through. Furthermore, the expectations of these populations for quality of life, especially in terms of environmental amenities (as opposed to necessities for sustaining life, or carrying capacity) are and will remain considerably lower than those of industrialized country populations, at least for a generation or so. On the other hand, environmental quality necessary to sustaining productivity of the biosphere for human use is likely to be a more urgent requirement in developing countries than in developed countries.

What this means is that the concept of "environmentally sound technology" or even sustainable development may mean something quite different in developing and developed countries, with productivity considerations playing a larger role in developing countries and ethical, aesthetic, and psychological considerations playing a larger role in the industrialized countries. Since a larger proportion of economic activities in developing countries than in industrialized countries are sensitive to the state of the environment, TA may be more of a *necessity* for sustainable economic growth in developing countries than it is for already industrialized countries, though the capacity for it is less. Whereas failure of foresight may lead to loss of quality of life in industrialized countries, the same lack of foresight could lead to social and economic catastrophe in a country at an earlier stage of development.

Increasingly, scientific matters are becoming dependent on government support and policies – and so are no longer the researchers' pre-serve. Science and technology have become both more noticeable be-cause of their repercussions and more open to scrutiny because of the public money invested in them. Public opinion is now more aware of government influence on the direction of national R&D efforts and of the role government can play in regulating technical change. Just as some national commissions have been set up in certain countries to examine the relationship between information technology and priv-acy, others have been created to discuss the ethical issues of bio-technological research. For example, in France, the mandate of the National Committee on Medical Bio-ethics is "to discuss the major moral problems raised by biological, medical and health research, whether it concerns the individual, specific social categories or society as a whole."

Many governments have taken on the new task of carrying out the research and analysis that is needed for technology assessment in the full sense, and in fact more energetically than has the private sector on its own with only market evaluation and competition as the spur. The aim is not only to foresee the consequences of technical changes without stifling innovation, but also to provide the basic understand-ing, procedures, and institutional mechanisms for regulating the con-ditions of competition between firms or between countries, taking into account the long-term effects of such changes.

In a period of deregulation, when the aim is to stimulate innova-tion, cut business costs, and lower taxes, it is tempting to equate the notion of less government interference with the legitimizing of tech-nological *laissez-faire*. It is true that new or tightened environmental protection, safety, and health regulations have altered the pace and direction of innovation in, for example, the chemical and pharma-ceutical industries; but whatever the drawbacks and shortcomings of the regulatory framework, technological *laissez-faire* would involve still greater disadvantages. Moreover, imposed standards prompt more research that leads to further innovations. Cultural necessity is no less the mother of invention than economic necessity: new regula-tions can sometimes stimulate rather than curb innovation.

The idea that society should exercise some control over the con-sequences of technical change stems from the very successes of the scientific and industrial enterprise, and the costs that these have sometimes entailed. The issue of government's responsibility as re-gards regulation of technical change is a thorny one; the criticisms levelled at government intervention tend either not to recognize or to minimize the responsibility. However, if the debate is about policy issues rather than technical matters, governments can help answer the key question of whether the cost-benefit ratio is or is not in the collective interest, and this is, for the developing countries as much as for the industrialized ones, what is at stake in the process of technology assessment.

References

1   Anderer, J., A. McDonald, and N. Nakicenovic. *Energy in a Finite World.* Report by the Energy Systems Program Group of the International Institute for Applied Systems Analysis. Cambridge, Mass.: Ballinger, 1981.

2   Bauer, R.A., with R.S. Rosenbloom and L. Sharpe. *Second Order Consequences: A Methodological Essay on the Impact of Technology.* Cambridge, Mass.: MIT Press, 1969.

3   Blisset, M., ed. *Environmental Impact Assessment.* New York: Engineering Foundation, 1976.

4   Brooks, H. *Science and Technology Policy for the 90's.* Paris: OECD, 1981.

5   ———. "The Role of International Research Institutions." In: H. Brooks and C.L. Cooper, eds. *Science for Public Policy.* New York: Pergamon Press, 1987, pp. 145–162.

6   Davis, K.S. "The Deadly Dust: The Unhappy History of DDT." *American Heritage* 22 (1971), no. 2: 44–47.

7   Dickson, D. *Alternative Technology and the Politics of Technical Change.* London: Fontana, 1974.

8   Dyson, F. "On the Hidden Costs of Saying No." *Bulletin of the Atomic Scientists* 31 (June 1975).

9   Hirschman, A. "The Principle of the Hiding Hand." *The Public Interest* (Winter 1967), no. 6: 10–25.

10   Laudan, L. *Progress and Its Problems: Towards a Theory of Scientific Growth.* Berkeley: University of California Press, 1977.

11   Mosteller, F. (Chairman, Committee for Evaluating Medical Technologies in Clinical Use). *Assessing Medical Technologies.* Washington, D.C.: National Academy Press, 1985.

12   Nichols, G. *Technology on Trial.* Paris: OECD, 1979.

13   OECD. *Technical Change and Economic Policy.* Paris: OECD, 1980.

14   ———. *New Technologies in the 1990s: A Socio-economic Strategy.* Paris: OECD, 1988.

15   Ozawa, C. *Recasting Science: Consensual Procedures in Public Policy Making.* Boulder, Colo.: Westview, 1991.

16   Parson, E.A., and W.C. Clark. "Learning to Manage Global Environmental Change: A Review of Relevant Theory." In: Judith Kildow, ed. *Global Environmental Problems and Solutions.* Cambridge, Mass.: MIT Press, 1992.

17   Rosenberg, N., and L.E. Birdzell. *How the West Grew Rich: The Economic Transformation of the Industrial World.* New York: Basic Books, 1986.

18   Rossini, F.A., A.L. Porter, P. Kelley, and D.E. Chubin. *Framework and*

Factors Affecting Integration within Technology Assessments. Atlanta: Georgia Institute of Technology, 1978.

19 Salomon, J.-J. *Prométhée empêtré: la résistance au changement technique*. Paris: Pergamon, 1981.

20 ———. *Le destin technologique*. Paris: Balland, 1992.

21 Utterback, J.M. "Innovation and Industrial Evolution in Manufacturing Industries." In: B.R. Guile and H. Brooks, eds. *Technology and Global Industry: Companies and Nations in the World Economy*. Washington, D.C.: National Academy Press, 1987.

22 Wad, A., and M. Radnor. "Technology Assessment: Review and Implications for Developing Countries." *Science Policy Studies and Documents* 63 (1984).

23 Wigglesworth, V.B. "DDT and the Balance of Nature." *Atlantic Monthly* 176 (1945): 107–113.

24 Wildavsky, A. "Public Policy." In: B.D. Davis, ed. *The Genetic Revolution: Scientific Prospects and Public Perceptions*. Baltimore: Johns Hopkins University Press, 1991.

25 Winner, L. "On Criticizing Technology." *Public Policy* (Winter 1972): 35–58.

26 Wynne, B. "The Rhetoric of Consensus Politics: A Critical Review of Technology Assessment." *Research Policy* 4 (1975).

Conclusion: Perspectives for the future

Francisco R. Sagasti

As we approach the twenty-first century the uncertain quest of mobilizing science and technology for development appears more elusive. The fragmentary character of our knowledge of the interactions between science, technology, and development and the fundamental changes these concepts are now experiencing make it difficult to derive authoritative and unambiguous conclusions from advances in the field. This task is made even more difficult by the constellation of political, economic, social, cultural, environmental, scientific, and technological changes now occurring, for they are continuously altering the ways in which the processes of knowledge generation and utilization interact with other spheres of human activity.

Nevertheless, even in this turbulent sea of change it is possible to focus on a few key findings and identify promising directions to be explored. Although we must be fully aware of the imperfect nature of our understanding of the interactions between science, technology, and development, the contributions to this volume show that it is possible to offer some conclusions and guidelines for strategy, policy, and decision-making.

First, advances in science and technology have created unprecedented opportunities for improvements in standards of living. During the past four centuries, the systematic process of subjecting abstract conceptions and propositions about the world to the test of empirical observations – which is the hallmark of modern science – has superseded other forms of knowledge generation. As a result, science-based technologies are steadily replacing or improving those that developed through trial and error, and modern science and tech-

511

nology have become an indispensable ingredient of whichever conception of development we may adhere to.

Second, we have become acutely aware that progress in material well-being for a growing fraction of the world's population coexists with stagnation and even deterioration in standards of living for the majority of poor people. Deprivation of food, health, education, and gainful employment besets a sizeable part of humanity, giving rise to new stresses on the environment that, in turn, undermine the basis for future development. The clash between rising aspirations and the reality of widespread poverty, largely triggered by growing awareness of the lifestyles of the affluent, has become a source of social tension, intolerance, and violence. Indeed, the conventional notion of "development," associated with relentless modernization, is being increasingly questioned on social and environmental grounds.

The combination of unprecedented opportunities brought about by advances in science and technology and enormous challenges in the developing regions highlights one of the fundamental paradoxes of our time: we have never had so much power to influence the course of civilization, to shape the way in which our species will evolve, and to create an ever-expanding range of opportunities for human betterment – but we remain unwilling or unable to use this new-found power to achieve our full potential as human beings. We are in the midst of a knowledge explosion that is dividing the world into fast-moving, rich societies that use knowledge effectively and slow-moving, poor societies that do not. The capacity to generate and utilize knowledge has become the key factor explaining differences in human progress at the end of the twentieth century.

As a consequence, regardless of the level of productive, scientific, or technological capabilities a country may have, it is now imperative to integrate science and technology considerations into the design of development strategies. The key for the vast majority of developing countries is not to produce knowledge, but to acquire, adapt, and use it effectively. This requires strategies, policies, and actions to obtain access to the growing international knowledge base, to develop complementary local science and technology capabilities, and to disseminate and use knowledge.

Different developing countries can achieve this in multiple and varied ways, but all of them can devise measures for mobilizing scientific and technological knowledge more effectively. The appropriate mix of strategy, policy coordination, and investment depends on a country's history and culture, resource endowments, quality of government, entrepreneurship, and external situation. For example, in low-income agrarian economies or primary commodity exporters, emphasis should be placed on increasing agricultural productivity and on producing primary commodities efficiently, while at the same time building the human capital and institutional base for a transition towards more knowledge-intensive industrial and service activities. Middle-income countries should focus on strengthening their limited scientific and technological capabilities to improve their access to foreign technology, and on adapting, diffusing, absorbing, and utilizing it more efficiently. The advanced developing countries should aim at closer integration with the global knowledge base, assimilating advanced technologies, and developing their own technologies in selected areas, tasks in which high-level human capital is crucial.

However, to appreciate fully the challenges and options available to different groups of developing countries it is necessary to develop new conceptual frameworks and habits of thought, as well as to reinterpret past experience in a rapidly changing context. Some elements of an agenda for improving our knowledge of the ways in which to effectively mobilize science and technology for development can be discerned from the discussions of the preceding chapters.

A first set of issues in the agenda is related to information and data gathering on science and technology activities and the way they interact with social and economic development. There is a need to construct indicators to reflect accurately the level of capabilities for generating and utilizing knowledge in developing countries, and also to gather information on the impact of policies and government interventions on the development of such capabilities. Comparative science and technology policy assessment efforts, pioneered in the OECD countries through the "Country Review" exercises that began in the 1960s, could be adapted to the situations that different groups of developing countries face in the 1990s. In addition to gathering much needed information, such exercises would help in spreading "best practice" in policy design and implementation.

The broad range of transformations in the institutional settings for conducting scientific and technological activities provides a second set of issues for research and study. The new patterns of interaction between the state, the market, and civil society – which find their concrete expression in the ways in which government agencies, private industry, non-governmental organizations, educational institutions, research centres, labour unions, professional associations, and similar entities relate to each other – are creating a richly interconnected environment for strategy design and policy implementation. It has now become necessary to map and understand better a host of new institutional arrangements for technology transfer and diffusion, research and development, and higher education, among many other fields, in order to evaluate their impact and consider their applicability. For example, in Latin America new partnerships between foreign firms, research centres, and community organizations have been forged for the evaluation of biodiversity in tropical forests (Costa Rica); co-operative arrangements between universities, consulting firms, and subsidiaries of transnational corporations have been established to develop and test software for business applications (Argentina); and government agencies have teamed up with transnational corporations in a joint venture to develop new products and provide industrial extension services (Chile).

The third set of issues arises from the need to fill knowledge gaps in important areas where opportunities may be lost unless timely action is taken. Some examples are international negotiations on intellectual and industrial property rights, in which the rules for access to the pool of international knowledge are being set; the design of mechanisms for international cooperation in science and technology, which acquires urgency because of the changes that international institutions are undergoing and the rising cost of building scientific and technological capabilities; and the exploitation of the window of opportunity offered to developing countries by the techno-economic paradigm that is emerging as a result of the spread of advances in information technology.

A fourth set of issues emerges out of the need to clarify the overall strategic options available to the different groups of developing countries, particularly with regard to their insertion into the global economy, the need to improve social conditions, the challenge of environmental sustainability, and the role that science and technology capabilities play in pursuing these strategic directions.

During the transition to the twenty-first century, no developing country can expect to remain isolated from global economic, social, and political forces. However, the persistent global inequalities in incomes and standards of living, the fall in the price of primary commodities and the deterioration of terms of trade, the burden of foreign debt, and the intensification of migratory currents indicate the non-viability of the current position of many developing countries in the world economy. As the globalization of trade and finance has

514

taken hold, during the 1980s and 1990s most developing nations have been undergoing a process of relative relaxation of their traditional ties to the international economy – from which they cannot extricate themselves. This period of a weakening of their close economic inter-relationships with the rest of the world is likely to be followed by a phase in which new links and patterns of strong interaction with the world economy will be formed.

There are degrees of freedom to influence the repositioning of the developing regions in the future international economy. But in a highly interconnected world, in which competitiveness will depend largely on the capacity to generate and utilize knowledge, scientific and technological capabilities will strongly influence the pattern of reinsertion that finally prevails when the period of relative flexibility and relaxation of linkages is over.

Growing social demands present a difficult challenge for developing countries. The expansion of modern productive activities is not likely to generate sufficient employment opportunities for an expanding labour force – at least for a considerable period, which in many developing countries may exceed a generation. It will therefore be necessary to explore strategic options that could improve social conditions through other, more direct, means.

One such option would provide low-cost, labour-intensive, basic social services using advances in information technology. The productivity of such services (basic education, preventive medicine, child care, nutritional extension, reforestation, maintenance of infrastructure, sanitation, waste disposal, personal services) does not depend, to a large extent, on the wage levels or on major investments in fixed assets. However, the organization of labour-intensive services creates problems of training, coordination, management, and administration. Traditional management and training techniques, largely developed before the widespread availability of microcomputers and the expansion of telecommunications, required a large bureaucracy and specialized administrative personnel.

These difficulties can be overcome through the use of advances in information technology (microcomputers, facsimile transmission, electronic and voice mail, computer databases, desktop publishing, video and sound recorders, interactive television, multimedia workstations, etc.), whose cost diminished rapidly during the 1980s. Hardware and software advances have revolutionized the practice of management and are transforming the nature and delivery of service activities. Added to the promotion of labour-intensive small-scale industries and to the expansion of construction activities, the provision of labour-intensive, high-technology, low-cost basic social services could absorb a sizeable portion of the growing labour force that cannot be employed in modern industry. This would set the stage for major improvements in social conditions, both through higher employment and through the direct benefits that the provision of such services bring to developing country populations.

Gaining access to environmentally friendly technologies developed elsewhere and developing solutions for environmental problems specific to developing countries are two of the main strategic options to meet the challenge of environmentally sustainable development. The United Nations Conference on Environment and Development (UNCED) held in Rio de Janeiro in July 1992, also known as the "Earth Summit," highlighted the commitment of political leaders to the preservation of the environment for future generations. Technology transfer and scientific research figured prominently in the UNCED negotiations, for it has become clear that the capacity to generate and utilize scientific and technological knowledge, often blended with traditional knowledge, holds the key to environmentally sustainable patterns of development.

Many environmental problems are specific to developing countries and require locally generated solutions (preservation of biodiversity, sustainable use of renewable resources, preventing soil erosion, improvements in water supply), while others can be addressed through the importation of technologies that reduce waste, cut energy consumption, and minimize pollution. As environmental sustainability has become a global concern, there is considerable scope for the design of strategic options that make use of the willingness of industrialized nations to provide development assistance for environmental preservation. Moreover, the preparation of "Environmental Action Plans" is now a condition for granting assistance from bilateral and multilateral sources such as the World Bank, the regional development banks, and the aid agencies of Japan, the United States, and Germany, among others.

These three items in the agenda of strategic options for developing countries – international insertion, social improvements, and environmental sustainability – pose a series of science and technology policy demands that must be met in light of the specific situations of developing countries. These range from maintaining an appropriate

516

policy environment (including macroeconomic stability, competitive pressures, innovation incentives) to establishing an adequate base of human resources (research scientists, managers, engineering and technical personnel, providers of extension services, skilled workers, knowledgeable observers of nature) and to designing policy interventions that focus on the effective acquisition and use of knowledge.

A fifth set of issues is related to governance and social guidance. Investments to create capacity in knowledge generation and utilization take a relatively long time to mature and require sustained support. Facing acute economic and social problems, most developing countries find themselves hard pressed to allocate resources to the uncertain task of building science and technology capabilities and forced to address urgent problems of immediate political concern. Creating and maintaining a social consensus that values science and technology and gives priority to the capacity to generate and utilize knowledge is one of the most urgent tasks for developing country leadership in the transition to the twenty-first century.

Scientific research and technological innovation flourish in open societies where there is a willingness to question orthodoxy and explore new ideas. These conditions are closely related to democratic institutions, where freedom of expression encourages the contrast of different viewpoints and opinions and fosters flexibility and adaptation. The extent to which democracy, development, and the capacity to generate and utilize knowledge are intertwined – and may have become inseparable – is one of the most important issues to elucidate in addressing the governance dimension of development strategies.

A final cluster of issues in the agenda for the future refers to the complex task of reinterpreting the meaning of and possible paths to development, linking them explicitly to the capacity to generate and utilize knowledge. Questions such as popular participation, institutional development, and social and political empowerment have begun to figure prominently as promising new avenues for development thinking. Each of these issues has important implications for the generation, dissemination, and utilization of knowledge, particularly for the rate and direction of technical change, the control of access to knowledge, and the interactions between modern and traditional knowledge.

As evidenced by the contributions to this volume, the repertoire of concepts and ideas in this field has increased significantly, although we are still a long way from having a comprehensive account of the ways in which knowledge, economic growth, and social improvements interact with each other and with the value and ethical dimensions of human progress. Such a reinterpretation should take into account the contemporary critique of modernity and the possible contribution of alternative traditions of knowledge production, particularly as social and environmental concerns move to centre stage in the developing world.

Contributors

JAN ANNERSTEDT is Associate Professor of Political Science at Roskilde University (Denmark) and Director of its international doctoral programme in Technology Policy, Innovation, and Socio-economic Development. He is also the founding Director of the Nordic Centre of Innovation in Lund (Sweden). He has been a consultant on science, technology, and innovation for international organizations such as the OECD, Unesco, UNCTAD, and UNIDO. While finishing his contribution to this volume, he was attached to the Research School of Social Sciences of the Australian National University in Canberra.

AJIT BHALLA is Chief of the Technology and Employment Division at the International Labour Office (ILO) in Geneva. He was previously Hallsworth Professorial Fellow in Economics, University of Manchester (UK); Visiting Research Associate at the Economic Growth Center, Yale University; Research Officer at the Institute of Economics and Statistics, Oxford University (UK); and Tutorial Fellow at the University of Delhi. Recent publications include *Technology and Employment in Industry* (Geneva: ILO, 1975, 1981, 1985); *New Technologies and Development* (Boulder, Colo.: Lynne Rienner, 1988); *Uneven Development in the Third World* (London: Macmillan, 1992).

HARVEY BROOKS is Professor of Applied Physics and Professor of Technology and Public Policy (Emeritus) at Harvard University. He is former Director of the Science, Technology, and Public Policy Program of the Kennedy School of Government at Harvard University. He served as a member of the US National Science Board and the President's Science Advisory Committee. He is a member of the National Academy of Sciences and of the National Academy of Engineering and former President of the American Academy of Arts and Sciences. He has published in the fields of nuclear engineering, underwater acoustics, and solid state physics, as well as in science and public policy.

JACQUES GAILLARD is Visiting Fellow at the Center for International Science and Technology Policy, George Washington University. He graduated from the Ecole

519

Supérieure d'Agriculture d'Angers and received his doctorate in science, technology, and society from the Conservatoire National des Arts et Métiers, Paris, in 1989. He worked as Scientific Secretary of the International Foundation for Science (IFS) in Stockholm from 1975 to 1985. During 1986–1991, he was Head of the Science, Technology, and Development Programme at the French Institute of Scientific Research for Development through Cooperation (ORSTOM). Recent publications include *Scientists in the Third World* (Lexington, Ky.: Univ. Press of Kentucky, 1991) and *Science Indicators for Developing Countries*, edited with Rigas Arvanitis (Paris: ORSTOM).

ANDREW JAMISON is Director of the graduate programme in Science and Technology Policy at the Research Policy Institute, University of Lund (Sweden). He studied the history of science at Harvard University before moving to Sweden in 1970. He has taught science and society courses at Lund, Copenhagen, and Gothenburg; he received a doctorate from the University of Gothenburg in the theory of science in 1983. His current research focuses on the cultural dimension of science and technology policy in developing countries. Among his recent publications are *Keeping Science Straight* (Gothenburg, 1988); *The Making of the New Environmental Consciousness*, co-author with Ron Eyerman and Jacqueline Cramer (Edinburgh, 1990); *Social Movements: A Cognitive Approach*, co-author with Ron Eyerman (Oxford, 1991).

JORGE KATZ is Professor of Industrial Economics, University of Buenos Aires. He graduated in economics from the University of Buenos Aires and received his Ph.D. from the University of Oxford. He has done extensive consulting on issues of industrial organization and technical change for ECLA, IDB, and other international agencies. His publications include *Technology Generation in Latin American Manufacturing Industries* (London: Macmillan, 1987); *Organización del Sector Salud: Puja Distributiva y Equidad*, edited with Alberto Muñoz (Buenos Aires: Centro Editor de América Latina, 1988); *El Proceso de Industrialización en la Argentina: Evolución Retroceso y Prospectiva*, edited with B. Kosacoff (Buenos Aires: Centro Editor de América Latina, 1989).

SANJAYA LALL is Lecturer in Development Economics at the Institute of Economics and Statistics and a Fellow of Green College at Oxford University. He studied economics in India and at Oxford University before joining the World Bank. He has taught at Oxford since 1968 and has also acted as a consultant for many international institutions, including the World Bank and a wide range of United Nations agencies. Among his many publications are *Foreign Investment, Transnationals and Developing Countries*, with Paul Streeten (1977), *The Multinational Corporation* (1980), *The New Multinationals* (1983), *Learning to Industrialize* (1987), and *Building International Competitiveness* (OECD, 1990). He recently co-edited *Current Issues in Development Economics* (Macmillan, 1991) and *Alternative Development Strategies for Sub Saharan Africa* (Macmillan, 1992).

NASSER PAKDAMAN teaches development economics at the Université de Paris VII. He studied in Teheran and Paris, and taught economics at the University of Teheran from 1967 to 1981, as well as at Princeton and Oxford universities. In addition to books and articles in Farsi, his main publications in French are: *Economie iranienne:*

analyse d'une économie sous-développée (Paris, 1968); *Bibliographie française de la civilisation iranienne*, 3 vols., in collaboration with A. Abolhamd (Teheran, 1970–1974); *Pour une histoire du développement: états, sociétés, développement*, with others (Paris, 1991). He is co-editor of *Çesmandaz*, a Farsi-language journal published in Paris, and is currently working on a study of the interaction between religious and economic factors in Iran.

AMITAV RATH is a Director of Policy Research International, a consultancy, and Visiting Professor at Carleton University, Ottawa. He studied at the Indian Institute of Technology, Kharagpur, and completed his gradute training at the University of California, Berkeley. He worked in the Science, Technology and Energy Policy Programme of the International Development Research Centre (IDRC) from 1982 to 1990 and was its Head for a number of years. He has taught at several universities in Canada, India, Jamaica, and the United States. His most recent publications include the special volume of *World Development* that he edited under the title "Science and Technology: Issues from the Periphery" (1990); *Technology and the International Agenda: Lessons for UNCED and Beyond*, co-author with B. Herbert-Copley (Ottawa: IDRC, 1992).

PAULO RODRIGUES PEREIRA is Adviser to the Brazilian Federal Secretariat of Science and Technology in Brasilia. He previously worked for the Brazilian Ministry of National Education and was Assistant Director for Scientific and Technological Cooperation at Digibras (the Brazilian state enterprise for computer science development). Prior to that he taught physics at the Universities of Rio de Janeiro, Brasilia, and Paraíba. He graduated in physics from the Swiss Federal Polytechnic (ETH) in Zurich and received his doctorate in science, technology, and society from the Conservatoire National des Arts et Métiers, Paris, in 1991.

IGNACY SACHS is Directeur d'études (Professor) at the Ecole des Hautes Etudes en Sciences Sociales in Paris, where he founded the International Research Centre on Environment and Development in 1973. He was Director of the Centre until 1985, since which time he has been Director of the Research Centre on Contemporary Brazil. He acted as a Special Adviser to the Secretary-General of the UN Conference on Human Environment in Stockholm in 1972 and of the UN Conference on Environment and Development in Rio in 1992. He is a founding member of the International Foundation for Development Alternatives. He studied at the Universities of Rio de Janeiro, Delhi (Ph.D., 1961), and the Central School of Planning and Statistics, Warsaw. His principal publications include *Development and Planning* (Cambridge: Cambridge University Press, 1987); *Studies in the Political Economy of Development* (Oxford: Pergamon, 1979); and *The Discovery of the Third World* (Cambridge, Mass.: MIT Press, 1976).

CÉLINE SACHS-JEANTET is responsible for coordinating the interministerial programme on the "University and the City" and is a researcher at the Fondation de la Maison des Sciences de l'Homme, in Paris. She has acted as a consultant to the World Bank, Unesco, and the French government. She studied at the University of Paris and the Massachusetts Institute of Technology (MIT), and received a doctorate in urban studies from the University of Paris XII in 1987. Her publications include *Sao*

Paulo. Politiques publiques et habitat populaire (Paris: Editions Maison des Sciences de l'Homme, 1990); and *Exploring the Human Dimensions of Development: A Review of the Literature* (Paris: Unesco).

FRANCISCO R. SAGASTI served from 1987 to 1991 as Chief of Strategic Planning for the World Bank. Earlier Dr. Sagasti served as an adviser to the Peruvian ministries of Foreign Affairs and Planning and Industry, and as a board member of an engineering design firm and professor at the Universidad del Pacífico in Lima. In 1979 he worked closely with the Secretary-General in the planning of the United Nations Conference on Science and Technology for Development and later chaired the UN Advisory Committee on Science and Technology for Development. From 1973 to 1978, Dr. Sagasti led an international study of S&T policies in 10 developing countries, in which more than 150 researchers in Africa, Asia, Latin America, and the Middle East participated. Dr. Sagasti has a Ph.D. in Social Systems Sciences from the Wharton School at the University of Pennsylvania, an M.Sc. in Industrial Engineering from Pennsylvania State University, and he earned two Industrial Engineering degrees from the Universidad Nacional de Ingeniería in Lima, Peru.

JEAN-JACQUES SALOMON is Director of the Centre Science, Technologie, Société at the Conservatoire National des Arts et Métiers, Paris, and President of the Collège de la Prévention des Risques Technologiques. Between 1963 and 1983 he set up and headed the Science and Technology Policy Division of the Organisation for Economic Co-operation and Development (OECD) in Paris. He has been visiting professor at several universities (MIT, Harvard, etc.) and visiting fellow at Clare Hall, Cambridge. He founded the International Council for Science Policy Studies, of which he was President after Derek de Solla Price, and was President of the Standing Committee for Social Sciences of the European Science Foundation. His principal recent publications include *Le destin technologique* (Paris: Balland, 1992); *L'écrivain public et l'ordinateur. Mirages du développement*, co-author with André Lebeau (Paris: Economica, 1988; English edition under the title *Mirages of Development* [Boulder, Colo.: Lynne Rienner, 1993]); and *Science, War and Peace*, editor (Paris: Economica, 1989).

HEBE VESSURI is Head of the Department of Science Studies at the Instituto Venezolano de Investigaciones Cientificas (IVIC) in Caracas. She writes on topics in the sociology and social history of twentieth-century Latin American science and on the interface between science and higher education. She has edited *La ciencia periférica. Ciencia y sociedad en Venezuela* (Caracas, 1984); *Ciencia académica en la Venezuela moderna* (Caracas, 1984); and *Las instituciones científicas en la historia de la ciencia en Venezuela* (Caracas, 1987). She is finishing a study of the origins of agricultural science in Venezuela and a book on scientific publishing in the periphery.

ATUL WAD is Director of Research, Technology, and Industrial Development Programs at International Business Development (IBD), Northwestern University, Evanston, Illinois. He is responsible for corporate and country projects involving technology sourcing, strategy, intelligence, and transfer, as well as industrial development. He has a Ph.D. in management and an undergraduate degree in mechanical engineering. He has been a visiting professor at the India Institute of Management in

Bangalore and a fellow at the Science Policy Research Unit at the University of Sussex. He has extensive consulting experience, and has also served for years as scientific officer with the UN Centre for Science and Technology for Development, where he was instrumental in developing the UN's Advanced Technology Alert System (ATAS). He is the author of numerous articles on science and technology for developing countries and of *Science, Technology, and Development* (Boulder, Colo.: Westview Press, 1988).

Name index

Subject index